The Cambridge Program for the Mathematics Test

Jerry Howett

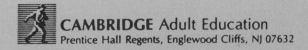

CAMBRIDGE Adult Education
Prentice Hall Regents, Englewood Cliffs, NJ 07632

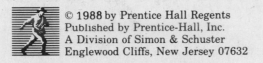 © 1988 by Prentice Hall Regents
Published by Prentice-Hall, Inc.
A Division of Simon & Schuster
Englewood Cliffs, New Jersey 07632

Printed in the United States of America

10 9 8 7 6 5

ISBN: 0-8428-8705-9

Prentice-Hall International (UK) Limited, *London*
Prentice-Hall of Australia Pty. Limited, *Sydney*
Prentice-Hall Canada Inc., *Toronto*
Prentice-Hall Hispanoamericana, S.A., *Mexico*
Prentice-Hall of India Private Limited, *New Delhi*
Prentice-Hall of Japan, Inc., *Tokyo*
Simon & Schuster Asia Pte. Ltd., *Singapore*
Editora Prentice-Hall do Brasil, Ltda., *Rio de Janeiro*

The Cambridge GED Program

Consulting Editors

Mary Ann Corley
Supervisor of Adult Education
Baltimore County Public Schools

Del Gratia Doss
Supervisor, Adult Basic Education
St. Louis, Missouri

Ron Froman
Administrator of Adult Education
Orange County Public Schools, Orlando, Florida

Dorothy Hammond
Coordinator
New York State Writing Project

Lawrence Levin
KNILE Educational & Training Association
Former Director ABE/ESL/HSE Services
New York City Board of Education

Noreen Lopez
Adult Educator
Illinois

Arturo McDonald
Assistant Superintendent—Adult Education
Brownsville, Texas

Cheryl Moore
Curriculum Director, Windham School System
Texas Department of Corrections

Carrie Robinson Weir
Director, Adult Education Resource Center
Jersey City State College

Harold Wilson
Director of Adult Basic Education
Indianapolis Public Schools

Jane Zinner
Director of Grants and Curriculum
Association for California School Administrators

Contributing Editors

Gloria Cohen, Ed.D.
Consultant, Curriculum & Gifted Education
New York Metropolitan Area

Carole Deletiner
Adult Basic Education Teacher
Formerly, New York City Technical College

Don Gerstein
Academic Educator
Wyoming State Penitentiary 1981–1986

Nathaniel Howard
Senior Training Representative
Consolidated Edison, New York City

Joan Knight
Former City-wide Supervisor of Staff Development
New York City Board of Education Adult Program

Bonnie Longnion
Dean of Continuing Education
North Harris County College, Kingwood, Texas

Joe Mangano
Research Associate
State University of New York at Albany

Ada Rogers
Adult Education GED Program
Broward County, Florida, School System

Ann Rowe
Education Specialist
New York State

Elois Scott
Reading Specialist
University of Florida, Gainesville

Stephen Steurer
Correctional Academic Coordinator
Maryland State Department of Education

Dr. Jay Templin
Professor of Biology
Widener University, Delaware Campus

Jeffrey Tenzer
Auxiliary Services for High Schools
New York City

The Cambridge GED Program

Executive Editor

Jerry Long

Senior Editor

Timothy Foote

Project Editors

James Fina
Diane Maass

Subject Editors

Jim Bedell
Diane Engel
Randee Falk
Scott Gillam
Rebecca Motil
Thomas Repensek

Art and Design

Brian Crede Associates
Adele Scheff
Hal Keith

Contents

Introduction   vii

Prediction

Instruction

Practice

Simulation

Answers and Solutions   309

Introduction

The following pages will introduce you to the Mathematics Test and to the organization of this book. You will read about ways you can use this book to your best advantage.

What Is The Mathematics Test?

The Mathematics Test of the GED Tests examines your ability to solve the kind of math problems you are likely to run into in your daily life.

What Kind of Questions Are on the Test?

When you take the test, you will answer questions based on brief passages and graphic material. Solving some of the problems will require working through two or more steps. For example, you may be given information about two prospective jobs with different pay and different annual raises. You also may be told the cost of transportation to each job. Then you may be asked to calculate which job would net more money at the end of two years. Answering that question would require performing several different operations.

Some questions will test your ability to use information presented in a graphic format. For example, you may be presented with a circle graph showing how a nursing home's budget is divided by percent. You then may be given the total budget in dollars and be asked to figure the dollar amount of one of the budget items.

If a question requires you to use a formula, the formula will be provided for you. You are required only to demonstrate that you can apply mathematical theory to solve the kinds of problems you may face in your work, as a consumer, or in your normal daily family life.

All the questions on the Mathematics Test are in multiple-choice format.

What Are the Problems Like?

Sometimes you will be given a short reading passage and then asked a question about it. For example, you may be given a list of the weights of several parcels, some in pounds and some in ounces. You then may be asked in what order the parcels should be stacked so that the heaviest ones are on the bottom and the lightest ones are on the top. Sometimes you will be given more information in a longer passage and be asked several questions.

About two-thirds of the problems are based on written passages. The other one-third are based on graphic material. You may be asked one or more questions about each of the pieces of graphic material.

One-fourth of the items do not require you to solve problems. Rather, they ask you to show how you would go about solving the problems. To get some items correct, you will have to recognize and answer that you are not given enough information to solve them.

The 56 items on the Mathematics Test cover the following content areas:

Arithmetic	
Measurement	30%
Number Relationships	10%
Data Analysis	10%
Algebra	30%
Geometry	20%

What You Will Find in This Book

This book gives you a four-step preparation for taking the Mathematics Test. The four steps are as follows:

Step One: Prediction

The Predictor Test is very much like the actual Mathematics Test but is only half as long. Taking the Predictor Test will give you an idea of what the real GED will be like. By evaluating your performance on the Predictor Test, you will get a sense of your strengths and weaknesses. This information will help you plan your studies accordingly.

Step Two: Instruction

The Instruction section is divided into seven chapters: Whole Numbers, Fractions, Decimals, Percents, Graphs, Algebra, and Geometry.

All but one of the chapters are divided into three levels. The levels are arranged so that the basic instruction is included in the lowest level, Level 1, of any chapter. Levels 2 and 3 include instruction on more and more complex applications of the basics. The first level in each chapter concentrates on how to calculate with the kind of numbers covered by the chapter or, in algebra and geometry, on basic principles. The second level of each chapter shows you how to apply calculations or basic principles in order to solve simple problems. The third level shows how to solve more complex problems.

Each level is divided into lessons. A lesson consists of instruction on one aspect of the material covered in that level. Each lesson ends with a brief exercise to help you test your knowledge of the material in the lesson. Throughout the instruction, most of the problems are not in multiple-choice format. This kind of practice should help you develop reliance on your ability to solve simple and complex problems.

Step Three: Practice

This section gives you valuable practice in answering the type of questions you will find on the actual Mathematics Test. There are two types of practice activity. The **Practice Items** are GED-like questions that are organized in three groups: arithmetic, algebra, geometry. The Practice Items allow you to test your understanding of one type of math at a time. The second type of practice, the **Practice Test,** is structured like the actual Mathematics Test. In the Practice Test, the types of math you need to use vary from item to item, just as on the real test. This section gives you an opportunity to practice a test similar to the GED.

Each practice activity contains 56 items, the same number of questions as are on the actual Mathematics Test. You can use your results to track your progress and to get an idea of how well prepared you are.

Step Four: Simulation

Finally, this book offers a simulated version of the Mathematics Test. It has been designed to be as close to the real test as possible. The number of questions, their level of difficulty, and their organization are the same as you will find on the actual test. You will have the same amount of time to answer the questions as you will have on the actual test. Taking the Simulated Test will help you find out how ready you are to take the real exam.

The Answers and Solutions

At the back of this book, you will find a section called Answers and Solutions. This contains the answers to all the questions in the Lesson Exercises, Level Previews and Reviews, Chapter Quizzes, Unit Test, Practice Items, the Practice Test, and the Simulated Test. The answer section is a valuable study tool: It not only tells you right answers, it also shows the process for solving each question successfully. You can benefit a great deal by consulting the answer section as soon as you complete an activity.

Using This Book

This book has been designed to give you several choices. Whether you are working with an instructor or alone, you can use this book to prepare for the Mathematics Test in the way that works best for you.

Take a Glance at the Contents

Before doing anything else, look over the Contents and get a feel for this book. You can compare the headings in the Contents with the descriptions found in the preceding pages. You also might want to leaf through the book to see what each section looks like.

Take the Predictor Test

Next, you will probably want to take the Predictor Test. As the introduction before the test explains, there is more than one way to take this test. Decide which way is best for you.

Your performance and score on the Predictor Test will be very useful to you as you work through the rest of this book. It will point out your particular strengths and weaknesses, which can help you plan your course of study.

Beginning Your Instruction

After you have analyzed your strengths and weaknesses, you are ready to begin instruction. The first step in instruction is to take the Level 1 Preview near the beginning of Chapter 1 of the Instruction section. If you score 80% or better on the Preview, go ahead to Level 1 of the Review. If you score 80% or better on the Level 1 Review as well, you can be satisfied that you understand the material in the first level. Go on to the Preview at the beginning of Level 2. You can continue skipping ahead in this fashion until you score below 80% on either a Preview or a Review. At that point, go back to the first lesson in the level and begin work.

As mentioned earlier, most chapters are divided into three levels: one on calculation, one on simple problems, and one on more complex problems. If you like, you can work through all the Level 1's, all the Level 2's, and all the Level 3's. Or you can work through each chapter in order—Level 1, Level 2, Level 3—before going on to the next chapter. Use the order for studying that is most interesting and effective for you.

Using the Practice Section

You will remember that the Practice Items in this section are divided into three sections: arithmetic, algebra, and geometry. This gives you another choice about how to use this book. You may complete the entire Instruction section before going on to the Practice section. Or, after having completed Chapters 1 through 5 of the Instruction section, you may wish to go on to the Practice Items that are concerned with arithmetic. Likewise, after completing Chapter 6 of the Instruction section, you could go on to the algebra Practice Items; and after completing Chapter 7 of the Instruction section, you could go on to the geometry Practice Items.

However, you should not do any work on the Practice Test until you have finished the entire Instruction section and worked through all the Practice Items.

Taking the Simulated Test

Finally, once you have completed the Instruction and Practice sections, you can take the Simulated Math Test. This will give you the most accurate assessment of how ready you are to take the actual test.

Try Your Best!

As you study the lessons and complete the activities and tests in this book, you should give it your best effort. Try to maintain a score of at least 80% correct as you work through the book. The Progress Charts will help you compare your work with this 80% figure. If you maintain 80% scores, you are probably working at a level that will allow you to do well on the GED test.

What Is The GED?

You are preparing for the GED Tests. The initials GED stand for General Educational Development. You may also have heard the tests referred to as the High School Equivalency Tests. The GED diploma is widely regarded as the equivalent of a high school diploma.

The GED Test is a way for millions of adults in the United States and Canada to get diplomas or certificates without returning to high school. Each year about half a million people take the GED tests.

Who Recognizes the GED?

The GED is recognized by employers, unions, and state and federal civil services. Many vocational institutes, colleges, and universities accept students who have obtained a GED. All fifty states and parts of Canada use the GED Test results to issue high school equivalency credentials. However, each state has its own standards for what constitutes a passing grade. For information on the requirements in your state, contact the High School Equivalency Program of the State Department of Education in your state's capital.

What Is Tested on the GED Test?

The material found on the GED is based on the subjects covered in most high schools around the country. Thus you will be learning about the subject areas that you would be most likely to study if you attended four years of high school. However, the focus of the GED is not on content but on *skills*. You will not have to memorize specific dates, names, and places. For example, whether you recall the day on which a battle was fought or the title of a novel is less important than whether you can read and understand a passage on history or literature.

In fact, there is a good chance that you have already been using many of the thinking skills that will be tested on the GED. For example, many people must do some writing in their lives, and the GED includes a test of writing skills. A lot of people read on the job or for pleasure, and reading skills are tested, too. Many people use basic mathematics for such things as figuring out a budget and doubling a recipe, and the GED includes a test of basic math skills.

So, instead of testing your memory, the GED will test your ability to get information and apply your thinking to that information.

How Is the GED Structured?

The GED is actually five separate tests. With one exception, the test is composed entirely of multiple choice questions. The one exception is the 200-word essay that you will be required to write as part of the Writing Skills Test.

The chart below describes the structure of the five tests:

Test #	Test Subject	Number of Items	Minutes Allowed
1	Writing Skills Part 1—Multiple Choice Part 2—Essay	55 1	75 45
2	Social Studies	64	85
3	Science	66	95
4	Interpreting Literature and the Arts	45	65
5	Mathematics	56	90

Now you are ready to begin preparing for the Mathematics Test. Good luck!

Prediction

Introduction

If you were going to take the GED test today, how do you think you would do? In which areas would you perform best, and which would give you the most trouble? The Predictor Test that follows can help you answer those questions. It is called a Predictor Test because your test results can be used to discover where your strengths and weaknesses are related to the actual Mathematics Test of the GED.

The Predictor Test is like the actual GED test in may ways. It will check your skills as you apply them to the kinds of problems you will find on the real test. The questions are like those on the actual test.

How to Take the Predictor Test

The Predictor Test will be most useful to you if you take it in a manner close to the way the actual test is given. If possible, you should complete it in one sitting with as little distraction as possible. Refer to the formulas on page 372 as often as you need to. So that you will have an accurate record of your performance, write your answers neatly on a sheet of paper or use an answer sheet provided by your teacher.

As you take the test, don't be discouraged if you find you are having difficulty with some (or even many) of the questions. The purpose of this test is to predict your overall performance on the GED and to locate your particular strengths and weaknesses. Relax; there will be plenty of opportunities to correct any weaknesses and retest them.

You may want to time yourself to see how long you take to complete the test. When you take the actual Mathematics Test, you will be be given 90 minutes. The Predictor Test is about half as long as the actual test, so if you finish within 45 minutes, you are right on target. At this stage, however, you shouldn't worry too much if it takes you longer.

When you are done, check your answers by using the answer section that begins on page 9. Put a check by each item you answered correctly.

How to Use Your Score

At the end of the test, you will find a Performance Analysis Chart. Fill in the chart; it will help you get a general idea about which areas you are more comfortable with and which give you the most trouble. As you work through the instruction in this book, you will complete several short exercises—Previews and Reviews—that will help you pinpoint your skill levels even more closely.

PREDICTOR TEST

Time: 45 minutes

Directions: *Before you begin to work through the lessons in this book, try the following test. Use any of the formulas on page 372 that you need. Choose the one best answer to each question. When you finish, check your answers. The answers include a guide to the sections of the book where the skills tested in each problem are presented.*

1. Simplify the expression $10^2 - 3^3$.

   (1) 11
   (2) 73
   (3) 91
   (4) 97
   (5) 127

2. Solve for m in the equation $9m - 4 = 7m + 18$.

   (1) 2
   (2) 7
   (3) 9
   (4) 11
   (5) 18

3. Below are the weights of five boxes.
   Box A—0.15 kg
   Box B—0.9 kg
   Box C—0.85 kg
   Box D—1.05 kg
   Box E—0.955 kg

   Which of the following sequences shows the boxes in order from heaviest to lightest?

   (1) C, B, A, E, D
   (2) E, D, A, C, B
   (3) A, C, B, E, D
   (4) D, B, C, A, E
   (5) D, E, B, C, A

4. In the diagram below, line r is parallel to line s, and $\angle a = 99°$. What is the measurement of $\angle c$?

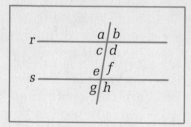

   (1) 9°
   (2) 81°
   (3) 89°
   (4) 101°
   (5) Not enough information is given.

5. Brian and Irene went out to dinner. The bill for food and drinks was $32. Which of the following tells, to the nearest dollar, the total amount they paid for dinner if they left a 12% tip?

   (1) $40
   (2) $38
   (3) $37
   (4) $36
   (5) $34

4

6. The diagram below shows the plan of a rectangular back yard. What is the diagonal distance in yards across the yard?

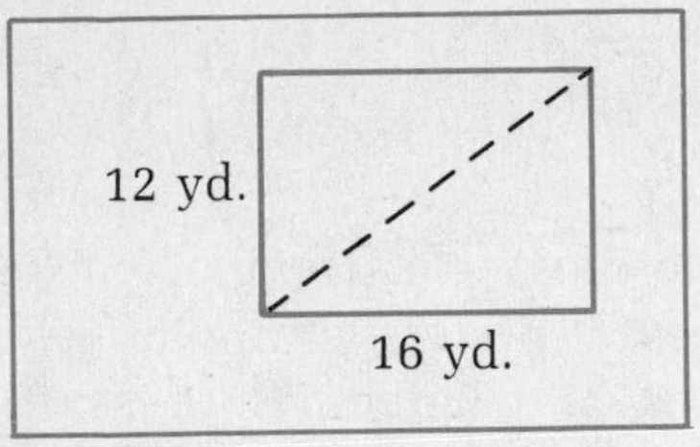

12 yd.

16 yd.

(1) 14
(2) 20
(3) 24
(4) 28
(5) 56

Items 7 and 8 are based on the following graph.

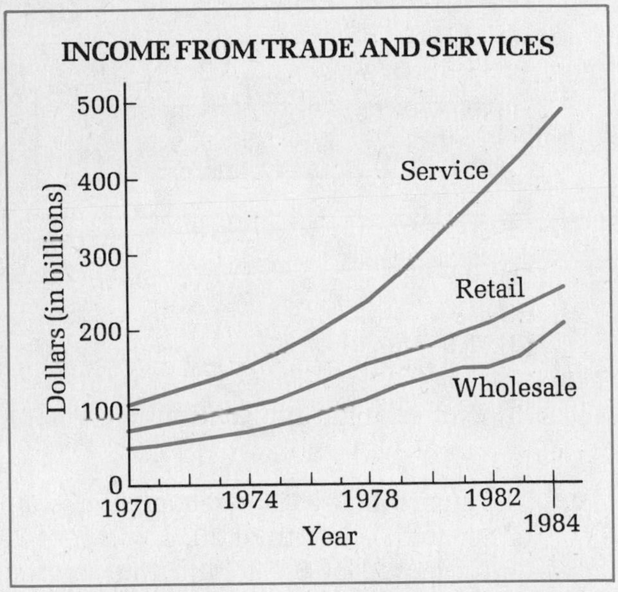

INCOME FROM TRADE AND SERVICES

Dollars (in billions)

Service

Retail

Wholesale

Year

7. The income from wholesale trade was about what fraction of the income from services in 1970?

(1) $\frac{1}{4}$
(2) $\frac{1}{2}$
(3) $\frac{2}{3}$
(4) $\frac{3}{4}$
(5) $\frac{5}{6}$

8. The total income produced from services in 1984 was about how many times the total income produced from services in 1970?

(1) $4\frac{1}{2}$
(2) 3
(3) $2\frac{1}{2}$
(4) 2
(5) $1\frac{1}{2}$

9. Following are the weights of the members of the Ramos family: Mr. Ramos, 195 pounds; Mrs. Ramos, 118 pounds; Susana, 84 pounds; and Alfredo, 156 pounds. What is the median weight in pounds of the members of the family?

(1) 137
(2) 140
(3) 156
(4) 166
(5) 174

10. Carol has a large bag that contains three glazed doughnuts, four cinnamon doughnuts, and five plain doughnuts. What is the probability that the first doughnut she takes from the bag will be cinnamon?

(1) $\frac{1}{4}$
(2) $\frac{1}{3}$
(3) $\frac{2}{5}$
(4) $\frac{1}{2}$
(5) $\frac{4}{5}$

11. What values of x will make the equation $x^2 - 2x - 24 = 0$ true?

(1) $+8$ and -3
(2) $+3$ and -8
(3) $+12$ and -2
(4) $+2$ and -12
(5) $+6$ and -4

Items 12 and 13 are based on the following information.

Max Reade bought 40 acres of farmland from Mr. Brown at a price of $800 an acre. Max made a down payment of 20% of the basic price of the land, and he financed the rest with a bank loan. Mr. Brown, the seller, had to pay $1600 to the real estate agent who handled the sale and $700 in legal fees.

12. How much was Max's down payment?

(1) $3200
(2) $4800
(3) $6400
(4) $5600
(5) Not enough information is given.

13. The real estate agent's fee was what percent of the basic price Mr. Brown asked for the land?

(1) 5%
(2) 7.5%
(3) 9%
(4) 12.5%
(5) Not enough information is given.

---

14. In a year, Tanya paid $124 in charges on purchases of $620 with her credit card. The charges represent what percent of the total purchases?

(1) 12.4%
(2) 18%
(3) 20%
(4) 50%
(5) 62%

15. David is twice as old as his daughter Catherine. David's wife Mary is three years younger than David. If x represents Catherine's age, which of the following expressions represents Mary's age?

(1) 2x − 3   (4) 3x + 2
(2) 2x   (5) 3x − 2
(3) 2x + 3

16. One week Stewart worked 35 hours at $8 an hour and 6 hours at $12 an hour. Which expression tells the amount, in dollars, that he made that week?

(1) 42(8 + 12)   (4) 35(8 + 12)
(2) 35(6) + 8(12)   (5) 35(8) + 6(12)
(3) 35(8)(6)(12)

17. Find the area of a triangle whose base measures 8 meters and whose height measures 6.5 meters. Express the answer to the nearest square meter.

(1) 52   (4) 21
(2) 29   (5) 13
(3) 26

18. The figure below is a centimeter scale. What is the distance in centimeters between points *A* and *C*?

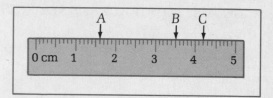

(1) 0.7
(2) 1.6
(3) 1.9
(4) 2.6
(5) Not enough information is given.

19. The figure below is a rectangular container with a length of 20, a width of 3, and a height of 0.75. Find the volume of the container.

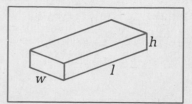

(1) 23.75
(2) 45
(3) 47.5
(4) 60
(5) Not enough information is given.

20. Colin drove east at an average speed of 55 mph. His brother Pete started at the same time and drove west at an average speed of 60 mph. Which of the following expressions shows the distance between them at the end of 2.5 hours?

(1) $(2.5 \times 55) + (2.5 \times 60)$
(2) $2.5 \times 55 \times 60$
(3) $2.5 \times 55 + 60$
(4) $(2.5 \times 55) - (2.5 \times 60)$
(5) $(55 + 60)(2.5 + 2.5)$

21. In Mr. Wilson's pasture there are 200 cows. 120 cows are Holsteins, and 80 are Jerseys. The cows pass one at a time through a door into the barn. What is the probability that a cow passing through the door will be a Jersey?

(1) $\frac{80}{120}$

(2) $\frac{80}{200}$

(3) $\frac{120}{200}$

(4) $\frac{(80 + 120)}{200}$

(5) $\frac{(120 - 80)}{200}$

22. Hank plans to build a new wall between the kitchen and living room of his house. He figures that he needs 6 sheets of drywall panels, 14 structural supports, one pound of nails, and a roll of drywall tape. Following are costs for materials.

Drywall panels: $8 each or 3 for $20.
Structural supports: $3 each.
Nails: $2 per pound.
Drywall tape: $4 per roll.

Which of the following tells the minimum total for materials?

(1) $ 49
(2) $ 88
(3) $ 96
(4) $108
(5) Not enough information is given.

23. Write an equation which expresses the following statement: One less than the quotient of a number divided by nine equals three.

(1) $9x + 1 = 3$ **(4)** $\frac{x}{9} - 1 = 3$
(2) $9x - 1 = 3$ **(5)** $\frac{x}{3} - 1 = 9$
(3) $\frac{x}{9} + 1 = 3$

24. In the triangle pictured below, $\angle A = \angle C$, and $\angle B = 44°$. Find the measurement of $\angle A$.

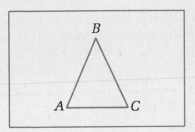

(1) 36°   **(3)** 68°   **(5)** 144°
(2) 44°   **(4)** 92°

25. Don and his assistant each worked 20 hours on a carpentry job. Don makes $5 an hour more than his assistant. Together they were paid $460 for the job. Find Don's hourly rate.

(1) $7   **(3)** $ 9   **(5)** $14
(2) $8   **(4)** $12

26. What is the slope of the line that passes through points C and D in the graph below?

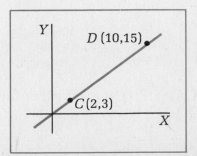

(1) $+\frac{3}{2}$ **(3)** $+\frac{4}{3}$ **(5)** $-\frac{4}{3}$

(2) $+\frac{3}{4}$ **(4)** $-\frac{3}{4}$

Item 27 refers to the following graph.

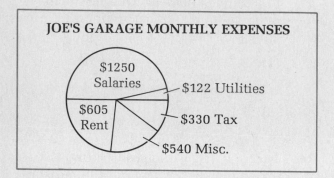

Item 28 refers to the following diagram.

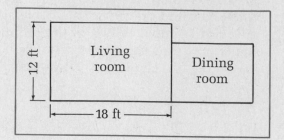

27. The graph represents the monthly expenses at Joe's Garage. Rent is approximately how many times as much as utilities?

(1) 2
(2) 3
(3) 5
(4) 8
(5) 10

28. The diagram shows the plan of a living room and a dining room. The ratio of the length to the width of each room is the same. The living room is 18 feet long and 12 feet wide. Which of the following do you need in order to find the *width* of the dining room?

(1) the perimeter of the living room
(2) the area of the living room
(3) the height of the ceilings
(4) the length of the dining room
(5) the volume of the living room

Answers are on pages 9–10.

Answers to the Predictor Test

1. (2) **73.** *Arithmetic/Applications*

$10^2 = 10 \times 10 = 100$
$3^3 = 3 \times 3 \times 3 = 27$
$100 - 27 = \mathbf{73}$

2. (4) **11.** *Algebra/Applications*

$$
\begin{array}{rcr}
9m - 4 = & & 7m + 18 \\
-7m & & -7m \\
\hline
2m - 4 = & & 18 \\
+4 & & +4 \\
\hline
\dfrac{2m}{2} = & & \dfrac{22}{2} \\
m = & & \mathbf{11}
\end{array}
$$

3. (5) **D, E, B, C, A.** *Arithmetic/Applications*

A = 0.15 = 0.150 kg
B = 0.9 = 0.900 kg
C = 0.85 = 0.850 kg
D = 1.05 = 1.050 kg
E = 0.955 = 0.955 kg

4. (2) **81°.** *Geometry/Skills*

$$
\begin{array}{r}
180° \\
- \ 99° \\
\hline
\mathbf{81°}
\end{array}
$$

5. (4) **\$36.** *Arithmetic/Applications*

$12\% = .12$

$$
\begin{array}{r}
\$32 \\
\times .12 \\
\hline
64 \\
3\ 2 \\
\hline
\$3.84
\end{array}
\qquad
\begin{array}{r}
\$32.00 \\
+ 3.84 \\
\hline
\$35.84 \text{ to nearest} \\
\text{dollar} = \mathbf{\$36}
\end{array}
$$

6. (2) **20.** *Geometry/Applications*

$c^2 = a^2 + b^2$
$c^2 = 12^2 + 16^2$
$c^2 = 144 + 256$
$c^2 = 400$
$c = \sqrt{400}$
$c = \mathbf{20}$

7. (2) $\frac{1}{2}$. *Graphs*

Income from wholesale trade in 1970 was about \$50 billion. Income from services in 1970 was about \$100 billion.

$\frac{50}{100} = \frac{1}{2}$

8. (1) $4\frac{1}{2}$. *Graphs*

Income from services went from about \$100 to about \$450 billion

$$
\begin{array}{r}
4\frac{50}{100} = \mathbf{4\frac{1}{2}} \\
100\overline{)450}
\end{array}
$$

9. (1) **137.** *Arithmetic/Applications*

In order: 84   118   156   195

Mean of middle two values:

$$
\begin{array}{r}
118 \\
+156 \\
\hline
274
\end{array}
\qquad
\begin{array}{r}
\mathbf{137} \\
2\overline{)274}
\end{array}
$$

10. (2) $\frac{1}{3}$. *Arithmetic/Applications*

$$
\begin{array}{ll}
\text{glazed} & 3 \\
\text{cinnamon} & 4 \\
\text{plain} & +5 \\
\hline
\text{total} & 12
\end{array}
\qquad
\begin{array}{ll}
\text{cinnamon} & \dfrac{4}{12} = \dfrac{1}{3} \\
\text{total} &
\end{array}
$$

11. (5) $x = +6$ **and** -4. *Algebra/Applications*

$x^2 - 2x - 24 = 0$

$(x - 6)(x + 4) = 0$

$$
\begin{array}{rcl}
x - 6 = 0 & \quad & x + 4 = 0 \\
+6 \quad +6 & & -4 \quad -4 \\
\hline
x = \mathbf{+6} & \text{and} & x = \mathbf{-4}
\end{array}
$$

12. (3) **\$6400.** *Arithmetic/Problem solving*

$20\% = .2$

$$
\begin{array}{r}
\$800 \\
\times \ 40 \\
\hline
\$32,000
\end{array}
\qquad
\begin{array}{r}
\$32,000 \\
\times \quad .2 \\
\hline
\mathbf{\$6400.0}
\end{array}
$$

13. (1) **5%.** Arithmetic/Problem solving

$$\frac{\text{fee}}{\text{land price}}\ \frac{1,600}{32,000} = \frac{1}{20}$$

$$\frac{1}{\overset{}{\underset{1}{20}}} \times \frac{\overset{5}{\cancel{100}}\%}{1} = 5\%$$

14. (3) **20%.** Arithmetic/Applications

$$\frac{\text{charges}}{\text{purchases}}\ \frac{124}{620} = \frac{1}{5}\quad \frac{1}{\underset{1}{\cancel{5}}} \times \frac{\overset{20}{\cancel{100}}\%}{1} = \mathbf{20\%}$$

15. (1) **2x − 3.** Algebra/Applications

Catherine's age = x

David's age   = 2x

Mary's age   = **2x − 3**

16. (5) **35(8) + 6(12).** Arithmetic/Problem solving

$$c = nr \qquad\qquad c = nr$$
$$c = 35 \times 8 + c = 6 \times 12$$

17. (3) **26.** Arithmetic/Applications

$$A = \tfrac{1}{2}bh$$
$$A = \tfrac{1}{2} \times 8 \times 6.5 = \mathbf{26}$$

18. (4) **2.6 cm.** Arithmetic/Applications

$$\begin{array}{r}\text{point } C \quad 4.2 \text{ cm}\\ \text{point } A \ \underline{-1.6}\\ \mathbf{2.6 \text{ cm}}\end{array}$$

19. (2) **45.** Arithmetic/Applications

$$V = lwh$$
$$V = 20 \times 3 \times 0.75 = \mathbf{45}$$

20. (1) **(2.5 × 55) + (2.5 × 60)** Arithmetic/Problem solving

$$\xleftarrow{\quad\text{Pete}\quad\quad\text{Colin}\quad}\rightarrow$$

$$d = rt \qquad\qquad d = rt$$
$$d = 60 \times 2.5 + d = 55 \times 2.5$$

21. (2) $\frac{80}{200}$. Arithmetic/Applications

$$\frac{\text{Jerseys}}{\text{cows}}\ \frac{80}{200}$$

22. (2) **$ 88.** Arithmetic/Problem solving

$$\frac{\text{drywall}}{}\ \frac{3}{20} = \frac{6}{x}\quad \text{boards}\ \begin{array}{r}14\\ \times 3\\ \hline 42\end{array}\quad \begin{array}{r}\$40\\ 42\\ 2\\ \underline{4}\\ \mathbf{\$88}\end{array}$$

$$3x = 120$$
$$x = \$40$$

23. (4) $\frac{x}{9} - 1 = 3$. Algebra/Applications

24. (3) **68°.** Geometry/Skills

$$\begin{array}{r}180°\\ \underline{-\ 44°}\\ 136°\end{array}\qquad \begin{array}{r}\mathbf{68°}\\ 2\overline{)136°}\end{array}$$

25. (5) **$14.** Algebra/Problem solving

assistant = x

Don   = x + 5

$$20x + 20(x + 5) = 460$$
$$20x + 20x + 100 = 460$$
$$40x + 100 = 460$$
$$\underline{\qquad -100 \quad -100}$$
$$\frac{40x}{40} = \frac{360}{40}$$
$$x \qquad = 9$$
$$x + 5 = 9 + 5 = \mathbf{14}$$

26. (1) $\frac{+3}{2}$. Geometry/Applications

$$C = (x_1, y_1) = (2, 3)$$
$$D = (x_2, y_2) = (10, 15)$$
$$m = \frac{y_2 - y_1}{x_2 - x_1} = \frac{15 - 3}{10 - 2} = \frac{12}{8} = \mathbf{\frac{+3}{2}}$$

27. (3) **5.** Arithmetic/Applications

Rent is approximately $600.
Utilities are approximately $120.

$$\begin{array}{r}\mathbf{5}\ \textbf{times}\\ 120\overline{)600}\end{array}$$

28. (4) **the length of the dining room.** Geometry/Applications

You are missing the corresponding part, the length, of the dining room.

PREDICTOR TEST
Performance Analysis Chart

Directions: Circle the number of each item you got correct on the Predictor Test. Count how many items you got correct in each row; count how many items you got correct in each column. Write the amount correct per row and column as the numerator in the fraction in the appropriate "Total Correct" box. (The denominators represent the total number of items in the row or column.) Write the grand total correct over the denominator **28** at the lower right corner of the chart. (For example, if you got 24 items correct, write 24 so that the fraction reads 24/**28**.) Item numbers in color represent items based on graphic material.

Item Type	Arithmetic (page 21)	Algebra (page 174)	Geometry (page 213)	TOTAL CORRECT
Skills			4, 24	/2
Applications	1, 3, 5, 7, 8, 9, 10, 14, 17, 18, 19, 21	2, 11, 15, 23	6, 26, 27, 28	/20
Problem Solving	12, 13, 16, 20, 22	25		/6
TOTAL CORRECT	/17	/5	/6	/28

The page numbers in parentheses indicate where in this book you can find the beginning of specific instruction about the various areas of mathematics you encountered on the Predictor Test.

Instruction

Mathematics

This section of the book is made up of seven chapters that can help you learn the things you need to know to pass the Mathematics Test.

The chapters (except Chapter 5) are divided into levels, which are in turn divided into individual lessons. You can think of the levels in each chapter as levels of difficulty. In each chapter, the lessons in Level 1 cover the basics. For example, in Level 1 in Chapter 1, the lessons cover how to calculate with whole numbers. In Chapter 2, the Level 1 lessons cover calculating with fractions. In Chapter 6, the Level 1 lessons cover the basics of algebra.

The lessons in each Level 2 are at the middle level of difficulty. To do the lessons in Level 2, you must already know how to do the kinds of problems covered in Level 1. In Level 2 you solve problems by applying the skills covered in Level 1. The lessons in each Level 3 are at the highest level of difficulty. Level 3 lessons present complex problems similar to the problems you will find on the GED.

Introduction

Two Methods for Using the Instruction

There are two methods for working through the math instruction.

Method 1. Work through the instruction section in order. Start with Level 1 in Chapter 1 and go straight through to the end of Level 3 in Chapter 7. As you work through each chapter, you will frequently tackle Level 3 problems. Then take the chapter test. This will give you practice with rather complex problems similar to those on the GED rather early in your period of instruction.

Method 2. Work through all the lessons in every Level 1. Then do the lessons in every Level 2. Finally, do the lessons in every Level 3. (Chapter 5 has no levels; it can be done when you are working through the Level 3's.) This way you can become comfortable with the basics of arithmetic, algebra, and geometry before you begin working with more complex problems.

The Progress Chart at the beginning of the instruction will make it easy for you to keep track of your work and record your performance on each lesson. There are many exercises and quizzes throughout the instruction, so you will have several opportunities to apply and test your understanding of the material you study. At the end of all the instruction is a test on all the material covered in the instruction.

Previews and Reviews. Each level begins with a Preview and ends with a Review. Previews and Reviews are made up of the kinds of problems covered in the levels they begin and end. Previews allow you to test your present skill in the material covered in the level. If you get 80% of the problems in a Preview correct, you may not have to work through that level. To be sure, try to solve the problems in the Review at the end of the level. If you get 80% correct again, you may feel comfortable in skipping the level. However, if you get fewer than 80% correct in either the Preview or the Review, you should study all the lessons in the level.

Decide whether you want to use Method 1 or Method 2 to work through the instruction. You may want to talk it over with a teacher. Then go to Level 1 in Chapter 1 and begin.

MATHEMATICS PROGRESS CHART

Directions: Use the following chart to keep track of your work. When you complete a lesson and have checked your answers to the items in the exercise, circle the number of questions you answered correctly. When you complete a Level Preview, Level Review, or Chapter Quiz, record your score on the appropriate line. The numbers in color represent scores at a level of 80% or better.

Lesson	Page	CHAPTER 1: Whole Numbers
		Level 1: Whole Number Skills
	21	Preview 1 2 3 4 5
1	22	Basic Operations:
		Addition and
		Subtraction 1 2 3 4 5 6 7 8 9 10
2	23	Basic Operations:
		Multiplication and
		Division 1 2 3 4 5 6 7 8 9 10
	25	Level 1 Review 1 2 3 4 5 6 7 8 9 10
		Level 2: Whole Number Applications
	26	Preview 1 2 3 4 5
1	26	Rounding Off 1 2 3 4 5
2	28	Distance and Cost
		Formulas 1 2 3 4 5 6 7 8 9 10
3	29	Powers 1 2 3 4 5
4	30	Square Roots 1 2 3 4 5
5	32	Perimeter 1 2 3 4 5 6 7 8 9 10
6	34	Area 1 2 3 4 5 6 7 8 9 10
7	37	Volume 1 2 3 4 5 6 7 8 9 10
8	39	Properties of
		Numbers 1 2 3 4 5 6 7 8 9 10
9	42	Mean and
		Median 1 2 3 4 5 6 7 8 9 10
10	44	Tables 1 2 3 4 5 6 7 8 9 10
	45	Level 2 Review 1 2 3 4 5 6 7 8 9 10
		Level 3: Whole Number Problem Solving
	46	Preview 1 2 3 4 5
1	47	Key Words 1 2 3 4 5 6 7 8 9 10
2	49	Multistep
		Problems 1 2 3 4 5 6 7 8 9 10
3	50	Multistep Area
		Problems 1 2 3 4 5 6 7 8 9 10
4	53	Estimating 1 2 3 4 5 6 7 8 9 10
5	54	Item sets 1 2 3 4 5 6 7 8 9 10
6	56	Setup Answers 1 2 3 4 5 6 7 8 9 10
	58	Level 3 Review 1 2 3 4 5 6 7 8 9 10
	59	Chapter 1 Quiz 1 2 3 4 5 6 7 8 9 10
		11 12 13 14 15 16 17 18 19 20

Lesson	Page	CHAPTER 2: Fractions
		Level 1: Fraction Skills
	62	Preview 1 2 3 4 5
1	63	Fractions 1 2 3 4 5 6 7 8 9 10
2	65	Interchanging Forms of
		Fractions 1 2 3 4 5 6 7 8 9 10
3	66	Adding and Subtracting
		Fractions 1 2 3 4 5 6 7 8 9 10
4	69	Multiplying and Dividing
		Fractions 1 2 3 4 5 6 7 8 9 10
	72	Level 1 Review 1 2 3 4 5 6 7 8 9 10
		Level 2: Fraction Applications
	72	Preview 1 2 3 4 5
1	73	Finding What Fraction
		One Number Is of
		Another 1 2 3 4 5
2	74	Word Problems with Basic
		Operations 1 2 3 4 5 6 7 8 9 10
3	76	Distance and Cost
		Formulas 1 2 3 4 5 6 7 8 9 10
4	76	Powers and Square Roots 1 2 3 4 5
5	77	Standard Units of
		Measurement 1 2 3 4 5 6 7 8 9 10
6	79	Perimeter, Area, and
		Volume 1 2 3 4 5
7	80	Comparing and Ordering
		Fractions 1 2 3 4 5
8	81	Tables 1 2 3 4 5 6 7 8 9 10
9	82	Ratio 1 2 3 4 5 6 7 8 9 10
10	83	Proportion 1 2 3 4 5
11	85	Proportion Word
		Problems 1 2 3 4 5 6 7 8 9 10
12	86	Probability 1 2 3 4 5 6 7 8 9 10
	88	Level 2 Review 1 2 3 4 5 6 7 8 9 10
		Level 3: Fraction Problem Solving
	89	Preview 1 2 3 4 5
1	89	Multistep Proportion
		Problems 1 2 3 4 5 6 7 8 9 10
2	91	Setup Answers 1 2 3 4 5 6 7 8 9 10
3	94	Mixed Units with
		Perimeter, Area, and
		Volume 1 2 3 4 5 6 7 8 9 10

1 | Whole Numbers

Objective

In this chapter, you will

- add, subtract, multiply, and divide whole numbers
- round off whole numbers
- use distance and cost formulas
- work with powers and square roots of whole numbers
- find the perimeter and area of plane figures; find the volume of solid figures
- apply the Commutative, Distributive, and Associative properties of numbers
- find the mean and the median of a set of numbers
- read tables
- understand key words
- solve multistep problems; solve word problems based on a reading passage
- estimate the best solution to a problem
- set up solutions to work problems

Level 1 Whole Number Skills

Preview

Directions: Solve each item.

1. What is the difference between 9000 and 496?

2. Find the product of 473 and 90.

3. Simplify $\frac{7635}{15}$.

4. Find the sum of 49, 207, 5653, and 28.

5. Divide 9879 by 21.

Check your answers. Correct answers are on page 309. If you have at least four answers correct, do the Level 1 Review on page 25. If you have fewer than four answers correct, study Level 1 beginning with Lesson 1.

Lesson 1   Basic Operations: Addition and Subtraction

You will use the four basic operations of addition, subtraction, multiplication, and division to solve many problems on the GED Test. Following are hints to help you avoid common mistakes with these operations.

Addition

The answer to an addition problem is called the **sum,** or total. When you set up an addition problem, be sure the numbers are lined up correctly. Put the units (or ones) under the units, the tens under the tens, the hundreds under the hundreds, and so on. Notice how the following example is set up.

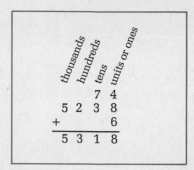

Example 1: 30,854 + 6672 + 53 = ?

Step 1. Line up the numbers. Put the units under the units, the tens under the tens, and so on.

Step 2. Add each column.

$$
\begin{array}{r}
30,854 \\
6,672 \\
+53 \\
\hline
37,579
\end{array}
$$

Subtraction

The answer to a subtraction problem is called the **difference.** When you take one number from another, be sure to put the larger number on top. Line up the numbers with units (or ones) under units, tens under tens, and so on. Then borrow carefully. It is a good idea to check subtraction problems. Add the answer to the lower number. The sum should equal the top number of the subtraction problem.

Look carefully at the borrowing in the next example.

Example 2: Find the difference between 8,249 and 50,600.

$$
\begin{array}{r}
5\ 0,6\ 0\ 0 \\
-\ 8,2\ 4\ 9 \\
\hline
\end{array}
\qquad
\begin{array}{r}
\overset{9}{}\\[-6pt]
\overset{4\ 10\ 5\ 10\ 10}{50,600} \\
-\ 8,249 \\
\hline
42,351
\end{array}
\qquad
\begin{array}{r}
8,249 \\
+42,351 \\
\hline
50,600
\end{array}
$$

Step 1. Put the larger number, 50,600, on top. Put units under units, tens under tens, and so on.

Step 2. Borrow and subtract.

Step 3. Check by adding the answer to the lower number.

Lesson 1   Exercise

Directions: *For items 1 and 2,* choose the correct setup for each problem. Then solve each problem.

1. Find the sum of 9015, 493, and 76.

   a.   9015   b.   9,015   c.   9015
            493   + 76,493   493
          + 76   + 76

2. What is the difference between 22,500 and 6,087?

   a.   22,500   b.   22,500   c.   6,087
        − 60,87   − 6,087   − 22,500

3. Take 763 from 5030.

4. What is the sum of 9704, 86, and 10,471?

5. Simplify 78 + 4,062 + 529.

6. Find the difference between 8,346 and 42,003.

7. Subtract 11,954 from 18,206.

8. Find the sum of 428, 61, 593, and 7.

9. Find the difference between 30,005 and 19,472.

10. What is the sum of 365 and 7,048?

Answers are on page 309.

Lesson 2   Basic Operations: Multiplication and Division

Multiplication

The answer to a multiplication problem is called the **product.** To find a product, put the number with more digits on top. When you multiply by the digit in the units place, be sure to start the **partial product** under the units. When you multiply by the digit in the next column moving left (the tens digit), be sure to start the partial product under the tens.

Example 1: Find the product of 876 and 204.

Step 1. Put 876 on top and find the partial products. Notice how the zero holds a place in the second partial product.

Step 2. Add the partial products.

$$
\begin{array}{r}
876 \\
\times\,204 \\
\hline
3\,504 \\
175\,20 \\
\hline
178{,}704
\end{array}
\left.\begin{array}{l} \\ \\ \end{array}\right\} \text{partial products}
$$

Division

The answer to a division problem is called the **quotient.** There are three common ways to write division problems. The problem "20 divided by 4 equals 5" can be written in any of the following ways.

$$20 \div 4 = 5 \qquad \frac{20}{4} = 5 \qquad 4\overline{)20}^{\,5}$$

    To find a quotient, repeat these four steps until you finish the problem: (1) divide, (2) multiply, (3) subtract, and (4) bring down the next number.

    It is important to line up division problems carefully. When you begin to find a quotient, you must have a digit in the answer for each remaining digit in the problem.

    To check a division problem, multiply the answer by the number you divided by. The product should equal the number you divided into. In the following example, be sure you understand how to get each number in the solution.

Example 2: Find the quotient of 4915 divided by 16.

Notice how the zero in the quotient holds a place. The zero shows that 16 does not divide into 11.

Notice that 16 does not divide evenly into 115 in the final step. The number left (3) is the **remainder.** When you check your answer, remember to add the remainder to the product of 16×307.

$$
\begin{array}{r}
307\ \text{r}3 \\
16\overline{)4915} \\
\underline{48} \\
11 \\
\underline{0} \\
115 \\
\underline{112} \\
3
\end{array}
$$

$$16 \times 307 = 4912$$
$$4912 + 3 = 4915$$

    Perhaps the most important tool in mathematics is the multiplication table. Decide for yourself whether your knowledge of the multiplication table is rusty. If it is, take the time now to practice the multiplication facts you are not sure about.

Lesson 2   Exercise

Directions: For items 1 and 2, choose the correct setup for each problem. Then solve each problem.

1. Simplify $\frac{9636}{12}$.

   a. 9636
      $\times\ \underline{\ 12}$

   b. $9636\overline{)12}$

   c. $12\overline{)9636}$

2. Find the product of 704 and 18.

   a. 704
      $\times\ \underline{\ 18}$

   b. $704\overline{)18}$

   c. $18\overline{)704}$

3. Find the product of 536 and 800.

4. Find the quotient of 7256 divided by 8.

5. Simplify $\frac{3588}{46}$.

6. Divide 3042 by 78.

7. What is the product of 34 times 230?

8. What is the product of 19 and 40,570?

9. 4424 ÷ 316 = ?

10. Find the quotient of 2392 divided by 523.

Answers are on page 309.

Level 1   Review

Directions: Solve each problem.

1. What is the sum of 890, 23, 4017, and 605?

2. Find the quotient of 11,043 divided by 12.

3. What is the difference between 12,050 and 9947?

4. Find the product of 76 and 308.

5. Simplify $\frac{24,768}{8}$.

6. What is the sum of 108,270, 33,580, and 6,095?

7. What is the product of 208 and 675?

8. Find the difference between 306,471 and 28,295.

9. Subtract 29,460 from 508,000.

10. Divide 25,728 by 32.

Check your answers. Correct answers are on page 309. If you have at least eight answers correct, go to the Level 2 Preview. If you have fewer than eight answers correct, go back to Lesson 1 and study Level 1.

Level 2   Whole Number Applications

Preview

Directions: Solve each problem.

1. Round off 496,273 to the nearest ten-thousand.

2. A pilot flew for seven hours at an average speed of 435 mph. How far did he fly?

3. What is the area of the figure pictured below?

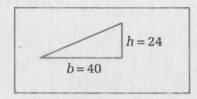

4. Jack sold five used cars one month. The prices were $8470, $1950, $6075, $3080, and $2155. What was the median price of the cars he sold that month?

5. Which of the following expressions is the same as $7(20 - 1)$?
   (1) $7 \times 20 - 1$
   (2) $20(7 - 1)$
   (3) $(7 \times 20) - (7 \times 1)$
   (4) $(20 \times 7) - (20 \times 1)$
   (5) $\dfrac{(20 - 1)}{7}$

Check your answers. Correct answers are on page 310. If you have at least four answers correct, do the Level 2 Review on pages 45–46. If you have fewer than four answers correct, study Level 2 beginning with Lesson 1.

Lesson 1   Rounding Off

Sometimes whole numbers are more exact than they need to be. For example, Mike weighs 178 pounds. We can say that he weighs about 180 pounds. 180 is 178 rounded off to the nearest ten. Rounding off is a way of making numbers easier to read and easier to use. You will use rounding off later when you learn to estimate answers.

To round off whole numbers, you must know the place value of every digit in a whole number. Below are the names of the first ten whole-number places.

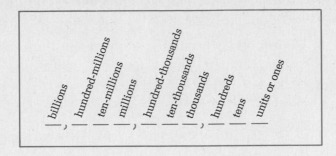

Follow these steps to round off a whole number.

1. Mark the digit in the place you want to round off to.

2. If the digit to the right of the marked digit is more than 4, add 1 to the marked digit.

3. If the digit to the right of the marked digit is less than 5, leave the digit you marked as it is.

4. Replace the digits to the right of the digit you marked with zeros.

Example 1: Round off 487 to the nearest ten.

Step 1. Mark the digit in the tens place, 8.   4<u>8</u>7

Step 2. Since the digit to the right of 8 is more   490
than 4, add 1 to 8. Put 9 in the tens place and
write 0 in the units place.

Example 2: Round off 196,275 to the nearest
ten-thousand.

Step 1. Mark the digit in the ten-thousands   1<u>9</u>6,275
place, 9.

Step 2. Since the digit to the right of 9 is more   200,000
than 4, add 1 to 9. To add 1 to 9, you must carry
1 over to the hundred-thousands column. Put
zeros to the right.

Lesson 1   Exercise

1. Round off each number to the nearest ten.
   78   164   3198   2433

2. Round off each number to the nearest hundred.
   847   1273   6580   351

3. Round off each number to the nearest thousand.
   3196   41,826   28,752   149,628

4. Round off each number to the nearest hundred-thousand.

   777,500   316,450   567,300   3,470,992

5. Round off each number to the nearest million.

   5,648,000   12,387,000   32,479,000   18,750,000

Answers are on page 310.

Lesson 2   Distance and Cost Formulas

You will use many formulas in this book. (There is a complete list of the formulas you will see on the GED Test on page 372.) A **formula** is a kind of shorthand mathematical instruction in which letters represent numbers. One useful formula is $d = rt$, where d = distance, r = rate, and t = time.

Two letters standing next to each other must be multiplied together. In words, the formula $d = rt$ means, "Distance is equal to the rate multiplied by the time." To use a formula, replace the letters with the numbers from a problem, and follow the instruction in the formula.

Example 1: Mike drove on a highway for five hours at an average speed of 55 mph. How far did he drive?

Step 1. Replace r with 55 and t with 5 in the formula $d = rt$.

$d = rt$
$d = 55 \times 5$

Step 2. Multiply 55 by 5. Notice that the distance is measured in miles.

$d = 275$ mi.

Another useful formula is $c = nr$, where c = total cost, n = number of units, r = rate or cost per unit. In words, the formula $c = nr$ means, "Cost is equal to the number of units multiplied by the rate."

Example 2: Ruth bought four cans of tuna at the cost of $1.29 per can. Find the total cost of the tuna.

Step 1. Replace n with 4 and r with $1.29 in the formula $c = nr$.

$c = nr$
$c = 4 \times \$1.29$

Step 2. Multiply $1.29 by 4.

$c = \$5.16$

Lesson 2   Exercise

Directions: Use the formula $d = rt$ or the formula $c = nr$ to solve each of the following problems.

1. Alfredo walked for three hours at a speed of 4 mph. How far did he walk?

2. The Johnsons drove for four hours at an average speed of 65 mph. How far did they drive?

3. A plane flew for five hours at an average speed of 475 mph. How far did the plane fly?

4. Sarah drove in city traffic for three hours at an average speed of 15 mph. How far did she drive?

5. Manny bought three shirts at $18 each. Find the total cost of the shirts.

6. At $3.60 per pound, what is the cost of 5 pounds of meat?

7. Find the cost of a dozen reams of typing paper at $6 a ream.

8. The Greenport Community Center bought 30 sets of desks and chairs at the rate of $65 a set. Find the total cost of the furniture.

9. Fontaine drove for six hours without stopping at an average speed of 55 mph. How many miles did she drive?

10. If milk is being sold at $1.51 a quart, how much will six quarts cost?

Answers are on page 310.

Lesson 3 Powers

A **power** is a product of a number multiplied by itself one or more times. 7^2 means, "Seven to the second power." The number 7 is the **base.** The number 2 is the **exponent.** The most common power is the power of two. When a number is raised to the second power, we say it is **squared.**
    Follow these steps to find a power.

1. Write the base as many times as the number of the exponent.
2. Multiply the base by itself.

Example 1: What is the value of 7^2?

Step 1. The exponent is 2. Write 7 two times. $7^2 = 7 \times 7$

Step 2. Multiply 7 by 7. $7^2 = 49$

Example 2: What is the value of 5^3?

Step 1. The exponent is 3. Write 5 three times. $5^3 = 5 \times 5 \times 5$

Step 2. Multiply 5 by 5 by 5. First find 5×5 $5^3 = 125$
= 25. Then find $25 \times 5 = 125$.

    There are some special cases with powers: 1 to any power = 1, a number to the first power is that number, and a number to the zero power = 1.

You can also combine powers. First find each power separately. Then add or subtract from left to right.

Example 3: What is the value of $3^4 - 6^2 + 2^3$?

Step 1. Find each power separately.

$$3^4 = 3 \times 3 \times 3 \times 3 = 81$$
$$6^2 = 6 \times 6 = 36$$
$$2^3 = 2 \times 2 \times 2 = 8$$

Step 2. Subtract 36 from 81. Then add 8.

$$81 - 36 = 45$$
$$45 + 8 = 53$$

Lesson 3   Exercise

Directions: Find the value of each of the following expressions.

1. $2^4 = ?$ $9^2 = ?$ $3^3 = ?$ $8^1 = ?$

2. $13^2 = ?$ $50^2 = ?$ $6^3 = ?$ $12^0 = ?$

3. $2^5 = ?$ $1^5 = ?$ $16^2 = ?$ $40^2 = ?$

4. $5^2 - 2^3 = ?$ $8^2 + 3^3 = ?$ $10^2 - 4^2 + 5^2 = ?$

5. $4^3 + 6^1 - 2^4 = ?$ $12^2 - 5^0 - 3^2 = ?$ $10^3 - 10^2 = ?$

Answers are on page 310.

Lesson 4   Square Roots

Roots are the opposite of powers. On the GED Test, you will have to solve problems with **square roots.** A square root is the opposite of a number to the second power. The sign for square root is $\sqrt{}$. To find a square root, find a number that when multiplied by itself gives the number inside the $\sqrt{}$ sign.

Example 1: Find the value of $\sqrt{36}$.

Ask yourself, "What number times itself is 36?" $\sqrt{36} = 6$
$6 \times 6 = 36$. So, 6 is the square root of 36.

Following is a list of common square roots.

$\sqrt{1} = 1$	$\sqrt{49} = 7$	$\sqrt{169} = 13$	$\sqrt{2500} = 50$
$\sqrt{4} = 2$	$\sqrt{64} = 8$	$\sqrt{196} = 14$	$\sqrt{3600} = 60$
$\sqrt{9} = 3$	$\sqrt{81} = 9$	$\sqrt{225} = 15$	$\sqrt{4900} = 70$
$\sqrt{16} = 4$	$\sqrt{100} = 10$	$\sqrt{400} = 20$	$\sqrt{6400} = 80$
$\sqrt{25} = 5$	$\sqrt{121} = 11$	$\sqrt{900} = 30$	$\sqrt{8100} = 90$
$\sqrt{36} = 6$	$\sqrt{144} = 12$	$\sqrt{1600} = 40$	$\sqrt{10,000} = 100$

You can find the square root of a number with a method that uses averages. Suppose that you did not know that $\sqrt{144} = 12$. When you divide a number by its square root, the answer equals the square root. ($144 \div 12 = 12$.) Guess an answer close to 144. A good guess is 10 because $10 \times 10 = 100$, which is fairly close to 144. Divide 144 by 10. $144 \div 10 = 14$ plus a remainder. Average the guess, 10, and the answer to the division problem, 14. $10 + 14 = 24$. $24 \div 2 = 12$, which is the correct answer.

Follow these steps to find the square root of a large number.

1. Guess an answer.

2. Divide the guess into the large number.

3. Average the guess and the answer to the division problem.

4. Check.

Example 2: Find the value of $\sqrt{1024}$.

Step 1. Guess. In the list of square roots, $\sqrt{900}$ = 30. 30 is too small, but it is easy to divide by.

$\sqrt{900} = 30$

Step 2. Divide 1024 by 30. Drop the remainder.

$$\begin{array}{r} 34 \\ 30\overline{)1024} \\ 90 \\ \hline 124 \\ 120 \end{array}$$

Step 3. Find the average of 30 and 34.

$$\begin{array}{r} 30 \\ +34 \\ \hline 64 \end{array} \qquad \begin{array}{r} 32 \\ 2\overline{)64} \end{array}$$

Step 4. Check. Multiply 32 by 32. 32 is the square root of 1024.

$$\begin{array}{r} 32 \\ \times 32 \\ \hline 64 \\ 96 \\ \hline 1024 \end{array}$$

When you use this method to find square roots, always guess a number that ends in zero. It is easier and faster to divide by these numbers. If the average is not the square root of the number, use the average as a new guess, and try again.

Lesson 4   Exercise

Directions: Find the value of each square root.

1. $\sqrt{289}$ = ? $\sqrt{784}$ = ? $\sqrt{1444}$ = ? $\sqrt{484}$ = ?

2. $\sqrt{1521}$ = ? $\sqrt{1849}$ = ? $\sqrt{529}$ = ? $\sqrt{2704}$ = ?

3. $\sqrt{2025}$ = ? $\sqrt{961}$ = ? $\sqrt{4761}$ = ? $\sqrt{3481}$ = ?

4. $\sqrt{3844} = ?$ $\sqrt{8649} = ?$ $\sqrt{5929} = ?$ $\sqrt{2401} = ?$

5. $\sqrt{4096} = ?$ $\sqrt{8836} = ?$ $\sqrt{6084} = ?$ $\sqrt{7056} = ?$

Answers are on page 310.

Lesson 5   Perimeter

There are three geometric figures which you will see often in this book: the rectangle, the square, and the triangle.

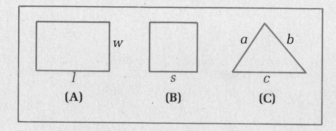

Figure A is a rectangle. A rectangle has four sides and four right angles. A right angle is a square corner. You will learn more about angles later. The sides across from each other are equal. The longer side is called the length. The shorter side is called the width. In Figure A, l is the length, and w is the width.

Figure B is a square. A square has four equal sides and four right angles. A square is referred to by its side. In Figure B, s stands for the length of the side.

Figure C is a triangle. A triangle has three sides. All three sides may be the same length. Two of the sides may be the same while the third is different. Or, all three sides may be different. In Figure C, the letters a, b, and c stand for the three sides.

Perimeter is a measure of the distance around a flat figure such as a rectangle, a square, or a triangle. To find the perimeter of a flat figure, add the measurements of each side. You can also use a formula to find the perimeter. Following are the formulas you will see on the GED Test for these three figures.

Perimeter (P) of

a rectangle $P = 2l + 2w$, where l = length and w = width
a square $P = 4s$, where s = side
a triangle $P = a + b + c$, where a, b, and c are the sides of the triangle

Perimeter is always measured in linear units, such as inches, feet, yards, or meters. In problems where no units are given, the perimeter is simply a number.

Example 1: Find the perimeter of a rectangle with a length of 20 feet and a width of 12 feet.

Step 1. Replace *l* with 20 and *w* with 12 in the formula for the perimeter of a rectangle.

$P = 2l + 2w$
$P = 2 \times 20 + 2 \times 12$

Step 2. Work out the formula. Do both multiplications before you add.

$P = 40 + 24$
$P = 64$ ft.

Example 2: Find the perimeter of the square pictured at the right.

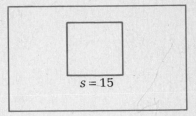

Step 1. Replace *s* with 15 in the formula for the perimeter of a square.

$P = 4s$
$P = 4 \times 15$

Step 2. Evaluate the formula. Since there are no units, the perimeter is simply 60.

$P = 60$

Example 3: Find the perimeter of the triangle pictured at the right.

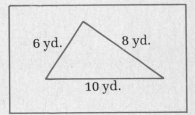

Step 1. Replace *a* with 6, *b* with 8, and *c* with 10 in the formula for the perimeter of a triangle.

$P = a + b + c$
$P = 6 + 8 + 10$

Step 2. Evaluate the formula. The unit of measurement is yards.

$P = 24$ yd.

Lesson 5   Exercise

Directions: Solve each problem.

1. Find the perimeter of a square with a side of 11 feet.

2. What is the perimeter of a rectangle with a length of 9 meters and a width of 7 meters?

3. Find the perimeter of the rectangle pictured at the right.

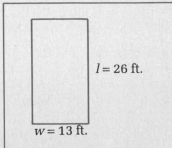

4. What is the perimeter of a square with a side of 20 inches?

5. Find the perimeter of the triangle pictured at the right.

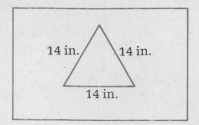

6. Find the perimeter of the square pictured at the right.

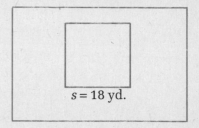

7. What is the perimeter of the rectangle pictured at the right?

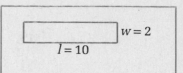

8. Find the perimeter of the triangle pictured at the right.

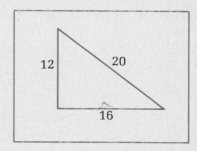

9. The sides of a triangle measure 7 centimeters, 8 centimeters, and 9 centimeters. Find the perimeter of the triangle.

10. What is the perimeter of the rectangle shown at the right?

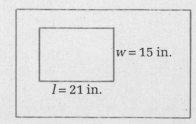

Answers are on page 311.

Lesson 6 Area

Area is a measure of the amount of space inside a flat figure. Area is always measured in square units, such as square inches, square feet, square yards, or square meters. You may see these units abbreviated with a second power. For example, ft.^{2} is the same as square feet.

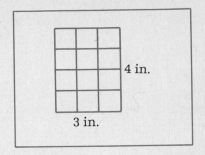

The rectangle shown at the right has an area of 12 in.[2] This means that there are 12 one-inch squares on the rectangle.

Following are the formulas for the areas of four of the figures you will encounter on the GED Test.

Area (A) of
a rectangle $A = lw$, where l = length and w = width
a square $A = s^2$, where s = side
a parallelogram $A = bh$, where b = base and h = height
a triangle $A = \frac{1}{2}bh$, where b = base and h = height

Example 1: Find the area of a rectangle with a length of 12 feet and a width of 9 feet.

Step 1. Replace l with 12 and w with 9 in the formula for the area of a rectangle.
$A = lw$
$A = 12 \times 9$

Step 2. Evaluate the formula. Notice that the area is measured in square feet.
$A = 108$ ft.[2]

Example 2: Find the area of the square pictured at the right.

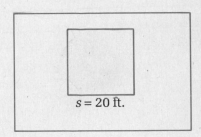

$s = 20$ ft.

Step 1. Replace s with 20 in the formula for the area of a square.
$A = s^2$
$A = 20^2$

Step 2. Evaluate the formula.
$A = 20 \times 20$
$A = 400$ ft.[2]

Example 3: Find the area of the parallelogram pictured at the right.

$h = 12$ in.
$b = 60$ in.

Step 1. Replace b with 60 and h with 12 in the formula for the area of a parallelogram.
$A = bh$
$A = 60 \times 12$

Step 2. Evaluate the formula.
$A = 720$ in.[2]

Example 4: Find the area of the triangle pictured at the right.

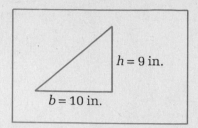

Step 1. Replace b with 10 and h with 9 in the formula for the area of a triangle.

$A = \frac{1}{2}bh$

$A = \frac{(10 \times 9)}{2}$

Step 2. Evaluate the formula.

$A = \frac{90}{2}$

$A = 45 \text{ in.}^2$

The height of the triangle in the last example was inside the triangle. The height of a triangle is always **perpendicular** to the base. This means that the height and base meet to form a square corner. Look carefully at each of the following triangles to see the positions of the height and the base.

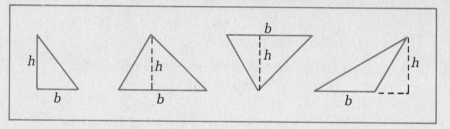

Lesson 6   Exercise

Directions: Solve each problem.

1. Find the area of a rectangle with a length of 16 inches and a width of 9 inches.

2. What is the area of the square shown at the right?

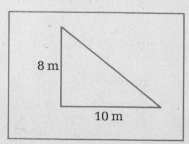

3. What is the area of the triangle shown at the right?

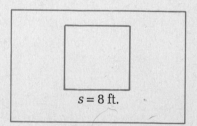

4. What is the area of a parallelogram with a base of 20 and a height of 14?

5. Find the area of a square with a side of 15 inches.

6. Find the area of a triangle with a base of 30 feet and a height of 18 feet.

7. What is the area of the rectangle pictured at the right?

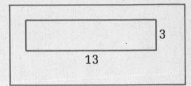

8. Find the area of the parallelogram pictured at the right.

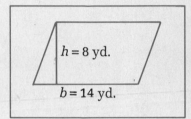

9. What is the area of the triangle shown at the right?

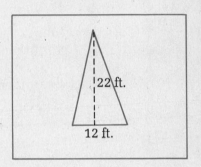

10. Find the area of a square with a side of 16 meters.

Answers are on page 311.

Lesson 7   Volume

Volume is a measure of the amount of space or the capacity inside a three-dimensional figure. The unit of measurement for volume is always cubic units, such as cubic inches or cubic meters. These units are sometimes abbreviated with exponents. Cubic feet is the same as ft.³ Two of the three-dimensional figures you may see on the GED Test are the cube and the rectangular container.

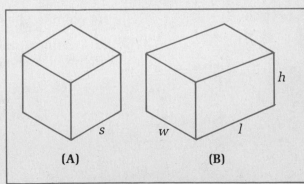

Figure A is a cube. A cube is a three-dimensional figure made up of right angles. Each edge or side has the same measurement. The cube is usually named by the length of its side.

Figure B is a rectangular container or rectangular solid. A rectangular solid is also made up of right angles, but the sides are not all the same. The different measurements are usually called the length, the width, and the height.

Following are the formulas for the volume of these two figures. Notice that for both the cube and the rectangular container you are multiplying the area of the plane figure by its height.

Volume (V) of a:
  cube $V = s^3$, where s = side
  rectangular container $V = lwh$, where l = length, w = width, and h = height

Example 1: Find the volume of a cube with a side of 2 feet.

Step 1. Replace s with 2 in the formula for the volume of a cube. $V = s^3$
$V = 2^3$

Step 2. Evaluate the formula. Notice that the volume is measured in cubic feet. $V = 2 \times 2 \times 2$
$V = 8$ ft.^{3}

Example 2: Find the volume of the rectangular container pictured at the right.

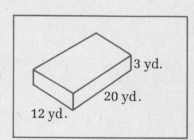

Step 1. Replace l with 20, w with 12, and h with 3 in the formula for the volume of a rectangular container. $V = lwh$
$V = 20 \times 12 \times 3$

Step 2. Evaluate the formula. $V = 720$ yd.^{3}

Lesson 7   Exercise

Directions: Solve each problem.
  1. Find the volume of a cube with a side of 6 feet.

  2. Find the volume of the rectangular container pictured at the right.

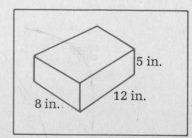

3. Find the volume of a rectangular solid with a length of 20 feet, a width of 5 feet, and a height of 6 feet.

4. What is the volume of the cube shown at the right?

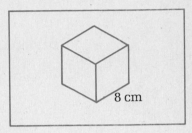

8 cm

5. Find the volume of a cube with a side of 1 yard.

6. What is the volume of a rectangular container which is 100 feet long, 24 feet wide, and 8 feet high?

7. Find the volume of the rectangular solid pictured at the right.

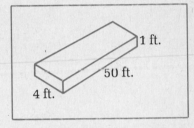

1 ft.

50 ft.

4 ft.

8. What is the volume of a cube with a side that measures 3 feet?

9. Find the volume of a rectangular container with a length of 18 yards, a width of 11 yards, and a height of 30 yards.

10. Find the volume of a cube with a side of 27 centimeters.

Answers are on page 311.

Lesson 8   Properties of Numbers

In this section, you will learn about three characteristics of mathematical operations. On the GED Test, you will see these properties applied to answer choices.

The **Commutative Property** states that when you add two numbers, you can add them in either order. The answers will be the same.

The commutative property for addition is $a + b = b + a$, where a and b are two numbers.

Example 1: Which expression below is equal to 12 + 14?

    (1) 12 × 14   **(4)** 12 − 14
    (2) 14 + 12   **(5)** 12(14)
    (3) 12 × 12

The correct choice is **(2)** 14 + 12. Both expressions equal 26.

This property also applies to multiplication. The commutative property for multiplication is $ab = ba$, where a and b are two numbers.

The commutative property is *not* true for subtraction or division.

Example 2: Which expression equals 15×3?
       (1) 3×15
       (2) $\frac{15}{3}$
       (3) $15 + 3$
       (4) 15×15
       (5) $3 + 15$

Choice **(1)** 3×15 is correct. Both expressions equal 45.

The **Associative Property** states that when you add three or more numbers, you can add them in any order.

The associative property for addition is $(a + b) + c = a + (b + c)$, where a, b, and c are three numbers.

The associative property also works for multiplication.

The associative property for multiplication is $(ab)c = a(bc)$, where a, b, and c are three numbers.

Notice in both cases how the parentheses are used to group numbers together.

The associative property is *not* true for subtraction or division.

Example 3: Choose the expression below which equals $7 (3 \times 10)$.
       (1) $7 + (3 \times 10)$
       (2) $(7 \times 3) + 10$
       (3) $7 + 3 + 10$
       (4) $(7 \times 3) \times 10$
       (5) $10 + 7 \times 3$

Choice **(4)** $(7 \times 3) \times 10$ is correct. $7(3 \times 10) = 7(30) = 210$. $(7 \times 3) \times 10 = (21) \times 10 = 210$. Both expressions equal 210.

The **Distributive Property** states that a number standing next to a set of parentheses must be multiplied by each number inside the parentheses.

The distributive property can be written two ways:
$a(b + c) = ab + ac$ or $a(b - c) = ab - ac$,
where a, b, and c are three numbers.

Example 4: Which expression is equal to $5(6 + 7)$?
       (1) $(5 \times 6) + 7$ **(4)** $5 + (6 \times 7)$
       (2) $(5 \times 6) + (5 \times 7)$ **(5)** $5 + (7 + 6)$
       (3) $5 \times 6 \times 7$

Choice **(2)** $(5 \times 6) + (5 \times 7)$ is correct. If you combine the numbers inside the parentheses in the expression $5(6 + 7)$, you get $5(13) = 65$. If you work out the two separate parts of $5 \times 6 + 5 \times 7$, you get $30 + 35 = 65$. Both expressions equal 65.

Lesson 8   Exercise

Directions: Choose the correct equivalent for each of the following expressions.

1. Which expression is equal to 9×6?

    (1) $9 + 6$ **(4)** $6(9 + 6)$
    (2) $9 - 6$ **(5)** 6×6
    (3) 6×9

2. Which of the following expressions equals $3 \times (12 \times 8)$?

    (1) $(3 \times 12) + 8$ **(4)** $(3 \times 12) + (3 \times 8)$
    (2) $(3 \times 12) \times 8$ **(5)** $(12 \times 8) + 3$
    (3) $3 + (12 \times 8)$

3. Which expression is the same as $(15 + 4) + 10$?

    (1) $15 \times 4 \times 10$ **(4)** $(15 + 4) \times 10$
    (2) $15(4 + 10)$ **(5)** $15 + (4 + 10)$
    (3) $4(15 + 10)$

4. Choose the expression which equals $7(8 - 1)$.

    (1) $(7 \times 8) - (7 \times 1)$ **(4)** $(7 \times 8) + (7 \times 1)$
    (2) $7 \times 8 \times 1$ **(5)** $(7 \times 8) + 1$
    (3) $(7 \times 8) - 1$

5. Which of the following expressions is the same as $20 + 30$?

    (1) 30×20 **(4)** $30 + 20$
    (2) $20 - 30$ **(5)** $60 - 30$
    (3) $20 + 20$

6. Which expression is equal to $(2 \times 19) + (2 \times 7)$?

    (1) $2 \times 19 \times 7$ **(4)** $10(2 + 7)$
    (2) $7(2 \times 19)$ **(5)** 14×19
    (3) $2(19 + 7)$

7. Choose the expression which equals $(6 + 2) + 12$.

    (1) $(6 \times 2) + 12$ **(4)** $6(2 + 12)$
    (2) $12 \times 2 \times 6$ **(5)** $6 + (2 + 12)$
    (3) $(12 \times 2) + 6$

8. Which of the following expressions equals $20(12 + 3)$?

    (1) $20 \times 12 + 3$ **(4)** $20 \times 12 \times 3$
    (2) $3 + 12 + 20$ **(5)** $\dfrac{(12 + 3)}{20}$
    (3) $(20 \times 12) + (20 \times 3)$

9. Bill drove in the morning for three hours at an average speed of 50 mph and again in the afternoon for 4 hours at the same speed. Which expression tells the total distance he drove?

    (1) $50 \times 3 \times 4$
    (2) $50 + (3 \times 4)$
    (3) $(50 + 3) + 50 + 4$
    (4) $50(3 + 4)$
    (5) $\dfrac{(3 + 4)}{50}$

10. Which expression tells the total value of five cartons of soup, each of which holds 24 cans which cost $.70 apiece?

    (1) $5(24 \times .70)$
    (2) $(5 \times 24) + .70$
    (3) $5 + 24 + .70$
    (4) $24(5 + .70)$
    (5) $5(24 + .70)$

Answers are on page 311.

Lesson 9   Mean and Median

There are two ways to find a "middle" value for a group of numbers. One way is called the **mean** or average. The other way is called the **median**. The mean is the more common method.

Follow these steps to find a mean of a group of numbers.

1. Add the numbers.
2. Divide by the number of numbers in the group.

Example 1: Sally works three days a week as a waitress. She made $24 on Thursday, $42 on Friday, and $48 on Saturday in tips. Find the mean amount of her tips for those days.

Step 1. Find the sum of the tips.

$$\begin{array}{r} \$\ 24 \\ 42 \\ +\ \underline{48} \\ \$114 \end{array}$$

Step 2. Divide the sum by the number of days, 3.

$$\begin{array}{r} \$\ 38 \\ 3)\overline{\$114} \end{array}$$

A median is the actual middle value of a group of numbers arranged in order.

Follow these steps to find a median of a group of numbers.

1. Put the numbers in order from smallest to largest.
2. The number in the middle is the median.
3. If there are two numbers in the middle, find the mean of those two numbers.

Example 2: Find the median for the following set of numbers: 287, 496, 317, and 409.

Step 1. Put the numbers in order from smallest to largest.   287   317   409   496

Step 2. Since the two numbers, 317 and 409, are in the middle, find the mean of 317 and 409. The median is 363.

$$\begin{array}{r} 317 \\ +409 \\ \hline 726 \end{array} \qquad \begin{array}{r} 363 \\ 2\overline{)726} \end{array}$$

Lesson 9   Exercise

Directions: Solve each problem.

1. In June, Al's electricity bill was $14. In July, it was $22. In August, it was $18. Find the mean electricity bill for those months.

2. What was the median electricity bill for the situation described in problem 1?

3. George Johnson weighs 185 pounds. His wife June weighs 138 pounds. Their daughter Jan weighs 97 pounds. Their son Joe weighs 88 pounds. What is the mean weight for the members of the Johnson family?

4. Deborah took five math tests. Her scores were 65, 86, 79, 92, and 88. Find her mean score for the five tests.

5. Paul is a traveling salesman. Monday he drove 284 miles; Tuesday, 191 miles; Wednesday, 297 miles; Thursday, 162 miles; and Friday, 256 miles. Find the mean distance he drove each day.

6. On Monday Juan bought 11 gallons of gas. On Wednesday he bought 9 gallons, and on Thursday he bought 13 gallons. Find the mean number of gallons he bought each time.

7. On Wednesday 213 people went to the Greenport Community Center Talent Show. Attendance on Thursday was 191; on Friday, 289; and on Saturday, 303. Find the mean attendance at the show each night.

8. Find the median attendance for the show described in problem 7.

9. Maxine priced cans of tuna at five different stores around town. She found the following prices for the same brand and can size: $1.29, $1.49, $1.16, $1.25, and $1.39. What was the median price for a can of tuna?

10. In a year Mr. Munro made $24,800. Mrs. Munro made $22,500. Their daughter Lee made $6,000, and their son Nick made $4,200. What was the mean income for each member of the Munro family?

Answers are on pages 311–312.

Lesson 10   Tables

A **table** is a set of numbers in rows or columns. A table is an orderly way of presenting detailed numerical information.

The table below tells how calories are used up during selected activities.

How Calories Are Used by a 155-Pound Person	
Activity	Calories Per Hour
Walking at 2 mph	140
Mopping floors	270
Splitting wood	450
Jogging at 5 mph	540

Example: How many more calories does a 155-pound person use in one hour of splitting wood than in one hour of walking at 2 mph?

Subtract the calorie amounts for the two activities. $450 - 140 = 310$

Lesson 10   Exercise

Items 1 to 10 are based on the following table.

	Per Capita Expenditures	Per Capita Taxes
New York	$1504	$1164
Massachusetts	1440	1137
Florida	856	694
Illinois	1068	800
Texas	898	705
California	1449	1098
U.S. average	1221	902

Source: U.S. Bureau of Census

1. Which state in the table had the highest per capita expenditures?

2. For the states shown in the table, what was the difference between the highest amount and the lowest amount in per capita expenditures?

3. Find the median per capita expenditures for the states shown in the table.

4. Texas was how far below the U.S. average in per capita expenditures?

5. Massachusetts was how far above the U.S. average in per capita expenditures?

6. What were the mean per capita taxes for the states shown in the table?

7. Capital City is in a state where the per capita expenditures were at the national average. The population of Capital City is 300,000. Find, to the nearest ten million, the amount spent by the state on Capital City.

8. Find the difference between the per capita expenditures and the per capita taxes in Florida.

9. Greenport, with a population of 15,000, is in a state where the taxes are at the U.S. average. To the nearest $100,000, how much did the people of Greenport pay altogether in state taxes?

10. For the states shown in the table, what is the difference between the highest and the lowest per capita state taxes?

Answers are on page 312.

Level 2   Review

Directions: Solve each problem.

1. Round off 2,386,475 to the nearest hundred thousand.

2. Find the cost of 12 gallons of gasoline at the price of $1.20 per gallon.

3. Simplify the expression $15^2 - 10^2 + 25^1$.

4. Find the value of $\sqrt{3481}$.

5. The sides of a triangle measure 8 meters, 11 meters, and 14 meters. Find the perimeter of the triangle.

6. What is the area of the rectangle pictured at the right?

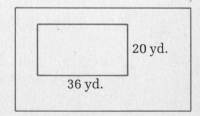

7. Find the volume of a rectangular container with a length of 25 feet, a width of 12 feet, and a height of 8 feet.

8. Ernie shipped packages with the following weights: 12 pounds, 36 pounds, 19 pounds, and 25 pounds. Find the mean weight of the packages.

9. What is the median weight of the packages in Problem 8?

10. Which expression below is the same as $(9 \times 15) + (9 \times 20)$?

(1) $(9 \times 15) + 20$ **(4)** $(9 + 15) + 20$

(2) $9(15 + 20)$ **(5)** $9(15 \times 20)$

(3) $9 + (15 \times 20)$

Check your answers. Correct answers are on page 312. If you have at least eight answers correct, go to the Level 3 Preview. If you have fewer than eight answers correct, go back to Lesson 1 and study Level 2.

Level 3   Whole Numbers Problem Solving

Preview

Directions: Solve each problem.

1. The town of Greenport received a $67,500 grant for a bilingual program in elementary schools. If the five elementary schools share in the money equally, how much will each school get?

2. Mark used 23 gallons of gasoline on a 437-mile trip. How far could Mark drive on one gallon of gasoline?

3. For four years the Millers made monthly mortgage payments of $350. Then they paid off $5000 and made $250 monthly payments for two more years. Altogether, how much did they pay on their mortgage?

4. The figure at the right shows the floor plan of an office. If carpet costs $25 a square yard, how much will it cost to carpet the office?

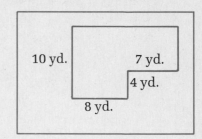

5. One week Max used 24 gallons of gasoline. The next week he used 8 gallons; the third week, 31 gallons; and the fourth week, 13 gallons. Which of the following expressions shows the average number of gallons of gasoline Max used each week?

(1) $\dfrac{(24 + 8 + 31 + 13)}{4}$

(2) $24 \times 8 \times 31 \times 13$

(3) $\dfrac{24 + 8}{2} + \dfrac{31}{3} + \dfrac{13}{4}$

(4) $(24 + 8) + (31 + 13)$

(5) $4(24 + 8 + 31 + 13)$

Check your answers. Correct answers are on pages 312–313. If you have at least four answers correct, do the Level 3 Review on pages 58–59. If you have fewer than four answers correct, study Level 3 beginning with Lesson 1.

Lesson 1   Key Words

Most of the problems that test whole-number skills on the GED Test are word problems. Here are some hints to help you identify the operations you will need to solve whole-number problems.

Addition problems are easy to recognize. *Sum, total, combined,* and *altogether* are key words that usually mean to add.

Subtraction problems are also easy to recognize. The word *difference* and the phrases *how much more?* and *how much less?* mean to subtract. The words *gross* and *net* often appear in subtraction problems. *Gross* refers to a total before any deductions are taken from it. *Net* is the amount that is left after the deductions are taken.

Example 1: Jeff gets a gross salary of $280 a week. His employer deducts $52 for taxes and pension. Find Jeff's net weekly salary.

$$\begin{array}{rl} \$280 & \text{gross salary} \\ -\ \ 52 & \text{deductions} \\ \hline \$228 & \text{net salary} \end{array}$$

The key words *product* and *times* do not often appear in multiplication problems. Watch for situations which mean to multiply. For example, you may be told information for one thing, such as the distance someone can drive on a gallon of gas. Then you may have to find information for several things, such as the distance the person can drive on several gallons.

Example 2: Celeste can drive her car for 18 miles (mi.) on one gallon of gas. How far can she drive with 12 gallons of gas?

$$\begin{array}{r} 18 \\ \times\,12 \\ \hline 36 \\ 18\ \ \\ \hline 216\ \text{mi.} \end{array}$$

You will rarely see the key word *quotient* in division problems, but the words *split* and *share* often mean to divide.

Example 3: Kim picked 85 pounds (lbs.) of apples. She wants to share the apples equally among herself and four friends. How many pounds will each person get?

Notice that Kim and her four friends are five people.

$$\begin{array}{r} 17\ \text{lbs.} \\ 5\overline{)85} \end{array}$$

You may be given information for several things, such as the price of several items. Then you will have to find the price of one item. This situation means to divide.

Example 4: Lydia paid $8 for 4 pounds of beef. Find the price of one pound.

Divide the total price by the number of pounds.

$$\begin{array}{r} \$2 \\ 4)\overline{\$8} \end{array}$$

You have learned to solve perimeter, area, and volume problems using formulas. However, you may not always see the key words in these problems. Remember that perimeter is a measure of the distance around a flat figure. A problem that asks you to find the amount of fencing needed to enclose a pasture is a perimeter problem.

Area is a measure of the amount of surface on a flat shape. A problem that asks you to find the amount of material needed to cover a flat surface is an area problem.

Volume is a measure of the capacity of a three-dimensional object. A problem that asks you to find the cubic feet of dirt that a truck can carry is a volume problem.

Lesson 1   Exercise

Directions: Solve each problem.

1. The town of Greenport spent $16,593,650 for police and fire protection in a year. The town spent $1,108,212 for health and welfare the same year. How much more did the town spend for police and fire protection than for health and welfare?

2. John, Tom, Bill, and Fred started a record store. At the end of a year, they had a profit of $27,936. If they shared the profit equally, how much did each person get?

3. Ruben Sutton made $16,456 last year. His wife Connie made $11,294. Their son Henry made $3,367 at his part-time job. Find the combined yearly income for the Sutton family.

4. It costs $1850 a year to educate one student at the Franklin School. There are 230 students in the school. Find the total cost for educating all the students at the school for a year.

5. In 1950, there were 14,273 people living in Elmford. In 1980, 8498 more people lived in Elmford than 1950. Find the population of Elmford in 1980.

6. Al wants to split a board 102 inches long into six equal pieces. Assuming no waste, how long will each piece be?

7. The swimming pool at the Greenport Community Center is 80 feet long, 20 feet wide, and 6 feet deep. Find the capacity of the pool in cubic feet.

8. Oregon became a state in 1859. For how many years had Oregon been a state in 1987?

9. A large farm in Texas is the shape of a square. Each side of the farm is 6 miles long. Find the distance around the farm.

10. In one month, the Simpsons spent $462 for utilities and mortgage payments, $436 for taxes and Social Security, $194 for car payments and gasoline, $323 for food, and $245 for everything else. How much did they spend altogether that month?

Answers are on page 313.

Lesson 2   Multistep Problems

Each problem in the last exercise required only one step to find a solution. Most problems are more complicated. There is no method that guarantees a successful solution to every word problem. Practice is probably the most helpful tool. A goal of this book is to give you a lot of practice with word problems.

Every problem in the next exercise requires at least two steps. Read each problem several times. Think about the operations you will use before you actually perform them. After you have solved each problem, think about your answer. Be sure your answers make sense.

Lesson 2   Exercise

Directions: Solve each problem.

1. Louise bought furniture priced at $650. She paid for the furniture in 15 equal payments of $52 each. How much more than $650 did she end up paying for the furniture?

2. Mr. Castro ordered three cases of tomato soup, four cases of chicken soup, and two cases of bean soup for his store. Each case contains 12 cans. Altogether, how many cans of soup did he order?

3. Every week Mark takes home $235, and his wife Heather takes home $240. Find their combined income for four weeks.

4. Joaquin's net monthly salary is $1300. He pays $290 a month for rent and $400 a month for child support. How much does he have left each month after these expenses?

5. For the Greenport Community Center, Phil bought 20 new baseball gloves at $25 each, eight baseball bats at $12 each, and six softballs at $8 each. Find the total cost of these items.

6. One week Winston worked for 35 hours at his regular wage of $6.50 an hour and for 8 hours overtime at $9.75 an hour. How much did he make altogether that week?

7. Adrienne lived for 18 months in an apartment. For the first year, she paid $260 a month rent. Then, for the last six months, she paid $290 a month. What was her average rent for the months she lived in the apartment?

8. Alex drove for four hours at an average speed of 55 mph and then for two hours at an average speed of 30 mph. Altogether, how far did he travel in those six hours?

9. The town of Greenport received $60,000 from the federal government and $45,000 from the state government for summer projects. Three projects shared the funds equally. How much did each receive?

10. Pat's gross monthly income is $1500. Her employer witholds $350 monthly. Find Pat's net salary for a year.

Answers are on page 313.

Lesson 3   Multistep Area Problems

You learned earlier to find the area of rectangles, squares, and triangles. You can use these skills to find the area of figures made up of two or more simple figures.

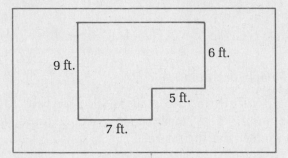

Example 1: Find the area of the figure shown above.

Step 1. Separate the figure into two rectangles. Then find the length of the missing sides. The figure below shows two rectangles labeled A and B. Notice that the left-side measurement, 9 ft., is equal to the sum of the two right-side measurements. The top measurement equals the sum of the two bottom measurements, 7 ft. and 5 ft.

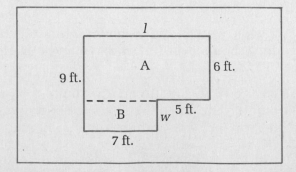

$w = 9 - 6 = 3$ ft.
$l = 7 + 5 = 12$ ft.

Step 2. Find the area of rectangle A.

$A = lw$
$A = 12 \times 6 = 72$ ft.²

Step 3. Find the area of rectangle B.

$A = lw$
$A = 7 \times 3 = 21$ ft.²

Step 4. Add the two areas.

$72 + 21 = 93$ ft.²

The figure in Example 1 can be separated in different ways. You can think of the figure as a large rectangle with a small rectangle cut from it.

Example 2: Find the area of the figure in Example 1 as the *difference* of two areas.

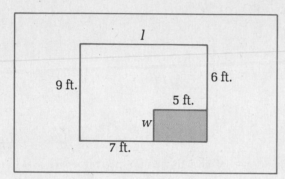

Step 1. The above figure shows the original figure as a large rectangle with the shaded rectangle cut from it. Again, find the length of the missing sides.

$l = 7 + 5 = 12$ ft.
$w = 9 - 6 = 3$ ft.

Step 2. Find the area of the large rectangle.

$A = lw$
$A = 12 \times 9 = 108$ ft.²

Step 3. Find the area of the small, shaded rectangle.

$A = lw$
$A = 5 \times 3 = 15$ ft.²

Step 4. Subtract the two areas.

$108 - 15 = 93$ ft.²

Lesson 3   Exercise

Directions: Solve each problem.

1. Find the area of the figure shown at the right.

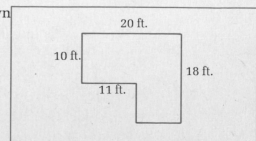

2. What is the perimeter of the figure in problem 1?

3. What is the area of the figure pictured at
    the right?

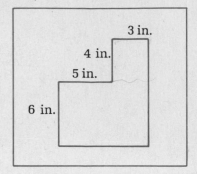

4. What is the perimeter of the figure in problem 3?

5. Find the area of the figure shown
    at the right.

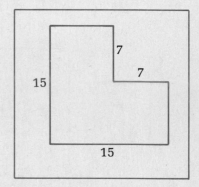

6. Find the perimeter of the figure in problem 5.

7. What is the area of the figure
    shown at the right?

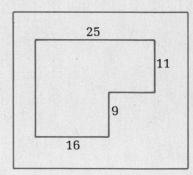

8. Find the perimeter of the figure in problem 7.

9. Find the area of the shaded part
    of the figure shown at the right.

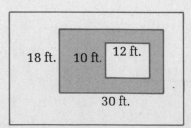

10. What is the area of the shaded portion of
    the figure pictured at the right?

    Answers are on page 314.

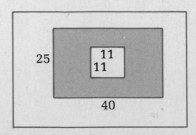

Lesson 4   Estimating

Estimating or **approximating** means making a good guess. Whenever you solve a word problem, think about the answer. Be sure it makes sense. You can often decide whether an answer makes sense if you first round off the numbers in the problem. Then solve the problem with these new round numbers. If your original answer is close to your rounded-off answer, your original answer probably makes sense.

The next exercise is similar to the exercise in Lesson 1. The choices which follow each question are to help you to think about an estimated answer.

Lesson 4   Exercise

Directions: Choose an approximate answer to each problem. Then find the exact answer for each problem.

1. The yearly budget for Capital City is $306,783,000. The yearly budget for Greenport is $15,487,000. How much bigger is Capital City's budget than Greenport's?

   (1) about $300 million   **(4)** about $150 million
   (2) about $250 million   **(5)** about $100 million
   (3) about $200 million

2. Three partners equally shared a $58,974 profit for their business. How much did each partner get?

   (1) a little less than $10,000   **(4)** a little less than $25,000
   (2) a little less than $15,000   **(5)** a little less than $30,000
   (3) a little less than $20,000

3. Linda's gross salary is $23,496. The total withheld from her salary is $3984. What is her net salary?

   (1) nearly $27,000   **(4)** nearly $12,000
   (2) nearly $25,000   **(5)** nearly $ 7,000
   (3) nearly $20,000

4. Jim Green made $29,260 last year. His wife Karen made $18,420. Their daughter Susan made $9455. Find their combined income.

   (1) a little less than $40,000   **(4)** a little less than $55,000
   (2) a little less than $45,000   **(5)** a little less than $60,000
   (3) a little less than $50,000

5. Richard paid $112 for four new work uniforms. How much did each uniform cost?

   (1) a little more than $15   **(4)** a little more than $30
   (2) a little more than $20   **(5)** a little more than $35
   (3) a little more than $25

6. Carlos cut an 87-inch-long copper pipe into three equal pieces. Find the length of each piece.

   (1) about 30 inches
   (2) about 25 inches
   (3) about 20 inches
   (4) about 15 inches
   (5) about 10 inches

7. It cost $2160 a year to educate a student at the Oak Street School. Find the cost of educating 30 students at the school for a year.

   (1) around $600,000
   (2) around $300,000
   (3) around $ 60,000
   (4) around $30,000
   (5) around $ 6,000

8. Ron is a truck driver. Monday he drove 328 miles; Tuesday, 217 miles; Wednesday, 421 miles; Thursday, 186 miles; and Friday, 313 miles. Find the total number of miles he drove that week.

   (1) about 900
   (2) about 1200
   (3) about 1400
   (4) about 1600
   (5) about 1800

9. Kevin makes $327.50 for a five-day work week. How much does he make in a day?

   (1) between $40 and $50
   (2) between $50 and $60
   (3) between $60 and $70
   (4) between $70 and $80
   (5) between $80 and $100

10. The floor of Roberto's basement is 62 feet long and 21 feet wide. What is the area of the floor in square feet?

    (1) between 600 and 800
    (2) between 800 and 1000
    (3) between 1000 and 1200
    (4) between 1200 and 1400
    (5) between 1400 and 1600

Answers are on page 315.

Lesson 5   Item Sets

You may see long reading passages on the GED mathematics test. You will have to solve several problems based on each of these passages. Use the paragraph below to study the examples which follow.

Hector is trying to decide which of two used cars to buy. The cars he likes are nearly identical. To pay for the car at Dealer A's shop, he must put down $500 cash and then make 24 equal monthly payments of $120 each. To pay for the car at Dealer B's, he does not have to make a down payment, but he must make 30 equal monthly payments of $115 each.

Example 1: What total amount would Hector have to pay for the car from Dealer A?

Multiply the monthly payments of $120 by the number of months, 24.

$120
× 24
480
240
$2880

Add the total monthly payments to the down payment.

$2880
+ 500
$3380

Example 2: Find the total price of the car at Dealer B's.

Multiply the monthly payment of $115 by the number of months, 30.

$115
× 30
$3450

Lesson 5   Exercise

Directions: *Items 1 to 4 are based on the following passage.*

Steve is a trainee on a new job. For six months he will make $950 a month. After the training period, he will make $550 a month more for one year. Then, if he does well on his performance, he will make another $250 a month for his second year as a regular employee.

1. How much will Steve make in his first full year as a regular employee?

2. How much will Steve make for his first 12 months at the job, including his training period?

3. What is Steve's average monthly income during his first 12 months at the job?

4. If Steve does well on the performance review at the end of his first year of regular employment, what will he make for his second year as a regular employee?

Items 5 to 10 are based on the following passage.

The yearly payroll for the eight people in management at Paulson's Plastics is $216,000. The payroll for the 35 laborers at the factory is $756,000. The ten people on the clerical staff make a total of $144,000 in a year.

5. What is the total number of people on the payroll at Paulson's?

6. What is the total yearly payroll at the factory?

7. Find the total monthly payroll at Paulson's.

8. What is the average yearly income of a laborer at Paulson's?

9. Find the average monthly income of a clerical person at the factory.

10. How much more is the average yearly income of a laborer at Paulson's than the average yearly income of a clerical person?

Answers are on page 315.

Lesson 6 Setup Answers

On some problems on the GED mathematics test, you may have to choose among a group of possible solutions. For these problems you are not looking for the exact answer but for the method to find the answer.

Parentheses are important in these setup solutions. Parentheses group together the amounts which must be combined first.

Example: On Friday 650 people attended the Greenport Fair. On Saturday 825 people were there, and on Sunday, 940. Everyone paid $5 to get into the fair. Which expression below tells the total receipts for those three days?

(1) $(5 \times 650) + 825 + 940$

(2) $\dfrac{(650 + 825 + 940)}{5}$

(3) $5(650 + 825 + 940)$

(4) $650 + 825 + (940 \times 5)$

(5) $650 + 825 + \left(\dfrac{940}{5}\right)$

Choice **(3)**, $5(650 + 825 + 940)$, is correct.

In choice **(1)**, only the Friday attendance is multiplied by the price of a ticket. In choice **(2)**, the entire attendance is divided by the price of a ticket. In choice **(4)**, only the Sunday attendance is multiplied by the price of a ticket. In choice **(5)**, only the Saturday attendance is divided by the price of a ticket.

Lesson 6 Exercise

Directions: Choose the correct answer to each problem.

1. Frank drove for two hours at 20 mph and for three hours at 60 mph. Which expression tells the total distance he drove?

(1) $2(20 + 60)$

(2) $3(20 + 60)$

(3) $5(20 + 60)$

(4) $(2 \times 20) + (3 \times 60)$

(5) $\dfrac{(20 + 60)}{5}$

2. Eva got scores of 80, 95, and 74 on three math quizzes. Which of the following expressions shows her mean score on the quizzes?

(1) $80 + 95 + 74 \div 3$

(2) $\dfrac{(80 + 95 + 74)}{3}$

(3) $\dfrac{80}{3 + 95 + 74}$

(4) $3(80 + 95 + 74)$

(5) $\dfrac{3}{(80 + 95 + 74)}$

3. Alberto bought four quarts of oil for $7.99 each and paid a total of $1.92 in tax. Which expression tells the amount he paid altogether for the oil?

    (1) $(4 \times 7.99) + 1.92$ **(4)** $4(7.99 - 1.92)$
    (2) $4 \times 7.99 \times 1.92$
    (3) $4(7.99 + 1.92)$ **(5)** $\dfrac{(7.99 + 1.92)}{4}$

4. For a play at the Greenport Community Center, members sold 350 tickets at $8 each and 425 tickets at $6 each. Which of the following expressions shows the total receipts for the tickets?

    (1) $(8 \times 6) + (350 + 425)$ **(4)** $(8 + 350) \times (6 + 425)$
    (2) $8 \times 6 \times 350 \times 425$ **(5)** $(8 + 6)(350 + 425)$
    (3) $(8 \times 350) + (6 \times 425)$

5. Denise's gross monthly income is $1800. Her monthly deductions are $360. Which expression shows her net yearly income?

    (1) $12(1800 + 360)$ **(4)** $12 \times 1800 \times 360$
    (2) $(12 \times 1800) + 360$
    (3) $12(1800 - 360)$ **(5)** $\dfrac{(1800 - 360)}{12}$

6. The elementary schools in Greenport had a joint festival for three nights. Receipts on Monday were $2500; on Tuesday, $4850; and on Wednesday, $4200. The money was shared equally by five schools. Which of the following expressions tells the amount each school received?

    (1) $\dfrac{(2500 + 4850 + 4200)}{5}$ **(4)** $\dfrac{(2500 + 4850 + 4200)}{3}$

    (2) $5(2500 + 4850 + 4200)$ **(5)** $\dfrac{5}{(2500 + 4850 + 4200)}$

    (3) $\dfrac{2500}{5 + 4850 + 4200}$

7. Mr. Vega ordered 20 boxes of shirts each containing 12 shirts for his store. He sent back two of the boxes because they were the wrong color. Which expression tells the total number of shirts he kept from that order?

    (1) $12 \times 20 \times 2$ **(4)** $(12 \times 20) - 2$
    (2) $12(20 - 2)$ **(5)** $12(20 + 2)$
    (3) $12 + (20 - 2)$

8. The length of a rectangle measures 100 feet, and the width measures 30 feet. Which of the following expressions shows the perimeter?

    (1) $2 \times 100 \times 30$ **(4)** $(2 \times 100) + (2 \times 30)$
    (2) 100×30 **(5)** $4(100 + 30)$
    (3) $(2 \times 100) - (2 \times 30)$

9. Roger rented a car for $35 a day for three days. Insurance cost $9 extra per day. Which of the following expressions shows the total amount Roger paid to rent the car, including the cost of insurance?

    (1) $3(35 - 9)$ **(4)** $\dfrac{(35 + 9)}{3}$
    (2) $3(35 + 9)$
    (3) $3 \times 35 \times 9$ **(5)** $(3 \times 35) + 9$

10. A rectangular container which measures 3 feet long, 2 feet wide, and 4 feet high is half full. Which of the following expressions shows the cubic dimensions of the contents of the container?

(1) $(3 \times 4 \times 2)2$

(2) $3 \times 2 \times 4 \times 2$

(3) $\dfrac{(3 \times 2 \times 4)}{2}$

(4) $3 \times 1 \times 4$

(5) $\dfrac{(3 \times 2 \times 4)}{3}$

Answers are on page 316.

Level 3   Review

Directions: Solve each problem.

1. Elton Electronics shipped 48 boxes which weighed 768 pounds altogether. If each box weighed the same, what was the weight of one box?

2. Gloria pays $315 a month for rent and $135 a month for car payments. She takes home $12,360 a year. How much does she have left over each year after the expenses of rent and car payments?

3. The boiler in the building where Louie is the superintendent uses 760 gallons of fuel oil each month for six months of the year, 1250 gallons a month for three months in winter, and 410 gallons a month for three months in summer. Find the total number of gallons used in a year.

4. Find the area of the figure pictured at the right.

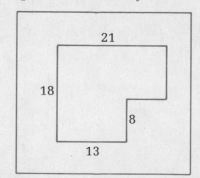

Items 5 through 8 are based on the following passage.

Paul wants to rent a car for four days. Dealer A charges $18 a day with no mileage charge. Insurance costs an additional $12.50 a day. Dealer B charges $12 a day and $.20 a mile. Insurance costs an additional $10.25 a day.

5. How much would a car from Dealer A cost for four days?

6. How much would a car from Dealer B cost for four days, excluding mileage?

7. If Paul rents from Dealer B and drives 100 miles, what will the cost of renting the car be?

8.   If Paul rents from Dealer B and drives 200 miles, what will the cost of renting the car be?

---

9.   Louise tutors four students in English. Her students are Mai Lee who is 33 years old, Carlos who is 25, Ludmilla who is 47, and Itzhak who is 19. Which expression gives the average age of Louise's students?

(1) $2(33 + 25) + 2(47 + 19)$

(2) $33 \times 25 \times 47 \times 19$

(3) $\dfrac{(33 + 25)}{2} + \dfrac{(47 + 19)}{2}$

(4) $\dfrac{(33 + 25 + 47 + 19)}{4}$

(5) $\dfrac{(33 + 25 + 47 + 19)}{\frac{1}{4}}$

10. Frank worked eight hours on Monday, six hours on Tuesday, and five hours on Wednesday painting a house. He made $12 an hour. Which expression tells the total amount he was paid for the job?

(1) $8(12 + 6 + 5)$

(2) $12(8 + 6 + 5)$

(3) $12 + 8 + 6 + 5$

(4) $(12 \times 8) + (6 \times 5)$

(5) $\dfrac{(8 + 6 + 5)}{12}$

Check your answers. Correct answers are on page 316. If you have at least eight answers correct, go to the Quiz. If you have fewer than eight answers correct, go back to Lesson 1 and study Level 3.

Chapter 1 Quiz

Directions: Solve each problem.

1.   Round off 986,548 to the nearest thousand.

2.   A trucker drove for six hours at an average speed of 48 mph. How far did he drive?

3.   At $29.50 a pair, what is the cost of a dozen pairs of sneakers?

4.   Simplify the expression $14^2 - 8^0 + 16^1$.

5.   What is $\sqrt{2401}$?

6.   Find the perimeter of the figure pictured at the right.

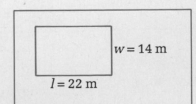

7. What is the area of the figure in Problem 6?

8. Find the volume of a cube with a side of 9 inches.

9. Lois is a real estate broker. In April she sold a house for $32,400. In May she sold another house for $46,600, and in June she sold a house for $29,300. Find the mean price of the houses she sold those months.

10. What was the median price of the houses in Problem 9?

Items 11 and 12 are based on the following table.

State	Per Capita Expenditures in a Year		
	Education	Welfare	Health
New York	$500	$392	$153
Illinois	371	278	69
California	553	397	90

Source: U.S. Bureau of the Census

11. How much more does California spend per capita than Illinois for education?
   - (1) $212
   - (2) $192
   - (3) $182
   - (4) $153
   - (5) $ 53

12. How much more does New York spend per capita on education, welfare, and health than Illinois?
   - (1) $129
   - (2) $198
   - (3) $229
   - (4) $243
   - (5) $327

13. Jack cut an 84-yard-long television cable into six equal pieces. How long was each piece?
   - (1) 20
   - (2) 18
   - (3) 16
   - (4) 14
   - (5) 12

14. Phil pays $112 a month on his car loan and $365 a month on his mortgage. What is the yearly total for these expenses?
   - (1) $4380
   - (2) $4770
   - (3) $5604
   - (4) $5724
   - (5) $5844

15. Find the total number of pounds in a shipment made up of eight cartons weighing 78 pounds each and five cartons weighing 112 pounds each.
   - (1) 1014
   - (2) 1099
   - (3) 1184
   - (4) 1232
   - (5) 1456

16. Find the area of the shaded part of the figure pictured at the right.

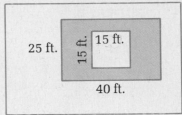

(1) 1000
(2) 940
(3) 775
(4) 725
(5) 375

Items 17 to 19 are based on the following passage.

On Monday morning Mike's odometer (mileage gauge) had a reading of 38,526. He drove for three days in the city where he lives, and he used a total of 16 gallons of gasoline. His odometer read 38,734 on Wednesday night. On Thursday and Friday he drove in the countryside, and he used another 14 gallons of gasoline. His odometer read 39,028 on Friday night.

17. How many miles did Mike get on a gallon of gasoline for the three days he drove in the city?

(1) 11
(2) 12
(3) 13
(4) 15
(5) 20

18. How many miles did Mike get on a gallon of gasoline for the two days he drove in the country?

(1) 18
(2) 21
(3) 24
(4) 25
(5) 26

19. Mike paid $1.15 a gallon for the gasoline he used. Altogether, how much did he spend on gasoline for those five days?

(1) $18.40
(2) $24.15
(3) $33.35
(4) $34.50
(5) $35.40

---

20. Sharon made $8.50 an hour for 35 hours and then $12.75 an hour for 6 hours of overtime work. Which expression tells the total amount she made?

(1) 41 × 8.50
(2) (35 × 12.75) + (6 × 8.50)
(3) 35(12.75 + 8.50)
(4) (35 × 8.50) + (6 × 12.75)
(5) 41(8.50 + 12.75)

Check your answers. Correct answers are on page 317. If you have at least sixteen answers correct, go to the next chapter. If you have fewer than sixteen answers correct, study the Whole Numbers chapter before you go on.

Chapter

2 | Fractions

Objective

In this chapter, you will

- Reduce fractions and raise fractions to higher terms
- Interchange forms of fractions
- Add, subtract, multiply, and divide fractions
- Find what fraction one number is of another
- Understand key words
- Solve distance and cost formulas containing fractions
- Work with powers and square roots of fractions
- Use and interchange standard units of measure
- Solve perimeter, area, and volume problems containing fractions
- Compare and order fractions
- Read tables containing fractions
- Set up ratios and proportions; solve proportion word problems
- Find the probability that an event will happen
- Solve multistep proportion problems
- Set up solutions to word problems
- Solve problems when extraneous information is given

Level 1   Fraction Skills

Preview

Directions: Solve each problem.

1. Reduce $\frac{24}{64}$ to lowest terms.

2. Find the sum of $4\frac{7}{20}$ and $9\frac{1}{2}$.

3. Find the difference between $9\frac{1}{2}$ and $4\frac{3}{8}$.

4. Find $12 \times 4\frac{1}{2} \times 2\frac{2}{3}$.

5. Divide $6\frac{1}{4}$ by $1\frac{7}{8}$.

Check your answers. Correct answers are on page 317. If you have at least four answers correct, do the Level 1 Review on page 72. If you have fewer than four answers correct, study Level 1 beginning with Lesson 1.

Lesson 1   Fractions

A **fraction** shows a part of a whole thing. Two days are two of the seven equal parts in a week. Two days are $\frac{2}{7}$ or two-sevenths of a whole week. Four cents are $\frac{4}{10}$ or four-tenths of a dime.

The top number of a fraction is called the **numerator.** It tells how many parts you have. The bottom number is called the **denominator.** It tells how many parts there are in the whole.

In the figure at the right, three parts are shaded out of a total of eight parts. We say that $\frac{3}{8}$ of the figure is shaded. The numerator 3 tells how many parts are shaded. The denominator 8 tells how many parts there are in the whole figure.

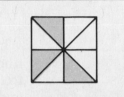

Fractions appear in three forms.

1. In a **proper fraction,** the numerator is less than the denominator. $\frac{2}{3}$, $\frac{7}{10}$, and $\frac{1}{50}$ are all proper fractions. A proper fraction does not have all the parts of the whole. The value of a proper fraction is always less than one whole.

2. In an **improper fraction,** the numerator is as big as or bigger than the denominator. $\frac{5}{5}$, $\frac{4}{3}$, and $\frac{24}{6}$ are all improper fractions. When the numerator and the denominator are the same, the value of an improper fraction is exactly one. $\frac{5}{5} = 1$. When the numerator is bigger than the denominator, the value of an improper fraction is more than one. Both $\frac{4}{3}$ and $\frac{24}{6}$ are more than one.

3. In a **mixed number,** a whole number stands beside a fraction. $1\frac{1}{4}$, $5\frac{2}{3}$, and $10\frac{5}{6}$ are all mixed numbers. The value of a mixed number is always more than one.

Reducing Fractions

Reducing means dividing both the numerator and the denominator by a number that goes into them evenly. Reducing changes the numbers in a fraction, but it does not change the value. $\frac{2}{4}$ of a dollar (two quarters) has the same value as $\frac{1}{2}$ a dollar (a fifty-cent piece).

To reduce a fraction, divide both the numerator and the denominator by a number that goes into them evenly. Then check to see if any other number will divide into both.

Example 1: Reduce $\frac{36}{48}$.

Step 1. Divide 36 and 48 by a number that goes evenly into both. 6 divides evenly into both 36 and 48. $\dfrac{36 \div 6}{48 \div 6} = \dfrac{6}{8}$

Step 2. Check to see if another number divides evenly into both 6 and 8. 2 goes evenly into $\dfrac{6 \div 2}{8 \div 2} = \dfrac{3}{4}$

both. Since no number besides 1 goes evenly into both 3 and 4, $\frac{3}{4}$ is reduced as far as it will go.

A fraction is in **lowest terms**, or simplest form, when it is reduced as far as it will go. The solutions to the next exercise show each fraction reduced in one step. You may have to use two or more steps to reduce some of the fractions.

Raising Fractions to Higher Terms

When you go on to study more mathematics, you will see that most operations have an opposite. The opposite of reducing fractions is raising fractions to higher terms. You will be increasing the numbers in both the numerator and the denominator, but the value of the fraction will stay the same. When fractions have the same value, they are called **equivalent fractions.** You will often raise fractions to higher terms when you add and subtract fractions. For example, you may want to change $\frac{3}{4}$ to 24ths. In this problem, you have a new denominator, 24, and you need to find the new numerator, which corresponds to 3.

To raise a fraction to higher terms, divide the smaller denominator into the larger denominator. Then multiply both the numerator and the denominator of the original fraction by the quotient.

Example 2: Raise $\frac{3}{4}$ to 24ths.

Step 1. Divide 4 into 24.

$$4\overline{)24} \quad \begin{array}{c} 6 \end{array}$$

Step 2. Multiply both 3 and 4 by 6. $\frac{3}{4}$ raised to 24ths is $\frac{18}{24}$.

$$\frac{3 \times 6}{4 \times 6} = \frac{18}{24}$$

To check a fraction you have raised to higher terms, reduce it. The reduced fraction should be the original fraction. For the example, reduce $\frac{18}{24}$ by 6. $\frac{18 \div 6}{24 \div 6} = \frac{3}{4}$

Lesson 1   Exercise

Directions: *For items 1 to 5, reduce each fraction to lowest terms.*

1. $\frac{5}{20} = ?$ $\frac{32}{56} = ?$ $\frac{25}{30} = ?$ $\frac{90}{200} = ?$

2. $\frac{40}{55} = ?$ $\frac{6}{18} = ?$ $\frac{45}{75} = ?$ $\frac{22}{24} = ?$

3. $\frac{24}{40} = ?$ $\frac{19}{38} = ?$ $\frac{60}{144} = ?$ $\frac{50}{1000} = ?$

4. $\frac{25}{45} = ?$ $\frac{48}{64} = ?$ $\frac{24}{72} = ?$ $\frac{9}{54} = ?$

5. $\frac{12}{60} = ?$ $\frac{20}{84} = ?$ $\frac{36}{76} = ?$ $\frac{68}{102} = ?$

For *items 6 to 10*, raise each fraction to higher terms by finding the missing numerator.

6. $\frac{9}{10} = \frac{?}{40}$ $\frac{5}{12} = \frac{?}{36}$ $\frac{2}{5} = \frac{?}{45}$ $\frac{2}{3} = \frac{?}{12}$

7. $\frac{4}{9} = \frac{?}{45}$ $\frac{3}{20} = \frac{?}{200}$ $\frac{3}{5} = \frac{?}{35}$ $\frac{9}{50} = \frac{?}{100}$

8. $\frac{7}{25} = \frac{?}{75}$ $\frac{1}{4} = \frac{?}{32}$ $\frac{2}{9} = \frac{?}{36}$ $\frac{5}{8} = \frac{?}{16}$

9. $\frac{6}{7} = \frac{?}{42}$ $\frac{7}{12} = \frac{?}{48}$ $\frac{5}{8} = \frac{?}{72}$ $\frac{1}{2} = \frac{?}{36}$

10. $\frac{7}{8} = \frac{?}{24}$ $\frac{3}{16} = \frac{?}{64}$ $\frac{4}{5} = \frac{?}{25}$ $\frac{4}{5} = \frac{?}{80}$

Answers are on pages 317–318.

Lesson 2   Interchanging Forms of Fractions

Changing Improper Fractions to Mixed Numbers

The answers to many fraction problems are improper fractions. These answers are easier to read if you change them to whole or mixed numbers.

To change an improper fraction, divide the numerator by the denominator, and write the answer as a whole number. The remainder, if any, is written as the numerator of the fraction with the same denominator. Check to see if you can reduce the remaining fraction.

Example 1: Change $\frac{32}{10}$ to a whole or mixed number.

Step 1. Divide 10 into 32.

$$\frac{32}{10} = 10\overline{)32} \quad \begin{array}{r} 3\frac{2}{10} \\ 30 \\ \hline 2 \end{array}$$

Step 2. Reduce the remaining fraction by 2. $3\frac{2}{10} = 3\frac{1}{5}$

Changing Mixed Numbers to Improper Fractions

When you multiply and divide fractions, you often have to change mixed numbers to improper fractions. This is the opposite of changing an improper fraction to a mixed number.

Follow these steps to change a mixed number to an improper fraction.

1. Multiply the denominator by the whole number.
2. Add the numerator.
3. Write the total over the denominator.

Example 2: Change $2\frac{3}{8}$ to an improper fraction.

Step 1. Multiply the denominator by the whole $8 \times 2 = 16$
number.

Step 2. Add the numerator to the product. $16 + 3 = 19$

Step 3. Write the total over the denominator. $2\frac{3}{8} = \frac{19}{8}$

Lesson 2   Exercise

Directions: *For items 1 to 4,* change each improper fraction to a whole or mixed number. Reduce each remaining fraction to lowest terms.

1. $\frac{11}{2} = ?$ $\frac{17}{3} = ?$ $\frac{25}{8} = ?$ $\frac{36}{9} = ?$

2. $\frac{52}{8} = ?$ $\frac{35}{10} = ?$ $\frac{28}{6} = ?$ $\frac{40}{12} = ?$

3. $\frac{36}{3} = ?$ $\frac{48}{5} = ?$ $\frac{17}{8} = ?$ $\frac{28}{19} = ?$

4. $\frac{7}{2} = ?$ $\frac{14}{10} = ?$ $\frac{38}{8} = ?$ $\frac{27}{4} = ?$

5. $\frac{54}{9} = ?$ $\frac{13}{2} = ?$ $\frac{16}{5} = ?$ $\frac{98}{15} = ?$

For items 6 to 10, change each mixed number to an improper fraction.

6. $3\frac{7}{10} = ?$ $2\frac{2}{3} = ?$ $7\frac{1}{2} = ?$ $1\frac{5}{6} = ?$

7. $10\frac{3}{8} = ?$ $2\frac{3}{4} = ?$ $7\frac{2}{5} = ?$ $12\frac{7}{20} = ?$

8. $4\frac{5}{8} = ?$ $6\frac{1}{2} = ?$ $9\frac{4}{5} = ?$ $1\frac{1}{3} = ?$

9. $5\frac{3}{5} = ?$ $8\frac{2}{3} = ?$ $3\frac{1}{2} = ?$ $9\frac{7}{8} = ?$

10. $1\frac{8}{9} = ?$ $4\frac{7}{10} = ?$ $5\frac{2}{7} = ?$ $7\frac{1}{8} = ?$

Answers are on page 318.

Lesson 3   Adding and Subtracting Fractions

The operations of adding and subtracting fractions are similar. For both operations, the denominators of the fractions you are working with must be the same. These operations are also usually written vertically. That is, the numbers you add or subtract are written in columns, one below the other. This is the method used most often in the United States. If you learned another method, you do not need to change. Simply compare your answers to the answers to these exercises.

Adding Fractions

To add fractions with the same denominators, first add the numerators. Then put the total over the denominator. Change an improper fraction answer to a mixed number and reduce.

Example 1: Find the sum of $\frac{3}{10}$ and $\frac{9}{10}$.

Step 1. Add the numerators. $3 + 9 = 12$. Then write the total over the denominator, 10.

$$\begin{array}{r} \frac{3}{10} \\ +\frac{9}{10} \\ \hline \end{array}$$

Step 2. Change $\frac{12}{10}$ to a mixed number and reduce.

$$\frac{12}{10} = 1\frac{2}{10} = 1\frac{1}{5}$$

When the fractions in an addition problem have different denominators, first find a **common denominator.** A common denominator is a number that every denominator in a problem divides into evenly. The lowest number that every denominator divides into evenly is called the **lowest common denominator,** or **LCD.**

Example 2: Add $\frac{3}{5}$ and $\frac{7}{10}$.

Step 1. Find the LCD. The lowest number both 5 and 10 divide into evenly is 10. Raise $\frac{3}{5}$ to 10ths.

$$\begin{array}{r} \frac{3}{5} = \frac{6}{10} \\ +\frac{7}{10} = \frac{7}{10} \\ \hline \frac{13}{10} = 1\frac{3}{10} \end{array}$$

Step 2. Add the new numerators. Put the total over 10, and change the improper fraction to a mixed number.

With mixed numbers, add the whole numbers separately. Then combine the whole number total with the fraction total.

Example 3: Add $2\frac{1}{2} + 3\frac{3}{4} + 4\frac{1}{6}$.

Step 1. Find the LCD. The lowest number that 2, 4, and 6 divide into is 12. Raise each fraction to 12ths.

Step 2. Add the new numerators and put the total over 12. Add the whole numbers. Then change $\frac{17}{12}$ to a mixed number, and add the mixed number to the whole number.

$$\begin{array}{r} 2\frac{1}{2} = 2\frac{6}{12} \\ 3\frac{3}{4} = 3\frac{9}{12} \\ +4\frac{1}{6} = 4\frac{2}{12} \\ \hline 9\frac{17}{12} = 10\frac{5}{12} \end{array}$$

Subtracting Fractions without Borrowing

To subtract fractions or mixed numbers, first change each fraction to a new fraction with a common denominator. Subtract the fractions and whole numbers separately.

Example 4: Subtract $4\frac{1}{2}$ from $5\frac{2}{3}$.

Step 1. Find the LCD. The LCD for 3 and 2 is 6. Raise each fraction to 6ths.

$$\begin{array}{r} 5\frac{2}{3} = 5\frac{4}{6} \\ -4\frac{1}{2} = 4\frac{3}{6} \\ \hline 1\frac{1}{6} \end{array}$$

Step 2. Subtract the numerators and put the difference over 6. Subtract the whole numbers and check to be sure the answer is reduced.

Subtracting Fractions with Borrowing

In problems where there is no top fraction to subtract from, you must borrow. To borrow means to take 1 from the whole number at the top. Rewrite (or rename) the 1 you borrow as a fraction. Remember that a fraction with the same numerator and denominator is equal to 1. The numerator and denominator should be the same as the denominator of the other fraction in the problem.

Example 5: Subtract $2\frac{3}{8}$ from 7.

Step 1. Borrow 1 from 7, and rewrite the 1 as $\frac{8}{8}$, since 8 is the LCD.

Step 2. Subtract the new fractions and the whole numbers. Be sure the answer is reduced.

$$
\begin{array}{rl}
7 = & 6\frac{8}{8} \\
-2\frac{3}{8} = & 2\frac{3}{8} \\
\hline
 & 4\frac{5}{8}
\end{array}
$$

In some subtraction problems, the fraction on top is not big enough to subtract the bottom fraction. To get a bigger fraction on top, borrow 1 from the whole number on top. Rewrite the 1 as a fraction. Then add this fraction to the old top fraction.

Example 6: Subtract $3\frac{4}{9}$ from $5\frac{1}{3}$.

Step 1. Find the LCD. The LCD for 3 and 9 is 9. Raise each fraction to 9ths.

Step 2. Since you cannot subtract $\frac{4}{9}$ from $\frac{3}{9}$, borrow 1 from 5, and rewrite the 1 as $\frac{9}{9}$. Add $\frac{9}{9}$ to $\frac{3}{9}$.

Step 3. Subtract the new fractions and the whole numbers. Be sure the answer is reduced.

$$
\begin{array}{rcccl}
5\frac{1}{3} = & 5\frac{3}{9} = & 4\frac{3}{9} + \frac{9}{9} = & 4\frac{12}{9} \\
-3\frac{4}{9} = & 3\frac{4}{9} & = & 3\frac{4}{9} \\
\hline
 & & & 1\frac{8}{9}
\end{array}
$$

Lesson 3   Exercise

Directions: Solve each problem.

1. $\frac{5}{8} + \frac{7}{8} = ?$ $3\frac{1}{6} + 8\frac{5}{6} = ?$

   $2\frac{3}{5} + 1\frac{4}{5} = ?$ $4\frac{11}{12} + 4\frac{7}{12} = ?$

2. $\frac{1}{4} + \frac{3}{5} = ?$ $\frac{5}{12} + \frac{3}{4} = ?$

   $8\frac{2}{5} + 1\frac{7}{15} = ?$ $6\frac{1}{2} + 5\frac{3}{8} = ?$

3. $\frac{7}{8} + \frac{2}{3} = ?$ $\frac{1}{6} + \frac{5}{9} = ?$

   $5\frac{3}{4} + 2\frac{2}{9} = ?$ $4\frac{3}{5} + 4\frac{1}{6} = ?$

4. $\frac{3}{10} + \frac{1}{2} + \frac{3}{4} = ?$ $\frac{3}{8} + \frac{2}{3} + \frac{1}{12} = ?$

   $2\frac{5}{6} + 5\frac{3}{8} + 3\frac{1}{4} = ?$ $1\frac{5}{9} + 4\frac{1}{2} + 3\frac{2}{3} = ?$

5. $\frac{11}{12} - \frac{7}{12} = ?$ $\frac{7}{8} - \frac{3}{8} = ?$

   $6\frac{9}{10} - 5\frac{7}{10} = ?$ $9\frac{15}{16} - 3\frac{3}{16} = ?$

6. $\frac{3}{4} - \frac{3}{8} = ?$ $\frac{4}{5} - \frac{3}{4} = ?$

   $7\frac{1}{2} - 2\frac{1}{5} = ?$ $4\frac{7}{9} - 1\frac{1}{6} = ?$

7. $7 - 3\frac{5}{8} = ?$ $9 - 4\frac{7}{12} = ?$

$8 - 2\frac{7}{10} = ?$ $12 - 5\frac{13}{16} = ?$

8. $5\frac{1}{6} - 2\frac{5}{6} = ?$ $9\frac{1}{3} - 3\frac{2}{3} = ?$

$8\frac{5}{12} - 7\frac{7}{12} = ?$ $4\frac{1}{5} - 1\frac{4}{5} = ?$

9. $6\frac{1}{5} - 3\frac{3}{4} = ?$ $7\frac{1}{2} - 1\frac{5}{8} = ?$

$9\frac{1}{3} - 5\frac{3}{4} = ?$ $2\frac{1}{2} - 1\frac{2}{3} = ?$

10. $8\frac{2}{3} - 1\frac{5}{8} = ?$ $5\frac{1}{4} - 2\frac{3}{4} = ?$

$3\frac{1}{9} - 1\frac{2}{3} = ?$ $5\frac{1}{8} - 4\frac{3}{8} = ?$

Answers are on pages 318–319.

Lesson 4   Multiplying and Dividing Fractions

The operations of multiplying and dividing fractions are similar. For both operations, the numbers you work with must be written in fraction form. This means that you must change every mixed number to an improper fraction. You must also write every whole number as a fraction with a denominator of 1. However, you do not have to find a common denominator for the fractions in these problems. Multiplying and dividing fraction problems are usually written horizontally. That is, the numbers you work with are written side by side.

Multiplying Fractions

The answers to fraction multiplication problems may seem strange. When you multiply whole numbers (except for 1 and 0), the answer is bigger than either of the two numbers you multiply. When you multiply fractions, the answer is smaller than either of the two fractions.

When you multiply fractions, you find a part of a part. If you multiply $\frac{1}{2}$ by $\frac{1}{2}$, you find $\frac{1}{2}$ of $\frac{1}{2}$. You know that $\frac{1}{2}$ of $\frac{1}{2}$ dollar is $\frac{1}{4}$ dollar. The product, $\frac{1}{4}$, is smaller than either of the fractions you multiplied.

To multiply fractions, multiply the numerators together, and multiply the denominators together. Check to see if the answer can be reduced.

Example 1: Find the product of $\frac{2}{5}$ and $\frac{1}{3}$.

Multiply the numerators together, and multiply $\frac{2}{5} \times \frac{1}{3} = \frac{2}{15}$
the denominators together. $\frac{2}{15}$ is reduced to low-
est terms.

Canceling is a way of making fraction multiplication problems easier. Canceling is similar to reducing. To cancel, divide a numerator and a denominator by a number that goes evenly into both of them.

Example 2: Multiply $\frac{3}{4} \times \frac{5}{6} \times \frac{4}{25}$.

Step 1. Cancel 3 and 6 by 3.

$$\frac{\overset{1}{\cancel{3}}}{4} \times \frac{5}{\underset{2}{\cancel{6}}} \times \frac{4}{25} =$$

Step 2. Cancel 4 and 4 by 4.

$$\frac{\overset{1}{\cancel{3}}}{\underset{1}{\cancel{4}}} \times \frac{5}{\underset{2}{\cancel{6}}} \times \frac{\overset{1}{\cancel{4}}}{25} =$$

Step 3. Cancel 5 and 25 by 5.

$$\frac{\overset{1}{\cancel{3}}}{\underset{1}{\cancel{4}}} \times \frac{\overset{1}{\cancel{5}}}{\underset{2}{\cancel{6}}} \times \frac{\overset{1}{\cancel{4}}}{\underset{5}{\cancel{25}}} =$$

Step 4. Multiply the numerators together, and multiply the denominators together.

$$\frac{\overset{1}{\cancel{3}}}{\underset{1}{\cancel{4}}} \times \frac{\overset{1}{\cancel{5}}}{\underset{2}{\cancel{6}}} \times \frac{\overset{1}{\cancel{4}}}{\underset{5}{\cancel{25}}} = \frac{1}{10}$$

Remember to write mixed numbers and whole numbers as improper fractions.

Example 3: Multiply $2\frac{1}{2}$ by 3.

Step 1. Change $2\frac{1}{2}$ and 3 to improper fractions.

$$2\frac{1}{2} \times 3 = \frac{5}{2} \times \frac{3}{1} =$$

Step 2. Since you cannot cancel, multiply the numerators together, and multiply the denominators together. Change $\frac{15}{2}$ to a mixed number.

$$\frac{5}{2} \times \frac{3}{1} = \frac{15}{2} = 7\frac{1}{2}$$

Dividing by Fractions

To divide by a fraction means to find out how many times a fraction goes into another number. For example, if you divide $\frac{1}{2}$ by $\frac{1}{4}$, you find out how many times $\frac{1}{4}$ goes into $\frac{1}{2}$. You know that $\frac{1}{4}$ dollar goes into $\frac{1}{2}$ dollar two times. Write this problem as $\frac{1}{2} \div \frac{1}{4}$. The $\div$ sign means "divided by."

To divide by a fraction, **invert** the fraction at the right, and follow the rules for multiplying fractions.

To invert means to rewrite a fraction with the numerator on the bottom and the denominator on the top. For example, when you invert $\frac{3}{4}$, you get $\frac{4}{3}$. $\frac{4}{3}$ is sometimes called the **reciprocal** of $\frac{3}{4}$.

Example 4: Divide $\frac{1}{2}$ by $\frac{1}{4}$.

Step 1. Invert the fraction you are dividing by. $\frac{1}{4}$ becomes $\frac{4}{1}$. Change the $\div$ sign to $\times$.

$$\frac{1}{2} \div \frac{1}{4} =$$

$$\frac{1}{2} \times \frac{4}{1} =$$

Step 2. Cancel 4 and 2 by 2. Then multiply the numerators and multiply the denominators.

$$\frac{1}{\underset{1}{\cancel{2}}} \times \frac{\overset{2}{\cancel{4}}}{1} = \frac{2}{1}$$

Step 3. Change $\frac{2}{1}$ to a whole number.

$$\frac{2}{1} = 2$$

Be sure to change mixed numbers or whole numbers to improper fractions.

Example 5: Divide 3 by $\frac{4}{5}$.

Step 1. Write 3 as an improper fraction with a denominator of 1. Invert $\frac{4}{5}$ and change the $\div$ sign to $\times$.

$$3 \div \frac{4}{5} =$$

$$\frac{3}{1} \times \frac{5}{4} =$$

Step 2. Multiply the numerators together and the denominators together. $\frac{3}{1} \times \frac{5}{4} = \frac{15}{4}$

Step 3. Change $\frac{15}{4}$ to a mixed number. $\frac{15}{4} = 3\frac{3}{4}$

Dividing by Mixed Numbers and Whole Numbers

Change every mixed number or whole number in a division problem to an improper fraction before you invert.

Example 6: Divide $\frac{3}{4}$ by 2.

Step 1. Rewrite 2 as an improper fraction with a denominator of 1.

$$\frac{3}{4} \div 2 =$$
$$\frac{3}{4} \div \frac{2}{1} =$$

Step 2. Invert $\frac{2}{1}$ to $\frac{1}{2}$. Change the $\div$ sign to $\times$, and multiply across.

$$\frac{3}{4} \times \frac{1}{2} = \frac{3}{8}$$

Example 7: Divide $4\frac{1}{2}$ by $2\frac{1}{4}$.

Step 1. Rewrite both $4\frac{1}{2}$ and $2\frac{1}{4}$ as improper fractions.

$$4\frac{1}{2} \div 2\frac{1}{4} =$$
$$\frac{9}{2} \div \frac{9}{4} =$$

Step 2. Invert $\frac{9}{4}$ to $\frac{4}{9}$. Change the $\div$ sign to $\times$. Cancel, and multiply across.

$$\frac{\overset{1}{\cancel{9}}}{\underset{1}{\cancel{2}}} \times \frac{\overset{2}{\cancel{4}}}{\underset{1}{\cancel{9}}} = \frac{2}{1}$$

Step 3. Change $\frac{2}{1}$ to a whole number. $\frac{2}{1} = 2$

Lesson 4   Exercise

Directions: Solve each problem.

1. $\frac{2}{3} \times \frac{4}{5} = ?$ $\frac{5}{8} \times \frac{3}{4} = ?$ $\frac{3}{4} \times \frac{1}{2} \times \frac{4}{5} = ?$

2. $\frac{9}{10} \times \frac{2}{3} = ?$ $\frac{3}{20} \times \frac{5}{12} = ?$ $\frac{2}{3} \times \frac{5}{6} \times \frac{7}{10} = ?$

3. $6 \times \frac{3}{4} = ?$ $\frac{5}{6} \times 15 = ?$ $8 \times \frac{11}{12} = ?$

4. $\frac{3}{4} \times 3\frac{1}{5} = ?$ $2\frac{1}{4} \times 1\frac{2}{3} = ?$ $5\frac{1}{3} \times 1\frac{5}{16} = ?$

5. $\frac{2}{3} \div \frac{4}{9} = ?$ $\frac{9}{10} \div \frac{2}{5} = ?$ $\frac{3}{4} \div \frac{1}{8} = ?$

6. $4 \div \frac{2}{3} = ?$ $8 \div \frac{4}{5} = ?$ $5 \div \frac{3}{4} = ?$

7. $1\frac{1}{9} \div \frac{5}{6} = ?$ $2\frac{5}{8} \div \frac{1}{4} = ?$ $6\frac{1}{2} \div \frac{3}{8} = ?$

8. $4\frac{1}{3} \div 5 = ?$ $10\frac{1}{2} \div 8 = ?$ $\frac{3}{8} \div 4 = ?$

9. $\frac{5}{9} \div 1\frac{1}{3} = ?$ $4 \div 1\frac{3}{4} = ?$ $\frac{3}{8} \div 2\frac{2}{3} = ?$

10. $1\frac{1}{4} \div 2\frac{1}{2} = ?$ $2\frac{3}{4} \div 1\frac{5}{8} = ?$ $7\frac{1}{2} \div 3\frac{3}{4} = ?$

Answers are on page 319.

Level 1   Review

Directions: Solve each problem.

1. Reduce $\frac{35}{84}$ to lowest terms.

2. Change $8\frac{4}{5}$ to an improper fraction.

3. Find the sum of $8\frac{3}{4}$ and $7\frac{9}{16}$.

4. Find the sum of $2\frac{3}{8}$, $4\frac{1}{2}$, and $6\frac{3}{5}$.

5. Subtract $3\frac{5}{12}$ from $7\frac{2}{3}$.

6. Find the difference between $8\frac{1}{5}$ and $3\frac{3}{4}$.

7. Find the product of $3\frac{3}{5}$ and 10.

8. Find $2\frac{1}{3} \times 4\frac{1}{8} \times \frac{1}{2}$.

9. Divide $6\frac{2}{3}$ by 8.

10. Find the quotient of $9\frac{1}{3}$ divided by $3\frac{1}{5}$.

Check your answers. Correct answers are on page 319. If you have at least eight answers correct, go to the Level 2 Preview. If you have fewer than eight answers correct, go back to Lesson 1 and study Level 1.

Level 2   Fraction Applications

Preview

Directions: Solve each problem.

1. Gordon received a shipment containing eight blue shirts, six white shirts, and six yellow shirts. Blue shirts made up what fraction of the shipment?

2. Miriam drove for $3\frac{1}{2}$ hours at an average speed of 52 mph. How far did she drive?

3. Change 40 months to years.

4. Find the area of a rectangle with a length of $6\frac{5}{8}$ inches and a width of 4 inches.

5. There are four women and six men on the steering committee for the Ninth Street Block Association. Find the ratio of the number of women to the total number of people on the steering committee.

Check your answers. Correct answers are on page 319. If you have at least four answers correct, do the Level 2 Review on page 88. If you have fewer than four answers correct, study Level 2 beginning with Lesson 1.

Lesson 1   Finding What Fraction One Number Is of Another

One of the most common applications of fractions skills is very simple. All you have to do is write a fraction and reduce it. Remember that the numerator tells the *part* and the denominator tells the *whole*.

Example 1: During a week in July, three workers at the Greenport post office were on vacation. Altogether, 24 people work at the post office. What fraction of the workers were on vacation that week?

Write a fraction where the numerator tells the number of workers on vacation and the denominator tells the total number of workers. Then reduce the fraction.

$$\frac{\text{part}}{\text{whole}} \quad \frac{3}{24} = \frac{1}{8}$$

In some problems, you will have to find the part that is referred to in the question.

Example 2: What fraction of the workers in Example 1 were at work during that week in July?

Step 1. To find the number of workers at work, subtract the number of workers on vacation from the total number of workers.

$$\begin{array}{ll} \text{total} & 24 \\ \text{on vacation} & -3 \\ \text{at work} & \overline{21} \end{array}$$

Step 2. Write a fraction where the numerator tells the number of workers at work and the denominator tells the total number of workers. Then reduce.

$$\frac{\text{part}}{\text{whole}} \quad \frac{21}{24} = \frac{7}{8}$$

In other problems, you will have to find the whole.

Example 3: Mary got two problems wrong and eight problems right on a quiz. What fraction of the problems did she get right?

Step 1. To find the total number of problems, add the number of problems Mary got right to the number she got wrong.

$$\begin{array}{ll} \text{right} & 8 \\ \text{wrong} & +2 \\ \text{total} & \overline{10} \end{array}$$

Step 2. Write a fraction where the numerator tells the number of problems Mary got right and the denominator tells the total number of problems. Then reduce.

$$\frac{\text{right}}{\text{total}} \quad \frac{8}{10} = \frac{4}{5}$$

Lesson 1   Exercise

Directions: Solve each problem. Be sure your answers are reduced.

1. Of the 15 students in Laura's GED class, 9 are men.
   a. Men are what fraction of the class?
   b. Women are what fraction of the class?

2. In the shop where Phil works there are 45 union members and 36 nonunion members.
   a. What fraction of the workers in the shop are union members?
   b. What fraction of the workers do not belong to the union?

3. Of the 33 cars in the lot at Dave's Auto Shop, 15 are new.
   a. What fraction of the cars in the lot are new?
   b. What fraction of the cars in the lot are used?

4. For his store, Felipe ordered 12 cases of cola, 2 cases of lime drink, and 6 cases of orange drink.
   a. What fraction of the cases were cola?
   b. What fraction of the cases were lime drink?
   c. What fraction of the cases were orange drink?

5. In the office where Kathleen works, 45 employees voted to join the union, 35 voted not to join, and 20 remained undecided.
   a. What fraction of the employees voted to join the union?
   b. What fraction voted against joining the union?
   c. What fraction remained undecided?

Answers are on pages 319–320.

Lesson 2   Word Problems with Basic Operations

In the Whole Numbers chapter, you learned some of the key words which tell you what operations to use. These key words apply to fraction problems as well.

Addition and subtraction problems are easy to recognize. Multiplication and division problems are sometimes more difficult. A fraction followed by the word *of* usually means to multiply.

Example 1: Tom's employer withholds $\frac{1}{5}$ of his salary for taxes and social security. Tom's gross salary is $1200 a month. How much does his employer withhold each month?

Find $\frac{1}{5}$ of $1200. Multiply $1200 by $\frac{1}{5}$. $\frac{1}{5} \times \frac{1200}{1} = \$240$

Be careful with the word *of*. It is a common word, and it does not always mean to multiply.

Example 2: The Smiths spend $\frac{1}{4}$ of their budget on rent and $\frac{1}{5}$ on utilities. These items make up what fraction of their budget?

In this problem, there is no number to multiply $\frac{1}{4}$ by. Simply add the two fractions.

$$\begin{aligned} \frac{1}{4} &= \frac{5}{20} \\ +\frac{1}{5} &= \frac{4}{20} \\ \hline & \frac{9}{20} \end{aligned}$$

Remember that in division problems the thing being divided must come first.

Example 3: Greg wants to split $3\frac{1}{2}$ pounds of roofing nails into four equal parts. How many pounds will be in each part?

Here the $3\frac{1}{2}$ pounds of nails are being divided. $3\frac{1}{2} \div 4 =$
Put $3\frac{1}{2}$ first and divide by 4.

$$\frac{7}{2} \div \frac{4}{1} =$$

$$\frac{7}{2} \times \frac{1}{4} = \frac{7}{8}$$

The word per, as in miles per gallon or miles per hour, can mean to divide.

Example 4: Geraldine drove for 84 miles in $1\frac{3}{4}$ hours. What was her average speed in miles per hour?

Divide the number of miles she drove by the $84 \div 1\frac{3}{4} =$
number of hours.

$$\frac{84}{1} \div \frac{7}{4} =$$

$$\overset{12}{\cancel{\frac{84}{1}}} \times \frac{4}{\cancel{7}} = 48 \text{ mph}$$

Lesson 2   Exercise

Directions: Solve each problem.

1. At the last election in Greenport, $\frac{3}{8}$ of the 9600 registered voters failed to vote. How many voters failed to vote?

2. How many of the registered voters described in problem 1 went to vote in that election?

3. Denise drove 209 miles on $9\frac{1}{2}$ gallons of gasoline. What was the average distance she drove on one gallon of gasoline?

4. From a total delivery of $2\frac{1}{2}$ tons of prepared concrete, Sal used $1\frac{9}{10}$ tons. How many tons of concrete were left?

5. How many $3\frac{1}{4}$-inch strips of wood can Phil cut from a piece that is 52 inches long?

6. A share of Acme, Inc., stock sold Wednesday afternoon for $24\frac{1}{4}$. By Thursday afternoon, the stock dropped $\frac{3}{8}$, and by Friday afternoon it dropped another $\frac{1}{8}$. Find the Friday afternoon price of the stock.

7. Joe bought a jacket, originally selling for $72, for $\frac{1}{4}$ off the original price. Find the sale price of the jacket.

8. The Greenport Development Corporation is selling off 24 acres of land in $\frac{3}{4}$-acre lots. How many lots can they get from 24 acres?

9. Mary makes $12.50 an hour working overtime. Friday night she worked $2\frac{1}{2}$ hours overtime. How much did she make for those hours?

10. Jeff works part-time at a gas station. Monday he worked $2\frac{3}{4}$ hours, Wednesday he worked $2\frac{1}{2}$ hours, and Friday he worked $3\frac{1}{4}$ hours. How many hours did he work altogether those three days?

Answers are on page 320.

Lesson 3   Distance and Cost Formulas

You may see fractions in problems that use the distance formula $d = rt$ or the cost formula $c = nr$.

Example: Chris walked for $\frac{3}{4}$ of an hour at a speed of 4 mph. How far did he walk?

Replace r with 4 and t with $\frac{3}{4}$ in the formula $d = rt$.

$$d = rt$$

$$d = \frac{\overset{1}{\cancel{4}}}{1} \times \frac{3}{\underset{1}{\cancel{4}}} = 3 \text{ mi.}$$

Lesson 3   Exercise

Directions: Solve each problem.

1.   Kate drove for $3\frac{1}{2}$ hours at an average speed of 40 mph. How far did she drive?

2.   A plane flew for $4\frac{1}{4}$ hours at an average speed of 520 mph. How far did the plane travel in that time?

3.   Find the cost of $\frac{3}{4}$ pound of ground beef at $2.80 per pound.

4.   What is the cost of $2\frac{1}{2}$ pounds of apples at the rate of $.84 a pound?

5.   How far can Anne walk in $1\frac{1}{2}$ hours if she averages 5 miles per hour?

6.   Find the cost of $9\frac{1}{2}$ feet of lumber at the cost of $.36 a foot.

7.   Paul rode his bike for $2\frac{3}{4}$ hours at an average speed of 22 mph. How far did he ride during that time?

8.   Find the cost of $2\frac{1}{4}$ pounds of wood screws at $.48 a pound.

9.   Find the cost of $1\frac{1}{4}$ pounds of cheddar cheese at $3.00 a pound.

10.   A plane flew $5\frac{3}{4}$ hours at 600 mph. How far did the plane fly?

Answers are on page 320.

Lesson 4   Powers and Square Roots

To raise a fraction to a power, remember to raise both the numerator and the denominator to the power.

Example 1: What is the value of $(\frac{3}{4})^2$?

Write $\frac{3}{4}$ two times and multiply across.

$$\left(\frac{3}{4}\right)^2 = \frac{3}{4} \times \frac{3}{4} = \frac{9}{16}$$

To find the square root of a fraction, remember to find the square root of both the numerator and the denominator.

Example 2: What is the value of $\sqrt{\frac{25}{64}}$?

Find the square root of 25 and the square root of 64.

$$\sqrt{\frac{25}{64}} = \frac{5}{8}$$

Lesson 4   Exercise

Directions: Find the value of each of the following expressions.

1. $\left(\frac{1}{2}\right)^2 = ?$ $\left(\frac{1}{9}\right)^2 = ?$ $\left(\frac{2}{3}\right)^3 = ?$ $\left(\frac{1}{10}\right)^4 = ?$

2. $\left(\frac{3}{10}\right)^3 = ?$ $\left(\frac{5}{6}\right)^2 = ?$ $\left(\frac{7}{12}\right)^2 = ?$ $\left(\frac{3}{4}\right)^3 = ?$

3. $\left(\frac{1}{6}\right)^4 = ?$ $\left(\frac{1}{2}\right)^5 = ?$ $\left(\frac{1}{8}\right)^3 = ?$ $\left(\frac{2}{7}\right)^3 = ?$

4. $\sqrt{\frac{25}{36}} = ?$ $\sqrt{\frac{4}{49}} = ?$ $\sqrt{\frac{1}{81}} = ?$ $\sqrt{\frac{9}{100}} = ?$

5. $\sqrt{\frac{64}{81}} = ?$ $\sqrt{\frac{1}{144}} = ?$ $\sqrt{\frac{9}{16}} = ?$ $\sqrt{\frac{36}{49}} = ?$

Answers are on page 320.

Lesson 5   Standard Units of Measurement

The list below gives units of measurement for length, weight, time, and liquid measure. These are the units commonly used in the United States and in Great Britain, where they are called Imperial units. Abbreviations for each unit are in parentheses. The list also tells how many smaller units each bigger unit is equal to. Memorize any units you do not already know.

Length

1 foot (ft.) = 12 inches (in.)
1 yard (yd.) = 3 ft. or 36 in.
1 mile (mi.) = 5280 ft. or 1760 yd.

Weight

1 pound (lb.) = 16 ounces (oz.)
1 ton (t.)   = 2000 lb.

Time

1 minute (min.) = 60 seconds (sec.)
1 hour (hr.)   = 60 min.
1 day (da.)   = 24 hr.
1 week (wk.)   = 7 da.
1 year (yr.)   = 365 da. or 12 months (mo.) or 52 wk.

Liquid Measure

1 pint (pt.)   = 16 oz.
1 quart (qt.)   = 2 pt. or 32 oz.
1 gallon (gal.) = 4 qt.

In some problems, you may need to change from a smaller unit to a larger unit. Divide the number of smaller units you have by the number of smaller units that make up one large unit. When you change from one unit of measurement to another, you will often use fractions.

Example 1: Change 18 months to years.

Divide 18 months by the number of months in one year, 12, and reduce the answer.

$$1\frac{6}{12} = 1\frac{1}{2} \text{ yr.}$$
$$12\overline{)18}$$
$$\underline{12}$$
$$6$$

In some problems, you may want to express an answer in more than one unit of measure.

Example 2: Change 18 months to years and months.

Divide 18 months by the number of months in one year, 12, and express the remainder as a number of months.

$$1 \text{ yr. } 6 \text{ mo.}$$
$$12\overline{)18}$$
$$\underline{12}$$
$$6$$

When you cannot divide by the number of smaller units in one large unit, simply reduce the problem.

Example 3: Change 8 inches to feet.

Since the number of inches in a foot, 12, does not divide into 8, write a fraction with 12 as the denominator, and reduce.

$$\frac{8}{12} = \frac{2}{3} \text{ ft.}$$

In other problems, you may want to change large units to smaller units. In this case, multiply the number of large units you have by the number of smaller units that make up one large unit.

Example 4: Change 10 feet to inches.

Multiply 10 feet by the number of inches in one foot, 12.

$$10 \times 12 = 120 \text{ in.}$$

Lesson 5   Exercise

Directions: Solve each problem.

1. a. Change 8 feet to yards.
    b. Change 8 feet to yards and feet.

2. a. Change 135 seconds to minutes.
    b. Change 135 seconds to minutes and seconds.

3. a. Change 10 quarts to gallons.
    b. Change 10 quarts to gallons and quarts.

4. a. Change 20 ounces to pounds.
    b. Change 20 ounces to pounds and ounces.

5. a. Change 20 months to years.
    b. Change 20 months to years and months.

For items 6 to 10, change each unit to the new unit indicated.

6. 10 mo. = ? yr.   45 min. = ? hr.   9 in. = ? ft.

7. 8 oz. = ? lb.   20 in. = ? yd.   2 qt. = ? gal.

8. 1800 lb. = ? t.   16 hr. = ? da.   1320 ft. = ? mi.

9. 6 gal. = ? qt. $3\frac{1}{4}$ lb. = ? oz.   5 da. = ? hr.

10. 4 yd. = ? in. $2\frac{1}{2}$ min. = ? sec. $1\frac{1}{2}$ t. = ? lb.

Answers are on pages 320–321.

Lesson 6   Perimeter, Area, and Volume

You may see fractions in problems where you have to find perimeter, area, or volume.

Example: What is the area of the triangle pictured at the right?

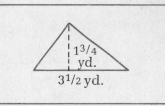

Step 1. Replace b with $3\frac{1}{2}$ and h with $1\frac{3}{4}$ in the formula $A = \frac{1}{2} bh$.

$A = \frac{1}{2} bh$

$A = \frac{1}{2} \times 3\frac{1}{2} \times 1\frac{3}{4}$

Step 2. Change the mixed numbers to improper fractions and multiply across.

$A = \frac{1}{2} \times \frac{7}{2} \times \frac{7}{4} = \frac{49}{16} =$

$3\frac{1}{16}$ yd.^{2}

Lesson 6   Exercise

Directions: Solve each problem.

1. a. What is the perimeter of the figure at the right?
    b. What is the area of the figure?

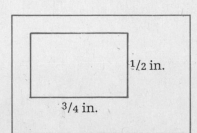

2. a. What is the perimeter of the figure pictured at the right?
    b. What is the area of the figure?

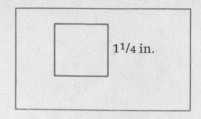

1¹/4 in.

3. a. Find the perimeter of the figure shown at the right?
    b. Find the area of the figure.

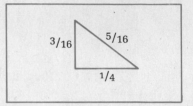

3/16   5/16

1/4

4. Find the volume of the figure at the right.

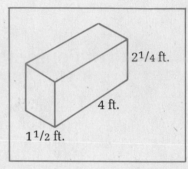

2¹/4 ft.

4 ft.

1¹/2 ft.

5. Find the volume of a cube whose side measures $4\frac{1}{2}$ feet.

Answers are on page 321.

Lesson 7   Comparing and Ordering Fractions

To compare the size of fractions, find a common denominator for each fraction you are comparing, and raise each fraction to higher terms with the new denominator.

Example: Arrange the following fractions in order from smallest to largest: $\frac{13}{20}, \frac{3}{5}, \frac{3}{4}$.

The LCD for 20, 5, and 4 is 20. Raise $\frac{3}{4}$ and $\frac{3}{5}$ to 20ths. Since $\frac{12}{20}$ is the smallest, $\frac{3}{5}$ comes first. $\frac{13}{20}$ is in the middle, and $\frac{3}{4}$ is largest.

$\frac{13}{20}$ $\frac{3}{5} = \frac{12}{20}$ $\frac{3}{4} = \frac{15}{20}$

In order: $\frac{3}{5}, \frac{13}{20}, \frac{3}{4}$

Lesson 7   Exercise

Directions: Solve each problem.

1. Find the larger fraction in each pair.

    a. $\frac{8}{15}$ or $\frac{1}{2}$ b. $\frac{2}{3}$ or $\frac{5}{9}$ c. $\frac{5}{8}$ or $\frac{4}{5}$ d. $\frac{5}{12}$ or $\frac{4}{9}$

2. Arrange in order from *smallest* to *largest*: $\frac{3}{5}$, $\frac{1}{2}$, $\frac{11}{20}$, $\frac{7}{10}$.

3. Arrange in order from *largest* to *smallest*: $\frac{5}{12}$, $\frac{1}{4}$, $\frac{5}{16}$, $\frac{1}{3}$.

4. Which is longer, $\frac{5}{8}$ inch or $\frac{9}{16}$ inch?

5. Bill wants to arrange the following packages in order from heaviest to lightest.

    Package A weighs $1\frac{1}{2}$ pounds.

    Package B weighs $1\frac{5}{8}$ pounds.

    Package C weighs $2\frac{3}{8}$ pounds.

    Package D weighs $2\frac{3}{16}$ pounds.

    Package E weighs $2\frac{1}{4}$ pounds.

    Which of the following sequences lists the packages in order from heaviest to lightest?
    (1) B, A, C, E, D
    (2) A, B, C, E, D
    (3) C, D, E, Λ, B
    (4) C, E, D, B, A
    (5) D, E, C, A, B

    Answers are on page 321.

Lesson 8   Tables

On the GED Test, you may be asked to answer fraction questions from the information given in a table.

Lesson 8   Exercise

Items 1 to 10 are based on the following table.

Wholesale Prices Per Pound		
	Imported	Domestic
Tomatoes	$.50	$ .60
Potatoes	.12	.24
Apples	.48	.64
Lemons	.30	.45
Lamb	.25	1.50
Beef	.45	.90

Source: U.S. Department of Agriculture

1. The price of a pound of imported lemons is what fraction of the price of a pound of domestic lemons?

2. The price of a pound of imported apples is what fraction of the price of a pound of domestic apples?

3. The price of a pound of imported potatoes is what fraction of the price of a pound of domestic potatoes?

4. The price of a pound of imported lamb is what fraction of the price of a pound of domestic lamb?

5. The price of a pound of imported tomatoes is what fraction of the price of a pound of domestic tomatoes?

6. The price of a pound of imported beef is what fraction of the price of a pound of domestic beef?

7. The price of a pound of imported lemons is what fraction of the price of a pound of imported tomatoes?

8. The price of a pound of imported apples is what fraction of the price of a pound of imported tomatoes?

9. The price of a pound of imported potatoes is what fraction of the price of a pound of imported apples?

10. The price of a pound of domestic beef is what fraction of the price of a pound of domestic lamb?

Answers are on page 321.

Lesson 9   Ratio

A **ratio** is a way of comparing two numbers. A ratio is very much like a fraction. The ratio of an inch to a foot is 1 to 12. The 1 stands for an inch, and 12 stands for the 12 inches in a foot. The 1 corresponds to the numerator, and the 12 corresponds to the denominator of a fraction.

You can write a ratio with the word *to*, with a colon (:), or as a fraction. The numbers in a ratio must be in the same order as the words in the problem. The ratio of an inch to a foot is 1 to 12, or 1:12, or $\frac{1}{12}$. Like fractions, ratios should be reduced to lowest terms.

Example 1: Sally takes home $800 a month, and she pays $200 for rent. What is the ratio of her rent to her take-home pay?

Step 1. Write a ratio that compares rent to take- $200:$800
home pay.

Step 2. Reduce the ratio to lowest terms.   1:4

In some problems, you may have to find one of the numbers in the ratio. Notice that ratio problems are similar to finding what fraction one number is of another.

Example 2: Mike's softball team won 12 games and lost 8. What is the ratio of the number of games the team won to the total number they played?

Step 1. Find the total number of games the team   won   12
played. Add the number they won to the num-   lost   + 8
ber they lost.   total   20

Step 2. Make a ratio of the number of games the   won:total
team won to the total number they played.   12:20

Step 3. Reduce the ratio to lowest terms.   3:5

Lesson 9   Exercise

Directions: Solve each problem.

1.   What is the ratio of 15 inches to a yard?

2.   Sandra's living room is 15 feet long and 12 feet wide. What is the ratio of the width to the length?

3.   In town, Louie gets 16 miles on a gallon of gasoline. In the country, he gets 24 miles per gallon. What is the ratio of his town gas mileage to his country gas mileage?

4.   Find the ratio of 4 ounces to a pint (1 pint = 16 ounces).

5.   The softball team at Ace Electronics has 20 members. Twelve of them are men. What is the ratio of the number of men to the number of women on the team?

6.   Jack bought a stereo for $450. He saved $150 by buying the stereo on sale. What was the ratio of the sale price to the original price?

7.   After paying $220 a month for rent, the Santiagos have $880 left for all other expenses. What is the ratio of rent to the total monthly take-home of the Santiagos?

8.   Dave weighed 280 pounds in March. By the following September, he had lost 70 pounds. Find the ratio of his March weight to his September weight.

9.   Denise has paid off $1500 on a $5000 car loan. What is the ratio of the amount she has already paid to the amount she still owes?

10.   On a test, Geraldine got 40 problems right and 16 problems wrong. What is the ratio of the number she got right to the total number of problems?

Answers are on page 321.

Lesson 10   Proportion

A **proportion** is a statement that says two ratios are equal. $\frac{6}{8} = \frac{3}{4}$ is a proportion. In a proportion, the **cross products** are equal. This means that the top of one side times the bottom of the other side is the same as the bottom of the first side times the top of the other. For the example, one cross product is $6 \times 4 = 24$. The other cross product is $8 \times 3 = 24$.

In many proportion problems, one number is missing. Usually a letter represents this number. To find the missing number in a proportion, write a statement which makes the cross products equal to each other. Then divide both sides of this statement by the number that is part of the cross product with the letter.

This is the first example of solving an equation in this book.

An **equation** is a statement that two amounts are equal.

Example 1: Solve for c in the proportion $\frac{5}{6} = \frac{30}{c}$

Step 1. Find the cross products. $5 \times c = 5c$ and $6 \times 30 = 180$. $\qquad \frac{5}{6} = \frac{30}{c}$

Step 2. Write an equation with the two cross products. $\qquad 5c = 180$

Step 3. Divide both sides of the equation by 5. The solution is 36. $\qquad \frac{5c}{5} = \frac{180}{5}$
$\qquad c = 36$

When the two ratios in a proportion are written with colons, first rewrite the ratios in fraction form.

Example 2: Solve for s in $s:9 = 5:6$.

Step 1. Rewrite the proportion in fraction form. $\qquad s:9 = 5:6$
$\qquad \frac{s}{9} = \frac{5}{6}$

Step 2. Write an equation with the cross products. $\qquad 6s = 45$

Step 3. Divide both sides of the equation by 6. The solution is $7\frac{1}{2}$. $\qquad \frac{6s}{6} = \frac{45}{6}$
$\qquad s = 7\frac{3}{6} = 7\frac{1}{2}$

Lesson 10   Exercise

Directions: Find the value of the letter in each proportion.

1. $\frac{c}{6} = \frac{7}{30}$ $\qquad \frac{9}{a} = \frac{3}{2}$ $\qquad \frac{5}{8} = \frac{m}{20}$ $\qquad \frac{4}{w} = \frac{18}{6}$

2. $\frac{9}{10} = \frac{27}{e}$ $\qquad \frac{x}{14} = \frac{4}{7}$ $\qquad \frac{1}{s} = \frac{8}{5}$ $\qquad \frac{8}{15} = \frac{r}{2}$

3. $3:n = 6:11$ $\qquad 2:5 = t:60$ $\qquad h:4 = 5:6$ $\qquad 9:2 = 1:d$

4. $6:n = 8:48$ $\qquad 3:5 = m:25$ $\qquad k:3 = 9:27$ $\qquad 4:16 = 8:p$

5. $7:35 = 2:r$ $\qquad a:5 = 20:100$ $\qquad 9:x = 1:2$ $\qquad 8:24 = y:3$

Answers are on pages 321–322.

Lesson 11   Proportion Word Problems

You can use proportion to solve many word problems. The key to solving proportion word problems is to label each number in the proportion. Remember that the parts of a proportion must correspond to each other. That is, if the top number at the left is measured in dollars, the top number at the right should also be measured in dollars.

Example 1: The ratio of lime to sand in a mixture of concrete is 1:3. Jeff is using 12 pounds of sand in the mixture. How much lime should he use?

Step 1. Make a proportion with the ratio of lime to sand on the left. On the right, put 12, the pounds of sand, on the bottom and put x, the pounds of lime, on top.

$$\begin{array}{c} \text{lime} \\ \text{sand} \end{array} \quad \frac{1}{3} = \frac{x}{12}$$

Step 2. Make an equation with the cross products.

$$3x = 12$$

Step 3. Divide both sides by 3. Jeff needs 4 pounds of sand.

$$\frac{3x}{3} = \frac{12}{3}$$
$$x = 4 \text{ lb.}$$

In some problems, you will not see the word *ratio*.

Example 2: If 12 cans of soda cost $4.50, how much do 30 cans cost?

Step 1. Make a proportion with the ratio of the number of cans to the cost. Put 30 on the top at the right. Put x in the place of the cost.

$$\begin{array}{c} \text{number} \\ \text{cost} \end{array} \quad \frac{12}{\$4.50} = \frac{30}{x}$$

Step 2. Make an equation with the cross products.

$$12x = 135.00$$

Step 3. Divide both sides by 12. The cost of 30 cans is $11.25.

$$\frac{12x}{12} = \frac{135.00}{12}$$
$$x = \$11.25$$

Lesson 11   Exercise

Directions: Solve each problem.

1. Six feet of lumber weigh 15 pounds. What is the weight of 16 feet of the same lumber?

2. The ratio of girls to boys at the Oakdale School is 5:4. There are 120 girls in the school. How many boys are there?

3. Carla makes $52.80 in 8 hours. How much does she make in 20 hours?

4. If 12 acres can yield 1440 bushels of corn, how many bushels of corn can a farmer expect from a 50-acre field?

5. The ratio of the width to the length of a snapshot is 3:5. If the snapshot is enlarged to be 15 inches wide, how long will it be?

6. Manny drove 72 miles in $1\frac{1}{2}$ hours. If he drives at the same average speed, how far can he go in 4 hours?

7. The scale on a map is $\frac{1}{2}$ inch = 15 miles. How far apart are two towns which are 2 inches apart on the map?

8. The ratio of blue to white paint in a certain mixture is 3:2. If Isabella uses 6 gallons of blue paint for the mixture, how many gallons of white paint should she use?

9. The ratio of wins to losses for the Greenport Grasshoppers baseball team was 3:5. The team won 15 games. How many games did they lose?

10. In a recent year, the divorce rate in the United States was 5 for every 1000 people. The population of Greenport is 15,000. If Greenport follows the national rate, how many divorces were there in Greenport that year?

Answers are on page 322.

Lesson 12   Probability

Probability tells the chance of an event happening. If you toss a coin, there are two possibilities. The coin may land showing its head, or it may land showing its tail. Probability is written like a fraction.

The numerator tells the number of possibilities that the event in question will happen. The denominator tells the total number of possibilities.

The probability that a coin will land showing its tail is $\frac{1}{2}$. The denominator, 2, means that there are a total of two possibilities when you toss a coin—heads or tails. The numerator, 1, means that there is just one possibility that the coin will land showing its tail.

Example 1: Tom has two white T-shirts, one gray T-shirt, one blue T-shirt, and one red T-shirt in a laundry bag. What is the probability that the first shirt Tom picks from the bag will be white?

Make a fraction with a numerator that tells the number of white T-shirts in the bag, 2, and a denominator that tells the total number of shirts in the bag, $2 + 1 + 1 + 1 = 5$.
$\frac{2}{5}$

Example 2: The first T-shirt Tom picked from the laundry bag was gray. What is the probability that the second shirt he picks will be white?

Make a fraction with a numerator that tells the number of white T-shirts left in the bag, 2, and a denominator that tells the total number of shirts left in the bag, $5 - 1 = 4$. Reduce $\frac{2}{4}$.
$\frac{2}{4} = \frac{1}{2}$

Lesson 12   Exercise

Directions: Solve each problem.

1. The Greenport Telephone book contains 7500 names. There are 15 Smiths listed in the phone book. What is the probability that in a random selection of Greenport telephone numbers, the first number will belong to a Smith?

2. Liz works as a cashier in a grocery store. At the end of the day, she had 12 quarters, 20 dimes, 18 nickels, and 30 pennies. She put all the coins in a bag.
   a. What is the probability that the first coin she takes from the bag will be a dime?
   b. What is the probability that that first coin she takes from the bag will be a penny?
   c. What is the probability that the first coin she takes from the bag will be either a quarter or a nickel?

3. Carlos received a shipment of sweaters to sell in his store. The shipment contained 10 small-size sweaters, 15 medium-size sweaters, and 8 large-size sweaters. What is the probability that the first sweater he takes from the box will be a medium size?

4. Stored in a locker at the Greenport Community Center are five hardballs, seven softballs, and three tennis balls.
   a. What is the probability that the first ball someone grabs from the locker will be a tennis ball?
   b. Frank began to take balls from the locker. The first ball he took was a hardball, and both the second and third were softballs. What is the probability that the fourth ball he takes from the locker will be another hardball?

5. Max is stacking cans in the supermarket. A carton contains eight cans of tomato sauce and six cans of green beans.
   a. What is the probability that the first can Max takes from the carton will be a can of tomato sauce?
   b. In fact, the first can Max took was tomato sauce, and the second was green beans. What is the probability that the third can Max takes will be a can of green beans?

6. José Acevedo, his wife Beatrice, and their son Felipe each bought a ticket for a chance to win a color TV. Altogether 540 tickets were sold.
   a. What is the probability that José will win the television?
   b. What is the probability that one of the Acevedos will win the television?

7. There are 100 marbles in a jar. Half of them are green, one quarter are yellow, and one quarter are red. If Maria takes a marble from the jar, what is the probability that she will pick a red one?

8. In problem 7, if the first five marbles Maria took from the jar were either green or yellow, what is the probability that the next marble she takes from the jar will be red?

9. If there are 1000 tickets for a chance to win a trip to Hawaii, and the Simpsons buy 20 chances, what is the probability that one of the Simpsons will win?

10. If there are three dimes, two nickels, and five quarters in a drawer, what is the probability that Fred will pick a dime after picking two quarters on his first two tries?

Answers are on page 322.

Level 2   Review

Directions: Solve each problem.

1. Yuki spends $240 a month for rent. She has $960 left over each month for other expenses. Rent is what fraction of Yuki's take-home pay?

2. Calvin drove 90 miles in $2\frac{1}{4}$ hours. What was his average speed in miles per hour?

3. Find the cost of $1\frac{3}{4}$ pounds of cheese at $1.84 a pound.

4. Find the value of $(\frac{3}{8})^3$.

5. Change 12 ounces to pounds.

6. Find the area of a triangle with a base of 10 inches and a height of $5\frac{1}{4}$ inches.

7. Arrange the following fractions in order from smallest to largest: $\frac{1}{4}$, $\frac{4}{9}$, $\frac{1}{3}$, $\frac{5}{12}$.

8. Members of the Central County finance committee made telephone interviews to try to find out whether people want a new community center. 65 people said yes, 25 people said no, and 10 said that they had no opinion. What is the ratio of the number of people who want the new community center to the total number of people who were interviewed?

9. The ratio of defective plastic bottles to the total number produced at Paulson's Plastics is 3:100. The factory produces 20,000 bottles a day. According to the ratio, how many of the bottles produced each day are defective?

10. A bag contains 5 green marbles, 7 blue marbles, and 3 black marbles. The first two marbles Chris took from the bag were blue. The third was green. What is the probability that the fourth marble Chris picks up will be black?

Check your answers. Correct answers are on pages 322–323. If you have at least eight answers correct, go to the Level 3 Preview. If you have fewer than eight answers correct, go back to Lesson 1 and study Level 2.

Level 3   Fraction Problem Solving

In this section of the Fractions chapter, you will learn to handle problems that look much like those on the GED Test. To find out whether you need to work through this section, try the following challenge.

Preview

Directions: Solve each problem.

1. The ratio of wins to losses for Kevin's basketball team was 8:5 last season. The team lost 15 games. How many games did they play altogether?

2. Of the total of 280 customers who bought something last week at Irene's Clothing Store, the ratio of the customers who paid cash to the total number of customers was 4:7. How many of the customers paid by some means other than cash?

3. Which of the following expressions gives the solution to 9:x = 17:20?
   (1) $9 \times 20 \times 17$
   (2) $\dfrac{(9 \times 17)}{20}$
   (3) $\dfrac{(17 \times 20)}{9}$
   (4) $\dfrac{(9 + 17)}{20}$
   (5) $\dfrac{(9 \times 20)}{17}$

4. What is the volume in cubic feet of a rectangular container which is 2 yards long, $1\frac{1}{2}$ feet wide, and 8 inches high?

5. Find the total cost of ten sheets of $\frac{3}{4}$-inch wallboard at the price of $12.50 per sheet.

Check your answers. Correct answers are on page 323. If you have at least four answers correct, do the Level 3 Review on pages 96–97. If you have fewer than four answers correct, study Level 3 beginning with Lesson 1.

Lesson 1   Multistep Proportion Problems

In some proportion problems, you have to change the ratios before you can solve the problems. Study the next examples carefully. In each problem, notice how the parts of the proportion correspond to the question which is asked.

Example 1: The ratio of men to women working at Central Hospital is 2:9. Thirty men work at the hospital. How many people work there altogether?

Step 1. In the ratio, 2 stands for men and 9 stands for women. Since the problem asks for the total, add 2 and 9 to get the number which stands for the total number of people.

men   2
women   +9
total   11

Step 2. Make a proportion with the ratio of the number of men to the total number of workers. Put 30 on the top at the right, and put x in the place of the total.

men $\quad \dfrac{2}{11} = \dfrac{30}{x}$
total

Step 3. Make an equation with the cross products.

$2x = 330$

Step 4. Divide both sides by 2. The total number of workers is 165.

$x = 165$

Example 2: Sam got 4 problems right out of 7 on a test with 35 questions. How many questions did he get wrong?

Step 1. 4 stands for the number right, and 7 stands for the total. Since the problem asks for the number wrong, subtract 4 from 7 to find the number wrong.

total   7
right   −4
wrong   3

Step 2. Make a proportion with the ratio of the number wrong to the total. Put x in the place of the number wrong, and put 35 in the place of the total.

wrong $\quad \dfrac{3}{7} = \dfrac{x}{35}$
total

Step 3. Make an equation with the cross products.

$7x = 105$

Step 4. Divide both sides of the equation by 7. Sam got 15 questions wrong.

$x = 15$

Lesson 1   Exercise

Directions: Solve each problem.

1. The ratio of the number of problems Shirley got right to the number she got wrong was 7:2. There were 36 problems on the test. How many problems did she get right?

2. The ratio of the number of men to the total number of workers at the factory where Dan works is 3:5. Altogether, 60 people work at the factory. How many women work there?

3. The ratio of domestic cars to imported cars in the lot at Al's Auto Shop is 3:2. There are 42 domestic cars in the shop. How many cars are there altogether?

4. Out of every 500 parts which are produced at Southern Steel, Inc., 497 are good. If the factory produces 15,000 parts a day, how many of the parts are defective?

5. The ratio of the amount Lois spends on car payments to her total monthly income is 2:7. She spends $96 a month on car payments. How much does she have left over each month for other expenses?

6. The ratio of the number of people who said they approved of a tax increase to the total number who were interviewed was 5:8. One hundred twenty people said they did not approve of the tax increase. How many people approved?

7. The ratio of the number of men to the number of women who registered to run in the Capital City Marathon was 9:4. Three hundred sixty women registered for the marathon. How many people registered altogether?

8. In a recent interview in Greenport, seven out of ten people said they prefer coffee in the morning. The rest said that they prefer tea. Twenty-seven people said they preferred tea. How many people were interviewed?

9. The ratio of the number of people who said they approved of the policy initiative to the total number who were interviewed was 4:5. If 150 people said they did not approve, how many people approved?

10. The ratio of the number of problems Don got right on the test to the number he got wrong was 9:1. There were 120 problems on the test. How many problems did Don get right?

Answers are on page 323.

Lesson 2   Setup Answers

You may see setup solutions to some fraction problems on the GED Test. Remember that these solutions show a method instead of an exact solution.

Example 1: Which of the following expressions shows 12 gallons changed to quarts?

(1) $12 + 4$
(2) 12×4
(3) $\frac{12}{4}$
(4) $\frac{12}{16}$
(5) $4(12 \times 4)$

To change gallons to quarts, you must multiply the number of gallons by 4, which is the number of quarts in one gallon. Choice **(2)**, 12×4, is correct.

Example 2: Which of the following expressions shows the solution to the proportion 9:7 = x:4?

(1) $\frac{(9 \times 7)}{4}$

(2) $\frac{(7 \times 4)}{9}$

(3) $\frac{(9 \times 4)}{7}$

(4) $9 \times 4 \times 7$

(5) $7(9 \times 4)$

Rewrite the proportion in fraction form. Choice **(3)**, $\frac{(9 \times 4)}{7}$, is correct. It shows the cross product $\frac{9}{7} = \frac{x}{4}$ of 9 and 4 divided by the other number in the proportion.

Lesson 2   Exercise

Directions: Choose the correct answer to each problem.

1. Change 20 inches to feet.
   (1) 20 + 12
   (2) 20 − 12
   (3) 20 × 12
   (4) $\frac{20}{12}$
   (5) 12 × 20

2. Change $4\frac{1}{2}$ pounds to ounces.
   (1) $16 + 4\frac{1}{2}$
   (2) $\frac{(4\frac{1}{2})}{16}$
   (3) $16 \times 4\frac{1}{2}$
   (4) $\frac{(4\frac{1}{2})}{12}$
   (5) $4\frac{1}{2} \times 12$

3. Fred bought six boards that were each 9 feet long. Express the total length of the boards in yards.
   (1) $\frac{(6 \times 9)}{36}$
   (2) $\frac{(6 \times 9)}{3}$
   (3) $3(6 \times 9)$
   (4) 3 + 6 + 9
   (5) 3(6 + 9)

4. Paul spliced together five lengths of wire that were each $2\frac{1}{2}$ feet long. Assuming no waste, what was the total length of the wire in inches?
   (1) $12(5 \times 2\frac{1}{2})$
   (2) $\frac{(2\frac{1}{2} \times 12)}{5}$
   (3) $(5 \times 2\frac{1}{2} + 12)$
   (4) $\frac{(5 \times 2\frac{1}{2})}{12}$
   (5) $(5 \times 2\frac{1}{2}) + 12$

5. Which of the following expressions shows the solution to 12:x = 34:9?

   (1) $\dfrac{(12 \times 9)}{34}$

   (2) $\dfrac{(12 \times 34)}{9}$

   (3) $12 \times 34 \times 9$

   (4) $\dfrac{(9 \times 34)}{12}$

   (5) $\dfrac{(12 + 9)}{34}$

6. Gloria can type 90 words per minute. Which of the following expressions shows the number of minutes she needs to type a 500-word report?

   (1) 90×500
   (2) 60×500
   (3) $\dfrac{500}{90}$

   (4) $\dfrac{500}{60}$

   (5) $\dfrac{500}{(90 \times 60)}$

7. Janet types 75 words per minute. How many minutes does she need to type a ten-page report if each page has an average of 275 words?

   (1) $10 \times 275 \times 75$

   (2) $\dfrac{75}{(10 \times 275}$

   (3) $\dfrac{(75 \times 275)}{10}$

   (4) $\dfrac{(10 \times 275)}{75}$

   (5) $75(10 \times 275)$

8. Which of the following expressions shows the solution to x:12 = 15:32?

   (1) $\dfrac{(15 \times 32)}{12}$

   (2) $\dfrac{(15 \times 12)}{32}$

   (3) $\dfrac{(12 \times 32)}{15}$

   (4) $12 \times 15 \times 32$
   (5) $32(15 \times 12)$

9. Kate's printer types 85 words per minute. How many words can it type in an hour?

   (1) 85×60
   (2) $85 + 60$
   (3) $\dfrac{85}{60}$

   (4) $85 - 60$
   (5) $\dfrac{60}{80}$

10. How many hours are required for the printer in problem 9 to type a report that contains 25,000 words?

    (1) $25,000 \times 85 \times 60$
    (2) $\dfrac{(25,000 \times 85)}{60}$
    (3) $\dfrac{(25,000 \times 60)}{85}$

    (4) $\dfrac{(25,000)}{(85 \times 60)}$
    (5) $\dfrac{25,000}{85}$

Answers are on page 323.

Lesson 3   Mixed Units with Perimeter, Area, and Volume

When you find perimeter, area, or volume, be sure that every dimension is in the same unit of measurement.

Example: Find the area in square inches of the rectangle pictured at the right.

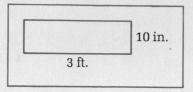

Step 1. Change 3 feet to inches.

3 ft. = 3 × 12 = 36 in.

Step 2. Replace *l* with 36 and *w* with 10 in the formula for the area of a rectangle.

$A = lw$
$A = 36 \times 10 = 360$ in.

Lesson 3   Exercise

Directions: Solve each problem.

1. Find the area in square feet of the rectangle pictured at the right.

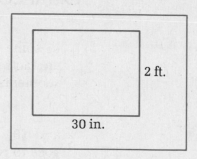

2. Find the perimeter in feet of a rectangle with a length of 2 yards and a width of 15 inches.

3. What is the area in square inches of a parallelogram with a base of one foot and a height of 9 inches?

4. Find the area in square feet of the square pictured at the right.

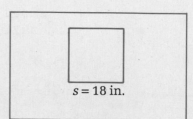

5. What is the area in square feet of the triangle pictured at the right?

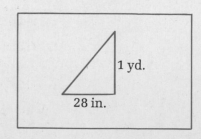

6. Find the volume in cubic feet of the rectangular container shown at the right.

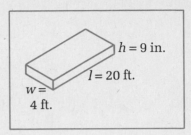

7. What is the volume in cubic yards of a rectangular solid which is 12 feet long, 6 feet wide, and 5 feet high?

8. What is the volume in cubic feet of a container which is 80 inches long, 2 feet wide, and 3 yards high?

9. Find the volume in square feet of a cube whose side measures 18 inches.

10. Find the volume in cubic yards of a container which measures 24 inches long, one foot wide, and a yard high.

Answers are on pages 323–324.

Lesson 4   Extraneous Information

In some problems, there is extraneous information. This means that there are facts which you do not need in order to solve the problems. Read word problems carefully to find this kind of unnecessary information.

Example: Mike takes home $1400 a month. He pays $350 a month for rent, and $84 a month for car payments. What fraction of his monthly income goes to rent?

Step 1. The question asks you to compare Mike's rent to his monthly income. The $84 car payment is extraneous information.

$84 is unnecessary.

Step 2. Make a fraction with the rent on top and the total monthly income on the bottom. Then reduce the fraction.

rent $\dfrac{350}{1400} = \dfrac{1}{4}$
total

Lesson 4   Exercise

Directions: For each problem, first identify any extraneous information. Then solve each problem.

1. Harold drove for $6\frac{1}{2}$ hours at an average speed of 48 mph. He used 24 gallons of gasoline. How far did he drive?

2. At the Third Street Day Care Center there are 32 boys and 28 girls. Altogether, 48 of the children have received vaccinations. Girls make up what fraction of the total number of children at the day care center?

3. A board 12 feet long and 10 inches wide weighs $3\frac{1}{2}$ pounds per linear foot. Find the total weight of the board.

4. Carmen bought $4\frac{1}{4}$ pounds of beef at $3.80 a pound and $5\frac{3}{4}$ pounds of chicken at $1.89 a pound. Find the total weight of her purchases.

5. On a test, Rick got 6 problems wrong and 54 problems right. Of the total number of problems, he guessed the answers to 8 of them. What fraction of the problems did he get right?

6. Geraldine drove 248 miles. She used 16 gallons of gasoline and paid $1.12 per gallon. On an average, how far did she drive on one gallon of gasoline?

7. The basement of Colin's house is 60 feet long, 20 feet wide, and $7\frac{1}{2}$ feet high. What is the area of the floor of the basement?

8. The ratio of gray paint to green paint in a certain color mixture is 2:5. Paul has 10 gallons of green paint and 3 gallons of paint thinner. How much gray paint does he need?

9. Alan can drive 25 miles on one gallon of gas. If he travels 175 miles at 50 mph, and pays $1.30 for a gallon of gas, how much will he spend on gas?

10. Bob's take-home pay is $1800 a month. Deductions from his gross monthly pay total $210. Bob's rent and payments for the month are $1200. What fraction of his pay does he have left over each month?

Answers are on page 324.

Level 3   Review

Directions: Solve each problem.

1. At the factory where Patty works, the ratio of management to workers is 3:14. Including management, there are 340 people working at the factory. How many people work in management?

2. The ratio of the number of girls to the number of boys at the Morningside Day Care Center is 5:4. There are 84 boys enrolled at the center. How many children are there altogether?

3. At Ace Products, the ratio of the amount of freight sent by rail to the total amount of freight that is sent is 2:9. All freight that does not go by rail goes by truck. In March, Ace Products shipped 140 tons of freight by truck. Find the weight of freight that was shipped by rail that month.

4. Which of the following expressions gives the solution to 11:16 = x:4?

(1) $\frac{(4 \times 11)}{16}$ (3) $16 \times 4 \times 11$ (5) $\frac{(16 \times 4)}{11}$

(2) $\frac{(16 \times 11)}{4}$ (4) $\frac{4}{(11 \times 16)}$

5. Bicycling burns off about 450 calories per hour. Which of the following expressions tells the number of hours required to burn off a pork chop which contains 310 calories and a glass of milk which contains 165 calories?

    (1) $\dfrac{(310 + 450)}{165}$ **(4)** $\dfrac{(165 \times 450)}{310}$

    (2) $310 + 165 + 450$

    (3) $\dfrac{(310 + 165)}{450}$ **(5)** $\dfrac{(310 \times 165)}{450}$

6. The workers at Green Gardens, Inc., can make 20 lawn mowers an hour. Which of the following expressions tells the total number of lawn mowers they can make working eight hours a day for five days a week?

    (1) $\dfrac{(20 \times 8)}{5}$ **(4)** $20 \times 8 \times 5$

    (2) $\dfrac{(20 \times 5)}{8}$ **(5)** $\dfrac{(20 + 8)}{5}$

    (3) $\dfrac{(5 \times 8)}{20}$

7. Find the area in square feet of the figure pictured at the right.

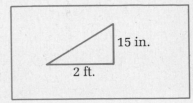

8. Find the volume in cubic feet of a cube with a side that is 18 inches long.

9. The Cruz family spends $360 a month for food and $240 a month for rent. What fraction of their total monthly income of $960 goes for food?

10. His employer deducts $\frac{1}{5}$ of Carlos's gross salary for taxes and $\frac{1}{10}$ for a savings fund. Carlos's gross salary is $2400 per month. How much does his employer deduct each month for taxes?

Check your answers. Correct answers are on page 324. If you have at least eight answers correct, go to the Quiz. If you have fewer than eight answers correct, go back to Lesson 1 and study Level 3.

Chapter 2 Quiz

Directions: Solve each problem.

1. A typist working in private industry makes an average of $14,200 a year. A typist working for the federal government makes $\frac{3}{4}$ of that amount. Find the average salary of a typist working for the federal government.

2. Donna weighs 120 pounds, and her husband Steve weighs 180 pounds. Donna's weight is what fraction of Steve's weight?

3. What is the value of $2\frac{1}{4}$ ounces of gold if gold is worth \$384 an ounce?

4. How far can a plane fly in $\frac{3}{4}$ of an hour if its average speed is 460 mph?

5. What is the value of $\sqrt{\frac{49}{100}}$?

6. Change 40 ounces to pounds.

7. Find the area of the figure pictured at the right.

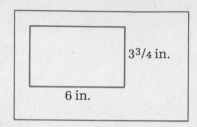

3 3/4 in.

6 in.

8. Following are the lengths of five pieces of wood.
   A—2 feet 9 inches
   B—37 inches
   C—$33\frac{1}{4}$ inches
   D—1 yard
   E—$2\frac{1}{2}$ feet

   Which of the following sequences shows the boards in order from shortest to longest?

   (1) A, E, D, C, B
   (2) E, A, C, D, B
   (3) D, E, A, C, B
   (4) C, B, A, E, D
   (5) C, B, D, A, E

9. On the shelves at Sal's grocery there are 80 quarts of whole milk out of a total of 128 quarts of milk products. What is the ratio of the number of quarts of whole milk to the number of quarts of other milk products?

10. The ratio of the amount of flour to the amount of sugar in a certain recipe is 5:2. If a cook uses 8 cups of flour, how many cups of sugar should he use?

For items 11 to 20, choose the correct answer.

11. A box contains 15 green tennis balls, 12 orange tennis balls, and 9 white tennis balls. What is the probability that the first ball someone picks from the box will be green?

   (1) $\frac{2}{5}$ **(4)** $\frac{5}{12}$

   (2) $\frac{3}{5}$ **(5)** $\frac{7}{12}$

   (3) $\frac{4}{5}$

12. In fact, the first tennis ball Alan took from the box in problem 11 was orange, and the next two he took were green. What is the probability that the fourth tennis ball he takes will be white?

(1) $\frac{9}{11}$ **(4)** $\frac{2}{3}$

(2) $\frac{8}{11}$ **(5)** $\frac{1}{3}$

(3) $\frac{3}{11}$

13. So far, the town of Greenport has raised $120,000 for harbor improvements. The ratio of the amount they have raised to the total amount they need is 3:5. How much do they still need to raise?

    (1) $200,000   **(4)** $72,000
    (2) $180,000   **(5)** $48,000
    (3) $80,000

14. For a circus performance, the ratio of the number of children's tickets which were sold to the number of adults' tickets was 5:2. 630 children's tickets were sold. Find the total number of tickets, including both children's and adults', which were sold.

    (1) 882   **(4)** 1890
    (2) 1260   **(5)** 2205
    (3) 1470

15. Which of the following expressions is a solution to 3:18 = 25:x?

    (1) $3 \times 25 \times 18$ **(4)** $\dfrac{18}{(3 \times 25)}$

    (2) $\dfrac{(3 \times 18)}{25}$ **(5)** $\dfrac{(18 \times 25)}{3}$

    (3) $\dfrac{(3 \times 25)}{18}$

16. Faye types an average of 50 words per minute. Which of the following expressions tells the number of minutes Faye needs to type a six-page report if each page has an average of 380 words?

    (1) $6 \times 380 \times 50$ **(4)** $\dfrac{(6 \times 50)}{380}$

    (2) $\dfrac{(6 \times 380)}{50}$ **(5)** $\dfrac{(50 \times 380)}{6}$

    (3) $\dfrac{380}{(6 \times 50)}$

17. Find the area in square feet of the figure shown at the right.

    (1) 120   **(4)** $6\frac{1}{4}$
    (2) 90   **(5)** 5
    (3) 10

    $s = 30$ in.

18. Find the volume in cubic feet of a rectangular container which has a length of 8 feet, a width of one yard, and a height of 3 inches.

    (1) 192   **(4)** 6
    (2) 72   **(5)** 3
    (3) 24

19. The Millers made a down payment of $6000 on a new house, and they paid the real estate broker $2400. The total price of the house was

$40,000. The down payment was what fraction of the price of the house?

(1) $\frac{3}{5}$

(2) $\frac{3}{20}$

(3) $\frac{3}{50}$

(4) $\frac{21}{50}$

(5) $\frac{21}{100}$

20. John rode his motorcycle 304 miles in $6\frac{1}{2}$ hours on $9\frac{1}{2}$ gallons of gasoline. What average number of miles did he ride on one gallon of gasoline?

(1) 16

(2) 19

(3) 32

(4) 48

(5) 64

Check your answers. Correct answers are on page 325. If you have at least sixteen answers correct, go to the next chapter. If you have fewer than sixteen answers correct, study the Fractions chapter before you go on.

3 Decimals

Objective

In this chapter you will

- Read and write decimals
- Add, subtract, multiply, and divide decimals
- Interchange decimals and fractions
- Round off decimals
- Solve word problems containing decimals
- Work with powers and square roots of decimals
- Use metric units of measurement
- Solve perimeter, area, and volume problems containing decimals
- Find the circumference and area of a circle; find the volume of a cylinder
- Compare and order decimals
- Read metric scales and tables containing decimals
- Sketch travel problems to find distance
- Find whether all the information necessary to solve a problem is given

Level 1   Decimal Skills

Preview

Directions: Solve each problem.

1. Write eighteen and twenty-three thousandths as a mixed decimal.

2. What is the sum of .409, .28, and .7?

3. What is the difference between .82 and .197?

4. What is the product of 4.5 and .26?

5. Change $\frac{4}{15}$ to a decimal.

Check your answers. Correct answers are on page 325. If you have at least four answers correct, do the Level 1 Review on pages 110–111. If you have fewer than four answers correct, study Level 1 beginning with Lesson 1.

Lesson 1   Decimals

A **decimal,** like a fraction, shows a part of some whole thing. Decimals are different from fractions in two ways. One difference is that the denominators of decimals are not written. The other difference is that only certain numbers—10, 100, 1000, etc.—can be decimal denominators.

Decimal denominators get their names from the number of **places** to the right of the decimal point. The decimal point itself does not take up a decimal place. Below is a list of place names. To the left of the point are the first four whole-number places. To the right of the point are the first six decimal places. Notice how the decimal place names all end in *-ths*. Be sure you know these place names before you go on.

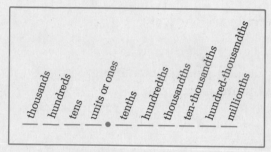

The most common application of decimals is our money system, which you have already used in this book. $.19 is a decimal. It represents 19 of the 100 pennies in a dollar. $.19 is the same as $\frac{19}{100}$ of a dollar.

A **mixed decimal** has a whole number to the left of the decimal point and a decimal fraction to the right. $2.35 is a mixed decimal. It represents two whole dollars and $\frac{35}{100}$ of another dollar.

Remember that a decimal gets its name from the number of places to the right of the decimal point. The number 386.4 has one decimal place because there is only one digit to the right of the point. The number 0.28 has two decimal places because there are two digits to the right of the point.

Zeros often cause trouble in decimals. The decimal .0807 has two zeros. These zeros keep 8 in the hundredths place and 7 in the ten-thousandths place. The decimal .0310 also has two zeros. However, the zero to the right of 1 is unnecessary. .0310 is the same as .031.

A decimal with no whole number is often written with a zero in the units place. Both 0.6 and .6 are acceptable.

Lesson 1   Exercise

Directions: *For items 1 and 2, tell the number of decimal places.*

1. a. 0.27   b. 21.486   c. 6.007
    d. 0.09546   e. 1240.06   f. 250.2

2. a. .37   b. 0.6134   c. .0034
    d. 1064.22   e. 9.1   f. .00604

For items 3 to 5, eliminate unnecessary zeros.

3.　a. 20.0670　　b. .4090　　　c. 028.70
　　 d. 01.20800　 e. 003.6　　　f. 04.500

4.　a. 03.040　　 b. .40　　　　c. .6070
　　 d. 0.003　　　e. .10700　　 f. .590

5.　a. 100.01070　 b. 0.39　　　 c. 74,000.10
　　 d. 6080.010　　e. 150.20　　 f. 19.20

Answers are on page 325.

Lesson 2　Reading and Writing Decimals

Remember that a decimal gets its name from the number of places to the **right** of the decimal point.

Reading Decimals

To read a decimal, count the number of places to the right of the decimal point. The last decimal place to the right corresponds to the name of the decimal. In mixed decimals, read the decimal point as the word "and."

Examples:

0.13　　is read "thirteen hundredths" because the number has two decimal places.

8.007　　is read "eight and seven thousandths" because the number has three decimal places.

0.0006　is read "six ten-thousandths" because the number has four decimal places.

Writing Decimals

When you write decimals, decide how many decimal places you need. Use zeros in places that are not filled in.

Examples:

Four thousandths = .004 or 0.004
　　because thousandths need three places. Notice how zeros fill the first two decimal places.
Sixteen and two hundredths = 16.02
　　because hundredths need two decimal places. Notice how a zero holds the first decimal place.

Lesson 2   Exercise

Directions: *For items 1 to 5*, write in the blank beside each number in the left-hand column the letter from the right-hand column which corresponds to each decimal or mixed decimal.

1. ____ 9.03   a. ninety and three hundredths

2. ____ 0.7   b. two and eight thousandths

3. ____ 0.0015   c. seven tenths

4. ____ 2.008   d. nine and three hundredths

5. ____ 90.03   e. fifteen ten-thousandths

For items 6 to 10, write each number as a decimal or a mixed decimal.

6. seven tenths

7. thirty-six thousandths

8. five hundred nineteen ten-thousandths

9. eight hundred fifty-four hundred-thousandths

10. three thousand seven hundred sixty-eight millionths

Answers are on page 325.

Lesson 3   Adding and Subtracting Decimals

Adding and subtracting decimals are easy operations. Be sure to line up each problem carefully.

To add decimals, line up the numbers with the decimal points under each other.

Example 1: Find the sum of .17, .9, and .256.

Step 1. Line up the numbers with decimal points under each other. Do not confuse the period at the end of the sentence with a decimal point.

$$\begin{array}{r} .17 \\ .9 \\ +.256 \\ \hline 1.326 \end{array}$$

Step 2. Add each column. 6 is the only digit in the thousandths column, and 7 and 5 are the only digits in the hundredths column. Notice how the sum of the tenths column carries over to the units.

When you add whole numbers with decimals or mixed decimals, put a point to the right of each whole number.

Example 2: Find the sum of 18, 2.35, and .482.

Step 1. Put a decimal point to the right of the whole number 18, and line up the numbers with the decimal points under each other.

$$\begin{array}{r} 18. \\ 2.35 \\ + \ \ .482 \\ \hline 20.832 \end{array}$$

Step 2. Add each column.

To subtract decimals, put a decimal point to the right of any whole number. Then line up the numbers with the decimal points under each other. Use zeros to give each number the same number of decimal places.

Example 3: Subtract .036 from .09.

Step 1. Line up the numbers with the decimal points under each other. Then put a zero to the right of .090 to give each number the same number of places.

$$\begin{array}{r} .090 \\ - .036 \\ \hline .054 \end{array}$$

Step 2. Borrow and subtract.

Example 4: Take .48 from 2.

Step 1. Put a decimal point to the right of 2, and line up the numbers with the decimal points under each other. Then put two zeros to the right of 2. to give each number the same number of places.

$$\begin{array}{r} 2.00 \\ - .48 \\ \hline 1.52 \end{array}$$

Step 2. Borrow and subtract.

Lesson 3   Exercise

Directions: Solve each problem.

1. .36 + .5 + .607 = ?   .38 + .619 + .2 = ?
    .3 + .9 + .7 = ?   .006 + .05 + .8 = ?

2. 2.5 + 18 + 1.07 = ?   .506 + 3.1 + 9 = ?
    38 + 4.078 + .0195 = ?   9.1 + .87 + .143 = ?

3. 6 − 2.5 = ?   8 − .19 = ?   .3 − .258 = ?   5.9 − 2.114 = ?
4. .015 − .009 = ?   1 − .0865 = ?   9 − .32 = ?   .6 − .24 = ?

5. .0174 − .0162 = ?   4.83 − 1.7123 = ?
    .62 + 3.18 + .0132 = ?   .206 + 3.605 + 2.7 = ?

Answers are on pages 325–326.

Lesson 4   Multiplying and Dividing Decimals

When you multiply and divide decimals, you do not have to line up the numbers with the decimal points under each other. However, in both multiplying and dividing, you must be careful to be sure you have the correct number of decimal places in your answers.

Multiplying Decimals

To multiply numbers with decimals, first set up the numbers for easy multiplication. Then count the number of decimal places in each number. Put the total number of places from the two numbers in the answer.

Example 1: Find the product of 3.17 and 8.

Step 1. Since 3.17 has more digits, put it on top and multiply.

Step 2. Count the decimal places in each number. 3.17 has two decimal places, and 8 has none. Put the total number of places, 2 + 0 = 2, in the answer.

```
  3.17 has 2 places
×    8 has 0 places
 25.36 has 2 places
```

In some problems you may have to add zeros to get enough decimal places in the answer. Note that multiplication by a number that is larger than 1 results in an answer that is larger than the number that was multiplied.

Example 2: Find the product of .06 and .4.

Step 1. Set up the problem and multiply.

Step 2. Count the decimal places in each number. .06 has two places, and .4 has one. Put the total, 2 + 1 = 3, in the answer. Put a zero to the left of 24 to get three decimal places.

```
  .06 has 2 places
× .4 has 1 place
 .024 has 3 places
```

Note that multiplication by a number that is smaller than 1 gives an answer that is smaller than the number multiplied.

Dividing Decimals by a Whole Number

When you divide a decimal (or a mixed decimal) by a whole number, set up your problem carefully. Then bring the decimal point up into the answer above its position in the problem and divide.

Example 3: Divide 2.88 by 6.

Set up the problem. Then bring the decimal point up into the answer above its position in the problem and divide.

```
     .48
6)2.88
   2 4
   ──
     48
     48
```

In some problems you will have to put zeros in your answer.

Example 4: Divide .348 by 4.

Set up the problem. Then bring the decimal
point up into the answer above its position in
the problem and divide. Notice the zero above
the 3. The zero shows that 4 does not divide
into .3. The zero also puts 8 in the hundredths
place.

$$\begin{array}{r} .087 \\ 4\overline{)\,.348} \\ \underline{32} \\ 28 \\ \underline{28} \end{array}$$

Dividing by Decimals

Dividing by decimals is complicated. The idea is to change the problem into
a similar problem in which the divisor is a whole number.

To divide by a decimal, first change the divisor into a whole number
by moving the decimal point to the right end of the divisor. Next, move the
decimal point in the dividend (the number you are dividing into) the same
number of places. If you move the point in the divisor one place, move the
point in the dividend one place. If you move the point in the divisor three
places, move the point in the dividend three places, and so on.

These steps are easier to understand with whole numbers. Look at the
problem 6 ÷ 2 = 3. The answer is the same if the decimal point is moved
one place to the right in both 6 and 2. The problem becomes 60 ÷ 20 = 3.
The problems are different, but the answers are the same.

Example 5: Divide 2.52 by .4.

Step 1. Set up the problem and make the divisor, .4, a whole number by moving the decimal
point one place to the right. Then move the
decimal point in the dividend the same number
of places, one.

$$.4\overline{)\,2.5.2}$$

Step 2. Bring the decimal point up into the
answer above its new position and divide.

$$\begin{array}{r} 6.3 \\ .4\overline{)\,2.5.2} \\ \underline{2\,4} \\ 1\,2 \\ \underline{1\,2} \end{array}$$

It is a good idea to check decimal division problems. Multiply the answer by the
decimal you divided by. The product should
equal the original dividend.

$$\begin{array}{r} 6.3 \\ \times\,.4 \\ \hline 2.52 \end{array}$$

When you divide a whole number by a decimal, you will have to add
zeros to the dividend in order to move the decimal point.

Example 6: Divide 45 by .05.

Step 1. Set up the problem and make the divisor, .05, a whole number by moving the decimal
point two places to the right. Then move the
decimal point in the dividend two places to the

$$.05\overline{)\,45.00.}$$

right. Remember that a whole number is understood to have a decimal point at the right end. Notice that we had to add two zeros to the dividend.

Step 2. Bring the decimal point up into the answer above its new position and divide.

$$.05\overline{)45.00.} \quad \begin{array}{r} 9\ 00. \\ \hline \end{array}$$

$$\begin{array}{r} 45 \\ \hline 0\ 00 \end{array}$$

To check the example, multiply the answer, 900, by the divisor, .05.

$$\begin{array}{r} 900 \\ \times\,.05 \\ \hline 45.00 = 45 \end{array}$$

Lesson 4 Exercise

Directions: Solve each problem.

1. $3.5 \times 7 = ?$ $.56 \times 8 = ?$

   $29 \times .04 = ?$ $.06 \times .5 = ?$

2. $.47 \times 16 = ?$ $.185 \times .4 = ?$

   $.82 \times 3.6 = ?$ $.59 \times .004 = ?$

3. $2.09 \times 30 = ?$ $.088 \times .52 = ?$

   $.0065 \times .6 = ?$ $215 \times .04 = ?$

4. $16.2 \div 6 = ?$ $4.32 \div 9 = ?$

   $24.195 \div 3 = ?$ $145.5 \div 15 = ?$

5. $.742 \div 14 = ?$ $47.5 \div 25 = ?$

   $1.152 \div 64 = ?$ $204.6 \div 31 = ?$

6. $128 \div .32 = ?$ $169 \div 2.6 = ?$

   $56 \div .08 = ?$ $108 \div 1.2 = ?$

7. $2.4 \div .008 = ?$ $261 \div .6 = ?$

   $465 \div 7.5 = ?$ $.312 \div .026 = ?$

8. $.63 \div .9 = ?$ $.252 \div .7 = ?$

   $.144 \div .03 = ?$ $.0318 \div .06 = ?$

9. $3.816 \div 3.6 = ?$ $.0192 \div .48 = ?$

   $.312 \div .052 = ?$ $.1875 \div .125 = ?$

10. $.304 \div .8 = ?$ $708 \div 1.5 = ?$

    $.6327 \div .703 = ?$ $.928 \div .8 = ?$

Answers are on page 326.

Lesson 5   Interchanging Decimals and Fractions

Later, especially when you work with percents, you will often need to change between decimals and fractions.

Changing Decimals to Fractions

To change a decimal to a fraction, write the digits in the decimal as the numerator. Write the name of the decimal—tenths, hundredths, thousandths, etc.—as the denominator. Then reduce the fraction.

Example 1: Change .08 to a fraction.

Write 08 as the numerator. The denominator is 100 because .08 has two places. Reduce $\frac{8}{100}$. $\frac{08}{100} = \frac{2}{25}$

Example 2: Change 9.6 to a mixed number.

Write 9 as a whole number and 6 as the numerator. The denominator is 10 because 9.6 has one decimal place. Reduce $9\frac{6}{10}$. $9\frac{6}{10} = 9\frac{3}{5}$

Changing Fractions to Decimals

The line in a fraction means "divided by." For example, $\frac{7}{20}$ means 7 divided by 20. To change a fraction to a decimal, divide the numerator by the denominator. Put a decimal point and zeros to the right of the numerator.

Example 3: Change $\frac{7}{20}$ to a decimal.

Divide 7 by 20. Put a decimal point and zeros to the right of 7. Bring the decimal point up into the answer.

$$
\begin{array}{r}
.35 \\
20\overline{)7.00} \\
\underline{6\ 0} \\
1\ 00 \\
\underline{1\ 00}
\end{array}
$$

With some fractions, the division will come out even with just one zero to the right of the decimal point. With other fractions the division will never come out even. You can stop after two decimal places and make a remainder with a fraction.

Example 4: Change $\frac{5}{6}$ to a decimal.

Divide 5 by 6. Put a decimal point and two zeros to the right of 5. Bring the decimal point up into the answer. Reduce $.83\frac{2}{6}$.

$$
\begin{array}{r}
.83\frac{2}{6} = .83\frac{1}{3} \\
6\overline{)5.00} \\
\underline{4\ 8} \\
20 \\
\underline{18} \\
2
\end{array}
$$

Lesson 5   Exercise

Directions: *For items 1 to 3,* change each decimal to a fraction and reduce.

1. .8 = ?   .04 = ?   .35 = ?   .005 = ?

2. .065 = ?   .0075 = ?   .002 = ?   .875 = ?

3. .6 = ?   .075 = ?   .006 = ?   .925 = ?

For items 4 and 5, change each mixed decimal to a mixed number and reduce.

4. 8.4 = ?   2.85 = ?   1.004 = ?   6.3 = ?

5. 10.125 = ?   3.009 = ?   4.80 = ?   12.16 = ?

For items 6 to 10, change each fraction to a decimal.

6. $\frac{9}{10}$ = ? $\frac{4}{5}$ = ? $\frac{9}{50}$ = ? $\frac{3}{8}$ = ?

7. $\frac{1}{2}$ = ? $\frac{2}{3}$ = ? $\frac{8}{25}$ = ? $\frac{1}{6}$ = ?

8. $\frac{2}{9}$ = ? $\frac{3}{4}$ = ? $\frac{5}{12}$ = ? $\frac{7}{8}$ = ?

9. $\frac{7}{10}$ = ? $\frac{5}{8}$ = ? $\frac{7}{12}$ = ? $\frac{1}{8}$ = ?

10. $\frac{4}{9}$ = ? $\frac{3}{5}$ = ? $\frac{5}{6}$ = ? $\frac{2}{50}$ = ?

Answers are on page 327.

Level 1   Review

Directions: Solve each problem.

1. Which of the following decimals is the same as 7.06?

    (1) seven and six tenths
    (2) seventy-six hundredths
    (3) seven and six hundredths
    (4) seventy-six thousandths
    (5) seven and six thousandths

2. Write sixty and twelve ten-thousandths as a mixed decimal.

3. What is the sum of .385, .6, and .09?

4. What is the difference between .058 and .0496?

5. What is the product of 12.8 and .35?

6. Divide 2.432 by 76.

7. Divide .658 by .07.

8. Change .625 to a fraction and reduce.

9. Change 4.32 to a mixed number and reduce.

10. Change $\frac{5}{16}$ to a decimal.

Check your answers. Correct answers are on page 327. If you have at least eight answers correct, go to the Level 2 Preview. If you have fewer than eight answers correct, go back to Lesson 1 and study Level 1.

Level 2   Decimal Applications

Preview

Directions: Solve each problem.

1. What is 0.2836 rounded off to the nearest thousandth?

2. To the nearest penny, what is the cost of 3.6 pounds of beef at $3.48 per pound?

3. What is the value of $(.5)^3$?

4. Change 655 millimeters to meters.

5. Find the area of a circle with a radius of 0.4 meters. Round off the answer to the nearest tenth.

Check your answers. Correct answers are on pages 327–328 If you have at least four answers correct, do the Level 2 Review on page 124. If you have fewer than four answers correct, study Level 2 beginning with Lesson 1.

Lesson 1   Rounding Off Decimals

The answers to many decimal problems have more places than they need. **Rounding off** makes decimals easier to read.
    Follow these steps to round off a decimal.

1. Mark the digit in the place you want to round off to.
2. If the digit to the right of the marked digit is more than 4, add 1 to the marked digit.
3. If the digit to the right of the marked digit is less than 5, leave the digit you marked as it is.
4. Drop the digits to the right of the digit you marked.

Example 1: Round off 3.18 to the nearest tenth.

Step 1. Mark the digit in the tenths place.   3.<u>1</u>8

Step 2. Since the digit to the right of 1 is more   3.2
than 4, add 1 to the marked digit and drop the
8.

Example 2: Round off 8.496 to the nearest hundredth.

Step 1. Mark the digit in the hundredths place.   8.4<u>9</u>6

Step 2. Since the digit to the right of 9 is more   8.50
than 4, add 1 to the marked digit and drop the
6. Notice that you must carry a digit over to the
tenths place.

Lesson 1   Exercise

Directions: *For items 1 and 2*, round off each number to the nearest tenth.

1. .634	.372	.08	5.16	0.3492
2. .15	.0764	4.81	.163	.74

For items 3 and 4, round off each number to the nearest hundredth.

3. .527	.483	2.019	8.296	0.9148
4. .796	.403	1.027	6.918	.093

For item 5, round off each number to the nearest thousandth.

5. .1386	.0577	1.7805	0.1052	6.4326

Answers are on page 328.

Lesson 2   Word Problems with Basic Operations

The next exercise gives you a chance to apply your decimals skills to some types of word problems you have already seen in this book. Division problems can be difficult. Remember that the number being divided must go inside the $\overline{)}$ sign.

Example 1: Tom wants to cut 13.5 yards of wire into six equal pieces. How long will each piece be?

The yards of wire are being cut. Put 13.5 inside
the $\overline{)}$ sign.

$$
\begin{array}{r}
2.25 \text{ yd.} \\
6\overline{)13.50} \\
\underline{12} \\
1\,5 \\
\underline{1\,2} \\
30 \\
\underline{30}
\end{array}
$$

Example 2: If a plane travels faster than the speed of sound for 3.5 hours
and flys 4690 miles, what is its average speed?

Use the formula $d = rt$. The distance is divided
by the time to get the rate.

$$
\begin{array}{r}
1340 \\
3.5\overline{)4690.0} \\
\underline{35} \\
119 \\
\underline{105} \\
140 \\
\underline{140} \\
00
\end{array}
$$

Lesson 2   Exercise

Directions: Solve each problem.

1. In 1980, the budget for services in Capital City was $12.5 million. By
    1985, the budget for services in Capital City increased $6.75 million
    over the 1980 budget. Find the amount of the 1985 budget for services.

2. Phil shipped four boxes. One box weighed 8 pounds; the second, 4.5
    pounds; the third, 7.65 pounds; and the fourth, 3.25 pounds. What was
    the mean weight of the boxes Phil shipped?

3. Before they went on vacation, the odometer in the Pagans' car read
    3789.6 miles. At the end of their vacation, the odometer read 4534.2
    miles. How many miles did they travel on their vacation?

4. If you multiply a batting average by the number of times at bat, you
    get the number of hits a player makes. Joe's batting average for the
    season was .268. He was at bat 45 times. How many hits did Joe make
    that season?

5. Sally bought 6.5 yards of lumber at a cost of $8.40 per yard. What was
    the total cost of the lumber?

6. Dan used 8.4 gallons of gas on a 200-mile trip. To the nearest gallon,
    find the average distance he drove on one gallon of gas.

7. It costs the Johnsons $0.036 for electricity to have their color television
    on for one hour. In a week the Johnsons watch television for 45 hours.
    How much does it cost them to watch television for a week?

8. At 10:00 A.M. Alfredo's temperature was 98.6°. At 12:00 noon it was

up 5.8°. By 2:00 P.M. it was down 3.9° from his noon temperature. What was Alfredo's temperature at 2:00 P.M.?

9. Steve drove for 6.25 hours at an average speed of 55 mph. To the nearest mile how far did he drive?

10. Nick makes $6.80 an hour. In a week he makes $238. How many hours does Nick work each week?

Answers are on page 328.

Lesson 3 Powers and Square Roots

To raise a decimal to a power, remember to count the total number of decimal places in the numbers you are multiplying.

Example 1: What is the value of $(.3)^2$?

Write .3 two times and multiply. Notice that the answer has two decimal places. $(.3)^2 = .3 \times .3 = .09$

To find the square root of a decimal, remember that the answer will have **half** as many places as the number you are finding the square root of.

Example 2: What is the value of $\sqrt{.0064}$?

The square root of 64 is 8. The square root of a four-place decimal has two places. Check by multiplying .08 by itself. $.08 \times .08 = .0064$. $\sqrt{.0064} = .08$

Lesson 3 Exercise

Directions: Find the value of each of the following expressions.

1. $(.5)^2 = ?$ $(.02)^3 = ?$ $(.4)^2 = ?$ $(.12)^2 = ?$

2. $(.07)^2 = ?$ $(.009)^2 = ?$ $(.1)^4 = ?$ $(1.5)^2 = ?$

3. $\sqrt{.16} = ?$ $\sqrt{.81} = ?$ $\sqrt{.0036} = ?$ $\sqrt{.0001} = ?$

4. $\sqrt{.0004} = ?$ $\sqrt{.0121} = ?$ $\sqrt{.000009} = ?$ $\sqrt{.0625} = ?$

5. $\sqrt{.0025} = ?$ $\sqrt{.0049} = ?$ $\sqrt{.64} = ?$ $\sqrt{.000144} = ?$

Answers are on page 328.

Lesson 4 Metric Measurement

Metric units of measurement are used in most countries outside the United States. One reason is its simplicity; there are only three basic units of measurement. Following are the three basic units of measure in the metric system.

1. The **meter** is the basic unit of length. A meter is a little longer than one yard.
2. The **gram** is the basic unit of weight. A gram is about $\frac{1}{30}$ of an ounce.
3. The **liter** is the basic unit of liquid measure. A liter is a little more than one quart.

To read metric measurements, learn the following prefixes.

kilo- (k)   means 1000.

deci- (d)   means .1 or $\frac{1}{10}$.

centi- (c) means .01 or $\frac{1}{100}$.

milli- (m) means .001 or $\frac{1}{1000}$.

Below is a list of the most common metric measurements. Abbreviations are in parentheses. Note that there is no s used in the abbreviation of a plural form. The list tells how many smaller units each bigger unit is equal to. Memorize any units you do not already know.

Length
1 kilometer (km) = 1000 meters (m)
1 meter   = 10 decimeters (dm)
1 meter   = 100 centimeters (cm)
1 meter   = 1000 millimeters (mm)

Weight
1 kilogram (kg) = 1000 grams (g)
1 gram   = 100 centigrams (cg)
1 gram   = 1000 milligrams (mg)

Liquid Measure
1 liter (l) = 10 deciliters (dl)
1 liter   = 100 centiliters (cl)
1 liter   = 1000 milliliters (ml)

To change from a **large** unit to a **small** unit, **multiply** by the number of small units that make up one large unit.

Example 1: Change 3.5 liters to milliliters.

Multiply 3.5 l by the number of milliliters in one liter, 1000.

$$\begin{array}{r} 3.5 \\ \times\,1000 \\ \hline 3500.0 = 3500 \text{ ml} \end{array}$$

To change from a **small** unit to a **large** unit, **divide** by the number of small units that make up one large unit.

Example 2: Change 85 centimeters to meters.

Divide 85 centimeters by the number of centimeters in one meter, 100.

$$100\overline{)85.00} = .85 \text{ m}$$

Lesson 4   Exercise

Directions: Solve each problem.

1. Change 2.5 meters to centimeters.
2. Change 3 kilograms to grams.
3. Change 4.8 liters to milliliters.
4. Change 6.5 kilometers to meters.
5. Change 1250 meters to kilometers.
6. Change 8 deciliters to liters.
7. Change 385 grams to kilograms.
8. Change 195 centimeters to meters.
9. Change 1680 milligrams to grams.
10. Change 1.6 liters to centiliters.

Answers are on page 328.

Lesson 5   Perimeter, Area, and Volume

You may see decimals in problems where you have to find perimeter, area, or volume.

Example: What is the area of the rectangle pictured at the right?

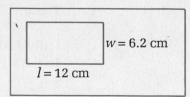

$w = 6.2$ cm
$l = 12$ cm

Step 1. Replace l with 12 and w with 6.2 in the formula $A = lw$.

$A = lw$
$A = 12 \times 6.2$

Step 2. Multiply across.

$A = 74.4$ cm

Lesson 5   Exercise

Directions: Solve each problem.

1. a. What is the perimeter of the figure at the right?
   b. What is the area of the figure?

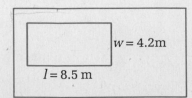

$w = 4.2$ m
$l = 8.5$ m

2. a. What is the perimeter of the figure
       pictured at the right?
    b. What is the area of the figure?

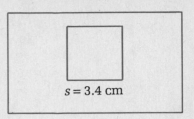

$s = 3.4$ cm

3. a. What is the perimeter of the figure
       shown at the right?
    b. What is the area of the figure?

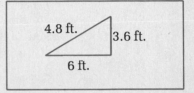

4.8 ft.   3.6 ft.

6 ft.

4. Find the area of the parallelogram at the
    right.

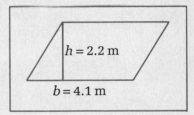

$h = 2.2$ m

$b = 4.1$ m

5. Find the volume of the figure at the right.

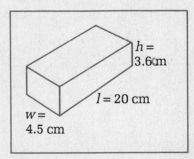

$h = 3.6$ cm

$l = 20$ cm

$w = 4.5$ cm

Answers are on page 328.

Lesson 6   Circumference and Area of Circles

So far in this book, you have found the perimeter and the area of figures
with straight sides. A **circle** is a curved line. Every point on the line is the
same distance from the center of the circle. Following are definitions of
some of the key words about circles.

The **diameter** is the distance across a circle. The diameter measures the widest part of a circle. The diameter always passes through the center.

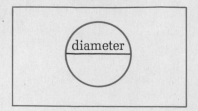

The **radius** is the distance from the center of the circle to the curved line that forms the circle. The radius is exactly one-half of the diameter. There are two formulas for the relationship between the diameter and the radius of a circle:

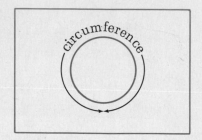

$d = 2r$ and $r = \frac{d}{2}$, where d stands for the diameter and r stands for the radius

The **circumference** is the distance around a circle. Circumference is the same as the perimeter of a circle.

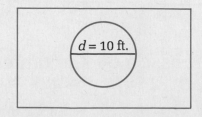

The Greek letter *pi*, or π, is the ratio of the circumference of a circle to its diameter. The value of π is usually given as 3.14.

The formula for the circumference (C) of a circle is:

$C = \pi d$, where $\pi = 3.14$ and d = diameter.

Remember that circumference is measured in units, such as inches, feet, yards, and meters.

The formula for the area (A) of a circle is:

$A = \pi r^2$, where $\pi = 3.14$ and r = radius.

Remember that area is measured in square units, such as square inches, square feet, square yards, and square meters.

Example 1: Find the circumference of the circle pictured at the right.

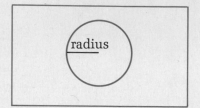

Step 1. Replace π with 3.14 and d with 10 in the formula for the circumference of a circle.

$C = \pi d$
$C = 3.14 \times 10$

Step 2. Evaluate the formula.

$C = 31.4$ ft.

Example 2: Find the circumference of the circle pictured at the right.

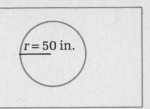

Step 1. Find the diameter of the circle.

$d = 2r$
$d = 2 \times 50$
$d = 100$ in.

Step 2. Replace π with 3.14 and d with 100 in the formula for the circumference of a circle.

$C = \pi d$
$C = 3.14 \times 100$

Step 3. Evaluate the formula.

$C = 314$ in.

Example 3: Find the area of the circle pictured at the right.

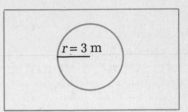

Step 1. Replace π with 3.14 and r with 3 in the formula for the area of a circle.

$A = \pi r^2$
$A = 3.14 \times 3^2$
$A = 3.14 \times 3 \times 3$

Step 2. Evaluate the formula.

$A = 28.26$ m^2

Lesson 6   Exercise

Directions: Solve each problem.

1. a. What is the radius of the circle pictured at the right?
    b. What is the circumference of the circle?
    c. What is the area of the circle?

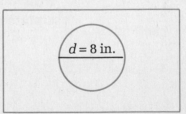

2. A circle has a radius of 30 feet.
    a. What is the diameter of the circle?
    b. What is the circumference of the circle?
    c. What is the area of the circle?

3. a. What is the diameter of the circle pictured at the right?

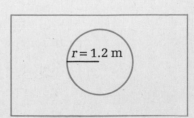

b. Find the circumference of the circle to the nearest tenth.
c. Find the area of the circle to the nearest tenth.

4. A circle has a diameter of 40 inches.
   a. Find the radius of the circle.
   b. FInd the circumference of the circle.
   c. Find the area of the circle.

5. A circle has a diameter of 2.8 inches. Use the improper fraction $\frac{22}{7}$ for the value of π.
   a. Find the circumference of the circle.
   b. Find the area of the circle.

Answers are on page 329.

Lesson 7   Volume of a Cylinder

You learned earlier that volume is a measure of the amount of space or the capacity inside a three-dimensional figure. So far, you have found the volume of figures with straight sides.

A **cylinder** is a figure shaped like a tin can. The top and the bottom of a cylinder are circles. The formula for the volume (V) of a cylinder is:

$V = \pi r^2 h$, where $\pi = 3.14$, r = radius, and h = height.

Remember that volume is measured in cubic units, such as cubic inches, cubic feet, cubic yards, and cubic meters.

Example: Find the volume of the cylinder pictured at the right.

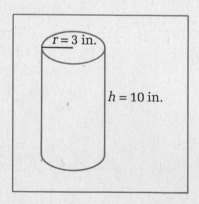

Step 1. Replace π with 3.14, r with 3, and h with 10 in the formula for the volume of a cylinder.

$V = \pi r^2 h$
$V = 3.14 \times 3^2 \times 10$

Step 2. Evaluate the formula.

$V = 3.14 \times 3 \times 3 \times 10$
$V = 282.6 \text{ in.}^3$

Lesson 7   Exercise

Directions: Find the volume of each of the cylinders pictured below.

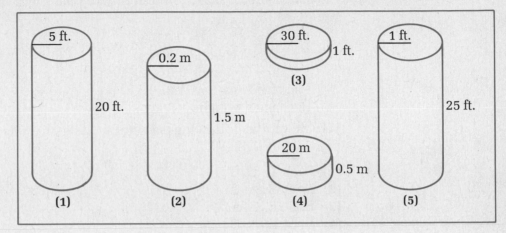

Answers are on page 329.

Lesson 8   Comparing and Ordering Decimals

To compare the size of decimals, give each decimal the same number of places by adding zeros. Then compare digits in each place, moving to the right, until you get to a place where the digits are not the same. The larger digit has the greater valuc.

Example: Arrange the following decimals in order from smallest to largest: 0.4, 0.404, 0.04.

Use zeros to give each decimal three decimal places. Since 0.040 is the smallest, 0.04 comes first. 0.400 or 0.4 is second, and 0.404 is largest.

$$0.4 \ \ \ = 0.400$$
$$0.404 = 0.404$$
$$0.04 \ \ = 0.040$$
In order: 0.04, 0.4, 0.404

Lesson 8   Exercise

Directions: Solve each problem.

1. Find the larger decimal in each pair.
    a. 0.056 or 0.05
    b. 0.19 or 0.2
    c. 1.08 or 1.082
    d. 0.075 or 0.57

2. Arrange in order from *smallest* to *largest*:
    0.021, 0.012, 0.21, 0.201.

3. Arrange in order from *largest* to *smallest*:
    0.38, 0.8, 0.083, 0.308.

4. Which is largest, 0.705 meter, 0.75 meter, or 0.075 meter?

5. Chester wants to arrange the following packages in order from heaviest to lightest.

    Package A weighs 0.65 kg.
    Package B weighs 1.05 kg.
    Package C weighs 0.065 kg.
    Package D weighs 1.65 kg.
    Package E weighs 1.5 kg.

    Which of the following sequences lists the packages in order from heaviest to lightest?

    (1) D, A, E, C, B
    (2) D, E, B, A, C
    (3) B, D, E, C, A
    (4) B, C, E, A, D
    (5) D, B, A, E, C

Answers are on page 329.

Lesson 9   Reading Metric Scales

A centimeter ruler is a tool for measuring length in the metric system. The longest lines on the ruler represent centimeters. The next longest lines represent half-centimeters (0.5 centimeters, or 5 millimeters). The shortest lines represent tenth-centimeters (0.1 centimeters, or 1 millimeter).

To measure a length with a centimeter ruler, determine how far from the left end a point is.

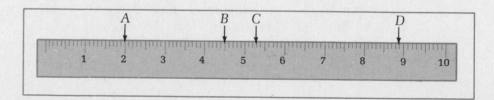

Example 1: How far from the left end of the 10-centimeter ruler pictured above is the point marked *A*?

A is at the second long line on the ruler.

A is 2 centimeters (20 millimeters) from the left.

Example 2: How far from the left end of the ruler is point *B*?

B is at the longest line between 4 and 5 centimeters.

B is 4.5 centimeters (45 millimeters) from the left.

Example 3: How far from the left end of the ruler is point C?

C is 3 short lines to the right of 5 centimeters.
C is 5.3 centimeters (53 millimeters) from the left.

Example 4: How far from the left end of the ruler is point D?

D is 9 short lines to the right of 8 centimeters.
D is 8.9 centimeters (89 millimeters) from the left.

Lesson 9   Exercise

Directions: *For items 1 to 10,* use the 10-centimeter ruler.

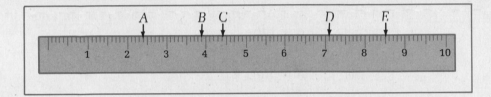

1. How far from the left end of the ruler is point A?

2. How far from the left end of the ruler is point B?

3. How far from the left end of the ruler is point C?

4. How far from the left end of the ruler is point D?

5. How far from the left end of the ruler is point E?

6. What is the distance between point B and point D on the ruler?

7. What is the distance between point A and point B?

8. What is the distance between point B and point C?

9. What is the distance between point C and point D?

10. What is the distance between point D and point E?

Answers are on page 329.

Level 2   Review

Directions: Solve each problem.

1. What is 6.2975 rounded off to the nearest hundredth?

2. A batting average is the number of hits a player gets, divided by the number of times at bat. Jeff got 17 hits out of 60 times at bat. Find his batting average to the nearest thousandth.

3. Celia gets an average of 22.5 miles on one gallon of gasoline. How far can she drive on 8.4 gallons of gas?

4. What is the value of $\sqrt{.0049}$?

5. Change 2.4 kilograms to grams.

6. Find the area of a triangle with a base of 5.4 meters and a height of 2.8 meters. Round the answer to the nearest tenth.

7. For the circle shown at the right, find the area to the nearest square inch.

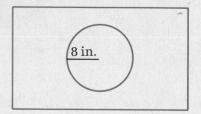

8. Find the volume of the cylinder pictured at the right.

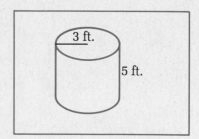

9. Arrange the following decimals in order from smallest to largest: 0.06, 0.066, 0.6, 0.065.

10. Point Y is how far from the left end of the centimeter scale shown below?

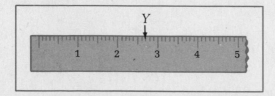

Check your answers. Correct answers are on pages 329–330. If you have at least eight answers correct, go to the Level 3 Preview. If you have fewer than eight answers correct, go back to Lesson 1 and study Level 2.

Level 3   Decimal Problem Solving

Preview

Items 1 and 2 are based on the following table.

Smokers and Nonsmokers (in millions)				
	Never Smoked	Used to Smoke	Smoke Now	Total Population
Male	28.8	22.2	27.6	78.7
Female	48.0	14.0	25.9	88.0

Source: U.S. National Center for Health Statistics

1. According to the table, how many more males smoke now than females? Choose from the following:

   (1) 1.3 million
   (2) 1.7 million
   (3) 2.7 million
   (4) 3.7 million
   (5) 8.2 million

2. The number of males who used to smoke is how many times as large as the number of females who used to smoke? Choose from the following:

   (1) more than $1\frac{1}{2}$ times
   (2) about 3 times
   (3) about 4 times
   (4) about $5\frac{1}{2}$ times
   (5) more than 6 times

3. Jack and Manny started driving at the same time and place and in the same direction. Jack drove for 2.5 hours at an average speed of 42 mph, and Manny drove for the same length of time at an average speed of 15 mph. How far apart were they at the end of 2.5 hours?

4. Two planes took off at the same time and place and headed in opposite directions. One plane flew east for 1.5 hours at an average speed of 350 mph, and the other flew west for 1.5 hours at an average speed of 420 mph. How far apart were they at the end of 1.5 hours?

5. Diane worked one week for 35 hours at her normal wage and six hours at her overtime rate of $6.80 an hour. How much did she make altogether that week?

(1) $159.80
(2) $219.30
(3) $238.00
(4) $278.80
(5) Insufficient data is given to solve the problem.

Check your answers. Correct answers are on page 330. If you have at least four answers correct, do the Level 3 Review on pages 131–132. If you have fewer than four answers correct, study Level 3 beginning with Lesson 1.

Lesson 1   Decimals and Tables

In the Whole Numbers Applications section of this book, you used tables to solve simple word problems. The numbers in tables are often expressed in decimals. For example, the table below tells the average number of hospital beds per 1000 people for several years.

Hospital Beds Per 1000 People						
	1960	1965	1970	1975	1980	1985
Rate per 1000	9.3	8.9	7.9	6.8	6.0	5.6*
* Estimate						

Source: American Hospital Association.

Example: According to the table, how many hospital beds did a city with a population of 500,000 have in 1975?

Step 1. The table tells the number of hospital beds per 1000 people. Divide the population by 1000.

$$1000 \overline{)500{,}000} \quad 500$$

Step 2. Multiply the 1975 rate, 6.8, by 500.

$$\begin{array}{r} 6.8 \\ \times 500 \\ \hline 3400.0 \text{ beds} \end{array}$$

Lesson 1   Exercise

Directions: *Items 1 to 5 are based on the above table.*

1. How many more hospital beds per 1000 people were there in 1960 than in 1985?

2. By how many beds per 1000 did the rate drop from 1965 to 1970?

3. The population of Greenport is 15,000. If Greenport followed the national average, how many hospital beds were there in Greenport in 1985?

4. The population of Capital City was 280,000 in 1980. If Capital City followed the national average, how many hospital beds were there in Capital City in 1980?

5. Which of the following statements describes the pattern for the rate of hospital beds per 1000 for the years shown in the table?
   (1) The rate first rose and then fell.
   (2) The rate first fell and then rose.
   (3) The rate gradually increased.
   (4) The rate gradually decreased.
   (5) No clear trend was discernible.

Items 6 to 10 are based on the following table.

Attendance at Selected Spectator Sports (in millions)			
	1975	1980	1984
Major League Baseball	30.4	43.7	45.2
National Football League	10.8	14.1	14.1
National Hockey League	9.5	10.5	11.4
Professional Basketball	7.6	10.7	11.1

Source: Statistical Abstracts of the United States.

6. For which year shown in the table was attendance at professional basketball games more than attendance at games of the National Hockey League?

7. Total attendance in 1984 at professional football, hockey, and basketball was how much less than attendance at major league baseball that year?

8. Attendance at major league baseball games was how much greater in 1980 than in 1975?

9. By how much did attendance at professional hockey games increase from 1975 to 1980?

10. Attendance at National Football League games in 1984 was about what fraction of attendance at major league baseball games that year? Choose from the following:
    (1) about $\frac{1}{4}$
    (2) about $\frac{1}{3}$
    (3) about $\frac{1}{2}$
    (4) about $\frac{2}{3}$
    (5) about $\frac{3}{4}$

Answers are on page 330.

Lesson 2   Sketching Travel Problems

You have already used the formula $d = rt$ to solve distance problems. In problems with more than one vehicle or traveler, it is useful to make a sketch of the trips. For the examples and the solutions to the next exercise, an arrow to the right means east. An arrow to the left means west. An arrow pointing up means north, and an arrow pointing down means south.

Example 1: Mark and Heather started hiking at the same time and the same place. Mark walked east at a rate of 5.5 mph, and Heather walked west at a rate of 4.5 mph. How far apart were they at the end of two hours?

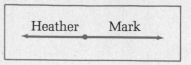

Step 1. Substitute each rate and a time of 2 hours in the formula $d = rt$.

$$d = rt \qquad d = rt$$
$$= 4.5 \times 2 \quad = 5.5 \times 2$$
$$= 9 \text{ mi.} \qquad = 11 \text{ mi.}$$

Step 2. Since they walked in opposite directions, the distance between them is the *sum* of the two distances.

$$9 + 11 = 20 \text{ mi.}$$

Example 2: Rick and Janina began driving along a highway at the same time and place and in the same direction. Rick drove at an average speed of 40 mph, and Janina drove at an average speed of 30 mph. How far apart were they after 1.5 hours?

Step 1. Substitute both rates and 1.5 hours into the formula $d = rt$. Notice that the arrows go in the same direction.

$$d = rt$$
$$= 40 \times 1.5 = 60 \text{ mi.}$$

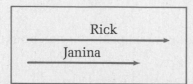

Step 2. Since they drove in the same direction, the distance between them is the *difference* between the two distances.

$$d = rt$$
$$= 30 \times 1.5 = 45 \text{ mi.}$$
$$60 - 45 = 15 \text{ mi.}$$

Lesson 2   Exercise

Directions: Solve each problem.

1. Pete drove west along a highway at an average speed of 55 mph. Lee began driving at the same time as Pete. She drove east at an average speed of 45 mph. How far apart were they after 0.5 hour?

2. Two friends began driving along a highway at the same time. One drove at an average speed of 40 mph, and the other at an average speed of 30 mph. How far apart were they at the end of 2.5 hours?

3. Two trains started at the same time, from the same place, on parallel tracks. One train traveled at an average speed of 60 mph. The other traveled at an average speed of 50 mph. How far apart were the two trains after 4.5 hours?

4. Miriam rode her bicycle west at an average speed of 12 mph. Her sister Marcia started riding her bicycle at the same time and rode east at an average speed of 10 mph. How far apart were they after 1.25 hours?

5. An express train started at 10:30 A.M. and traveled at an average speed of 50 mph. A local train started along the same track at 11:30 A.M. and traveled at an average speed of 30 mph. How far apart were the two trains at 1:00 P.M.?

6. Elizabeth drove her car at an average speed of 24 mph. John began bicycling at the same time and in the same direction as Elizabeth. He road at an average speed of 16 mph. How far apart were they at the end of 0.75 hour?

7. Two airplanes took off at the same time from the same place. One flew north at an average speed of 360 mph. The other flew south at an average speed of 480 mph. How far apart were the two planes at the end of 2.25 hours?

8. Stanton started bicycling at an average speed of 20 mph. Lil started hiking along the same road at the same time at an average speed of 4 mph. At the end of 3.5 hours, how much further ahead was Stanton?

9. Two delivery vans left from the distribution center at the same time, one traveling north at 55 mph and the other south at an average speed of 30 mph. How far apart were the two vans after 1.5 hours?

10. Juan drove east on a highway at an average speed of 40 mph for 4.5 hours. Carlos started driving at the same time as Juan. Carlos drove west for 3 hours at an average speed of 45 mph. Then he turned around and headed east at an average speed of 55 mph. How far apart were Juan and Carlos at the end of 4.5 hours?

Answers are on pages 330–331.

Lesson 3   Not Enough Information Is Given

On the GED Test and in the rest of this book, you will see multiple-choice questions where the last choice says, "Not enough information is given." This means that at least one number is missing and that there is no way to find a solution. To get practice with this kind of problem, try the next exercise.

Lesson 3   Exercise

Directions: For each problem tell what piece of information is missing. The first problem has been done as an example.

1. Guadalupe mailed four packages. Two of them weighed 4.75 pounds each, and a third weighed 5.6 pounds. Find the average weight of the four packages.

    [Answer: The weight of the fourth package is missing.]

2. Find the area of the triangle pictured at the right.

    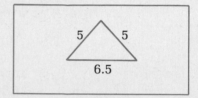

3. Mai Lee bought 6 pounds of chicken at $.89 a pound and 6 pounds of ground beef. What was the total cost of her purchases?

4. According to the table about attendance at spectator sports on page 127, how many of the spectators at National Football League games in 1980 were female?

5. According to the table about hospital beds per 1000 people on page 126, what was the average number of beds per hospital in 1975?

6. The Millers spend $280 a month for rent, $310 a month for food, and $89.90 a month for car payments. What decimal fraction of their income goes for these three expenses?

7. Manny worked 35 hours at his normal rate of $3.90 an hour and five hours at his overtime rate. How much did he make altogether that week?

8. Hans took 4.5 pounds of nails from a nail barrel to build the framing for a remodeled kitchen. Find the weight of the nails that remained in the barrel.

9. Mike uses about 30 gallons of gasoline a month to drive to and from work. He pays $.90 a gallon for the gasoline. Find Mike's total transportation costs, including parking, for a five-day work week.

10. The Jacksons' living room is 16 feet long and 12 feet wide. Find the volume of the living room in cubic feet.

Answers are on page 331.

Level 3   Review

Directions: *Items 1 to 4 are based on the following table.*

Per Capita Consumption of Selected Food Items (in pounds)						
Item	1960	1965	1970	1975	1980	1984
Pork	77.7	67.2	72.6	55.4	73.5	65.6
Fish	10.3	10.8	11.8	12.2	12.8	13.6
Cheese	8.3	9.6	11.5	14.3	17.6	21.7

Source: U.S. Department of Agriculture.

1. For which item shown in the table was the per capita consumption in 1984 less than the per capita consumption in 1960?

2. The per capita cheese consumption in 1984 was how many times as large as the per capita cheese consumption in 1960? Choose from the following:

    (1) about $1\frac{1}{2}$ times
    (2) more than $2\frac{1}{2}$ times
    (3) about 4 times
    (4) about 5 times
    (5) more than 6 times

3. By how many pounds did the per capita consumption of pork drop from 1970 to 1975?

4. In which year shown in the table did the consumption of cheese surpass the consumption of fish?

---

5. Phil hiked due north at a rate of 3.5 mph for 1.75 hours. Sue started at the same time and place and hiked due south at a rate of 4.5 mph. How far apart were they at the end of 1.75 hours?

6. An express train traveled for 2.25 hours at an average speed of 60 mph. A local train started at the same time as the express and traveled in the same direction on a parallel track at an average speed of 40 mph. How far apart were the trains at the end of 2.25 hours?

7. Sandy drove east for 3.5 hours at an average speed of 38 mph. Dick started at the same time as Sandy. Dick drove east for 2.5 hours at an average speed of 42 mph. Then he turned around and drove west for one hour at an average speed of 54 mph. How far apart were Sandy and Dick at the end of 3.5 hours?

8. What is the perimeter of a rectangle with a length of 6.25 m?

    (1) 18.75 m   **(4)** 6.25 m
    (2) 15 m   **(5)** Not enough information is given.
    (3) 12.5 m

9. George pays $2.75 a day to park at his job. To drive to his job, he uses 6.5 gallons of gasoline a week, which he buys for $0.90 a gallon. What are his total transportation costs, including parking, for one work week?

   (1) $ 8.60
   (2) $14.65
   (3) $19.60
   (4) $20.25
   (5) Not enough information is given

10. Cheryl bought 3 pounds of pork at $1.29 a pound and 4 pounds of beef. What was the total cost of her purchases?

    (1) $9.03
    (2) $7.87
    (3) $5.16
    (4) $3.87
    (5) Not enough information is given

Check your answers. Correct answers are on page 331. If you have at least eight answers correct, go to the Quiz. If you have fewer than eight answers correct, go back to Lesson 1 and study Level 3.

Chapter 3   Quiz

Directions: Solve each problem.

1. Write two hundred six and seven tenths as a mixed decimal.

2. What is 0.3962 rounded off to the nearest hundredth?

3. To the nearest penny, what is the cost of 6.5 gallons of gasoline at the rate of $0.895 per gallon?

4. Find the mean weight of the following parcels: 5.8 pounds, 2.75 pounds, 6.95 pounds, and 5.9 pounds.

5. From a chunk of cheddar cheese weighing 4.2 kilograms, Sam cut a piece weighing 1.95 kilograms. Find the weight of the remainder.

6. How much did Colin make for 7.5 hours of overtime if his overtime pay rate is $11.50 an hour?

7. A batting average is the number of hits a player gets divided by the number of times at bat. Andrew got 23 hits out of 75 times at bat. Find his batting average to the nearest thousandth.

8. What is the value of $(2.8)^2$?

9. Change 6.2 liters to milliliters.

10. Find the area of the rectangle pictured at the right.

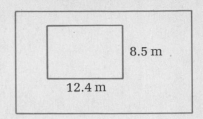

8.5 m

12.4 m

11. Find the circumference of a circle whose radius measures 0.75 meters.

12. What is the area, to the nearest tenth, of a circle with a radius of 1.6 meters?

13. Find the volume of a cylinder with a radius of 3 meters and a height of 10 meters.

14. Choose the sequence which shows the following lengths of wood in order from longest to shortest.

> A—1.4 m
> B—0.43 m
> C—1.3 m
> D—1.034 m
> E—0.34 m

(1) B, E, D, A, C   **(4)** B, A, E, C, D
(2) A, C, D, B, E   **(5)** C, D, B, E, A
(3) D, C, E, A, B

Items 15 to 18 are based on the following table.

Turkeys Raised in a Year (in millions)	
Arkansas	12.9
California	20.2
Minnesota	27.0
Missouri	13.0
North Carolina	29.4
Pennsylvania	6.8
Virginia	11.4
Wisconsin	7.1

Source: U.S. Department of Agriculture

15. Which of the states shown in the table raised the *third* largest number of turkeys?

(1) California
(2) Minnesota
(3) Missouri
(4) North Carolina
(5) Virginia

16. The total number of turkeys raised in the United States for the year shown in the table was 170.7 million. The four states which raised the most turkeys raised about what fraction of the national total?

(1) about $\frac{1}{10}$

(2) about $\frac{1}{4}$

(3) less than $\frac{1}{3}$

(4) over $\frac{1}{2}$

(5) Not enough information is given

17. Together, Pennsylvania and Wisconsin raised how many more turkeys than Arkansas?

(1) 3 million
(2) 2.5 million
(3) 2 million
(4) 1.5 million
(5) 1 million

18. For the states shown in the table, the state which raised the most turkeys raised about how many times as many turkeys as the state which raised the fewest?

(1) about $1\frac{1}{2}$ times
(2) less than 2 times
(3) about 3 times
(4) more than 4 times
(5) more than 10 times

___

19. At the same time, Pete drove west on Highway 40, and his son Chris drove east. Pete drove at an average speed of 55 mph for 3.5 hours. Chris drove for three hours at an average speed of 25 mph. Then he turned around and drove west for a half hour at an average speed of 40 mph. How many miles apart were Pete and Chris at the end of 3.5 hours?

(1) 287.5
(2) 247.5
(3) 212.5
(4) 180
(5) 117.5

20. Sal bought two shirts that cost $19.95 each. Find the total price of the shirts, including tax.

(1) $43.19
(2) $41.90
(3) $41.50
(4) $39.90
(5) Not enough information is given

Check your answers. Correct answers are on pages 331–332. If you have at least sixteen answers correct, go to the next chapter. If you have fewer than sixteen answers correct, study the Decimals chapter before you go on.

Chapter 4

Percents

Objective

In this chapter, you will

- interchange percents and decimals
- interchange percents and fractions
- find a percent of a number when a number is given
- find what percent one number is of another
- find a number when a percent of it is given
- solve word problems containing percents
- use the formula for interest
- solve multistep percent problems
- use percents in comparisons

Level 1   Percent Skills

Preview

Directions: Solve each problem.

1. Change the decimal .035 to a percent.

2. Change 84% to a fraction and reduce.

3. Change the fraction $\frac{7}{12}$ to a percent.

4. 120 is what percent of 48?

5. 75% of what number is 72?

Check your answers. Correct answers are on page 332. If you have at least four answers correct, do the Level 1 Review on pages 143–144. If you have fewer than four answers correct, study Level 1 beginning with Lesson 1.

Lesson 1   Interchanging Percents and Decimals

So far in this book you have used fractions and decimals as ways of describing parts of a whole. Percents are a third way to describe parts of a whole.

Percent means "out of 100." One whole is 100%, or 100 out of 100 parts. 1% means 1 out of 100 parts.

Percents look a lot like two-place decimals. Remember that two-place decimals are hundredths. 9% is the same as the decimal .09.

When you work with percent problems, you will usually change the percent in a problem to either a decimal or a fraction. The list below includes some of the most common percents along with the decimal form and the fraction form for each percent. Take the time now to memorize this chart.

Percent		Decimal		Fraction	Percent		Decimal		Fraction
25%	=	.25	=	$\frac{1}{4}$	20%	=	.2	=	$\frac{1}{5}$
50%	=	.5	=	$\frac{1}{2}$	40%	=	.4	=	$\frac{2}{5}$
75%	=	.75	=	$\frac{3}{4}$	60%	=	.6	=	$\frac{3}{5}$
$12\frac{1}{2}\%$	=	$.12\frac{1}{2}$	=	$\frac{1}{8}$	80%	=	.8	=	$\frac{4}{5}$
$37\frac{1}{2}\%$	=	$.37\frac{1}{2}$	=	$\frac{3}{8}$	10%	=	.1	=	$\frac{1}{10}$
$62\frac{1}{2}\%$	=	$.62\frac{1}{2}$	=	$\frac{5}{8}$	30%	=	.3	=	$\frac{3}{10}$
$87\frac{1}{2}\%$	=	$.87\frac{1}{2}$	=	$\frac{7}{8}$	70%	=	.7	=	$\frac{7}{10}$
$33\frac{1}{3}\%$	=	$.33\frac{1}{3}$	=	$\frac{1}{3}$	90%	=	.9	=	$\frac{9}{10}$
$66\frac{2}{3}\%$	=	$.66\frac{2}{3}$	=	$\frac{2}{3}$	$16\frac{2}{3}\%$	=	$.16\frac{2}{3}$	=	$\frac{1}{6}$
					$83\frac{1}{3}\%$	=	$.83\frac{1}{3}$	=	$\frac{5}{6}$

When you work with percents, you will usually change the percent in every problem to a decimal or to a fraction.

Changing Percents to Decimals

To change a percent to a decimal, move the decimal point two places to the *left,* and remove the percent sign.

Example 1: Change 45% to a decimal.

Move the decimal point two places to the left, and remove the percent sign. Notice that 45% is understood to have a decimal point at the right end.

$45\% = 45\% = .45$

Example 2: Change 80% to a decimal.

Move the decimal point two places to the left, and remove the percent sign. Notice that you can also drop the final zero. The zero does not affect the value of .8.

$80\% = 80\% = .8$

Example 3: Change 3.5% to a decimal.

Move the decimal point two places to the left, and remove the percent sign. Notice that you

$3.5\% = 03.5\% = .035$

must add a zero in order to get two places. This zero keeps 3 and 5 in their positions.

Changing Decimals to Percents

To change a decimal to a percent, move the decimal point two places to the *right,* and write a percent sign.

Example 4: Change .35 to a percent.

Move the decimal point two places to the right, and write a percent sign. You do not have to write the decimal point when it comes at the end.

$$.35 = .35 = 35\%$$

Example 5: Change 0.65 to a percent.

Move the decimal point two places to the right, and write a percent sign. You can drop the zero because it does not affect the value of 6 or 5.

$$.065 = .06\,5 = 6.5\%$$

Example 6: Change .4 to a percent.

Move the decimal point two places to the right, and write a percent sign. Here you must add a zero in order to get two places.

$$.4 = .40 = 40\%$$

Lesson 1　Exercise

Directions: *For items 1 to 5, change each percent to a decimal.*

1.	$75\% = ?$	$4\% = ?$	$62.5\% = ?$	$7\% = ?$
2.	$60\% = ?$	$150\% = ?$	$1\% = ?$	$300\% = ?$
3.	$8\frac{3}{4}\% = ?$	$12.6\% = ?$	$.6\% = ?$	$15\% = ?$
4.	$.1\% = ?$	$.25\% = ?$	$2.5\% = ?$	$200\% = ?$
5.	$12\% = ?$	$105\% = ?$	$9.7\% = ?$	$.2\% = ?$

For items 6 to 10, change each decimal to a percent.

6.	$.46 = ?$	$.08 = ?$	$.045 = ?$	$.08\frac{1}{3} = ?$
7.	$.9 = ?$	$.25 = ?$	$.005 = ?$	$.05 = ?$
8.	$.0825 = ?$	$.4 = ?$	$.675 = ?$	$1.2 = ?$
9.	$4.75 = ?$	$.625 = ?$	$8.0 = ?$	$.66\frac{2}{3} = ?$
10.	$.003 = ?$	$.80 = ?$	$6.35 = ?$	$.3\frac{2}{3} = ?$

Answers are on page 332.

Lesson 2 Interchanging Percents and Fractions

Percents are different from fractions in two ways. One difference is that 100 is the only denominator a percent can have. The other difference is that the denominator is not written. The percent sign, %, stands for the denominator 100, and the number before the percent sign is the numerator. 9% means 9 out of 100 equal parts. 9% is the same as the fraction $\frac{9}{100}$.

When you work with percent problems, you will change many of the percents to fractions.

Changing Percents to Fractions

To change a percent to a fraction, write the digits in the percent as the numerator, and write 100 as the denominator. Reduce the fraction to lowest terms.

Example 1: Change 16% to a fraction.

Write the digits in the percent as the numerator, and write 100 as the denominator. Reduce to lowest terms.

$$\frac{16}{100} = \frac{4}{25}$$

If the percent has a decimal in it, first change the percent to a decimal. Then change the decimal to a fraction.

Example 2: Change 3.5% to a fraction.

Step 1. Move the decimal point two places to the left.

$$3.5\% = 03.5 = .035$$

Step 2. Change .035 to a fraction and reduce.

$$\frac{35}{1000} = \frac{7}{200}$$

If the percent has a fraction in it, write the digits in the percent as the numerator, and write 100 as the denominator. Remember that the fraction line is an indication to divide. Divide the numerator by the denominator.

Example 3: Change $16\frac{2}{3}\%$ to a fraction.

Step 1. Write the digits in the percent as the numerator and 100 as the denominator.

$$16\frac{2}{3}\% = \frac{16\frac{2}{3}}{100}$$

Step 2. Divide the numerator by the denominator.

$$16\frac{2}{3} \div 100 =$$

$$\frac{50}{3} \div \frac{100}{1} =$$

$$\frac{\overset{1}{50}}{3} \times \frac{1}{\underset{2}{100}} = \frac{1}{6}$$

Changing Fractions to Percents

To change a fraction to a percent, find a fraction of 100%. In other words, multiply 100% by the fraction.

Example 4: Change $\frac{9}{20}$ to a percent.

Multiply 100% by the fraction.

$$\frac{9}{20} \times 100\% =$$

$$\frac{9}{20} \times \frac{\overset{5}{\cancel{100}}}{\underset{1}{1}} = 45\%$$

When you change a fraction to a percent, you may end up with a fraction in the answer. Then the percent can be written either with a decimal or with a fraction, whichever is easier to work with.

Example 5: Change $\frac{1}{9}$ to a percent.

Multiply 100% by the fraction.

$$\frac{1}{9} \times 100\% =$$

$$\frac{1}{9} \times \frac{100}{1} = 11\frac{1}{9}\%$$

Lesson 2   Exercise

Directions: For *items 1 to 5*, change each percent to a fraction and reduce.

1. $15\% = ?$ $85\% = ?$ $96\% = ?$ $60\% = ?$

2. $275\% = ?$ $8\% = ?$ $450\% = ?$ $42\% = ?$

3. $1.5\% = ?$ $4.8\% = ?$ $12.5\% = ?$ $6.25\% = ?$

4. $6\frac{1}{4}\% = ?$ $18\frac{3}{4}\% = ?$ $13\frac{1}{3}\% = ?$ $56\frac{1}{4}\% = ?$

5. $5\frac{5}{8}\% = ?$ $2.75\% = ?$ $80.5\% = ?$ $61\frac{1}{4}\% = ?$

For *items 6 to 10*, change each fraction to a percent.

6. $\frac{3}{5} = ?$ $\frac{12}{25} = ?$ $\frac{1}{2} = ?$ $\frac{2}{3} = ?$

7. $\frac{23}{100} = ?$ $\frac{3}{10} = ?$ $\frac{5}{8} = ?$ $\frac{3}{16} = ?$

8. $\frac{2}{7} = ?$ $\frac{5}{6} = ?$ $\frac{19}{20} = ?$ $\frac{4}{5} = ?$

9. $\frac{4}{5} = ?$ $\frac{3}{8} = ?$ $\frac{5}{7} = ?$ $\frac{8}{9} = ?$

10. $\frac{1}{6} = ?$ $\frac{3}{7} = ?$ $\frac{4}{15} = ?$ $\frac{1}{20} = ?$

Answers are on page 333.

Lesson 3   Finding a Percent of a Number

In the Fractions chapter, you learned that a fraction of a number means to multiply. A percent of a number also means to multiply. To multiply by a percent, first change the percent to either a decimal or a fraction. Then multiply by the decimal or the fraction.

With some problems, it is easier to change the percent to a decimal.

Example 1: Find 2.5% of 300.

Step 1. Change the percent to a decimal. $2.5\% = 02.5\% = .025$

Step 2. Multiply 300 by the decimal.

$$\begin{array}{r} .025 \\ \times\ 300 \\ \hline 7.500 = 7.5 \end{array}$$

With other problems, it is easier to change the percent to a fraction.

Example 2: Find $16\frac{2}{3}\%$ of 48.

Step 1. Change the percent to a fraction. $16\frac{2}{3}\%$ $16\frac{2}{3}\% = \frac{1}{6}$ is on the list you should have memorized.

Step 2. Multiply 48 by $\frac{1}{6}$. $\frac{1}{\underset{1}{6}} \times \frac{\overset{8}{48}}{1} = 8$

In some problems, you can change the percent to either a decimal or a fraction. Both methods are easy.

Example 3: Find 60% of 40.

Using a decimal:

Step 1. Change the percent to a decimal. $60\% = 60\% = .6$

Step 2. Multiply 40 by .6.

$$\begin{array}{r} 40 \\ \times\ .6 \\ \hline 24.0 = 24 \end{array}$$

Using a fraction:

Step 1. Change the percent to a fraction. $60\% = \frac{3}{5}$

Step 2. Multiply 40 by $\frac{3}{5}$. $\frac{3}{\underset{1}{5}} \times \frac{\overset{8}{40}}{1} = 24$

Lesson 3   Exercise

Directions: *For items 1 to 3, first change each percent to a decimal.*

1. 6% of 150 = ?   40% of 160 = ?

    5.4% of 80 = ?   0.8% of 50 = ?

2. 9% of 90 = ?   3% of 300 = ?

    125% of 36 = ?   1.9% of 200 = ?

3. 6.25% of 300 = ?   10.4% of 500 = ?

    12.8% of 80 = ?   .2% of 100 = ?

For items 4 to 6, first change each percent to a fraction.

4. $66\frac{2}{3}\%$ of 240 = ? $12\frac{1}{2}\%$ of 400 = ?

    $33\frac{1}{3}\%$ of 18 = ? $16\frac{2}{3}\%$ of 96 = ?

5. $12\frac{1}{2}\%$ of 24 = ? $37\frac{1}{2}\%$ of 1000 = ?

$62\frac{1}{2}\%$ of 48 = ? $83\frac{1}{3}\%$ of 120 = ?

6. $37\frac{1}{2}\%$ of 24 = ? $87\frac{1}{2}\%$ of 1200 = ?

$33\frac{1}{3}\%$ of 600 = ? $87\frac{1}{2}\%$ of 6000 = ?

For items 7 to 10, change each percent to either a decimal or a fraction.

7.   75% of 84 = ?   8.5% of 400 = ?

250% of 36 = ?   60% of 200 = ?

8.   20% of 200 = ?   150% of 1000 = ?

300% of 21 = ? $62\frac{1}{2}\%$ of 800 = ?

9.   90% of 130 = ?   50% of 28 = ?

25% of 116 = ?   35% of 260 = ?

10. 76% of 80 = ?   40% of 35 = ?

70% of 30 = ?   22% of 190 = ?

Answers are on pages 333–334.

Lesson 4   Finding What Percent One Number Is of Another

In some percent problems, you must find a percent as an answer. These problems are similar to the fraction problems in which you find what fraction one number is of another. When you find what percent one number is of another, you compare a *part* to a *whole*. First make a fraction with the part as the numerator and the whole as the denominator. Then change the fraction to a percent.

In these problems, the whole usually follows the word *of*.

Example 1: 18 is what percent of 24?

Step 1. Make a fraction with the part, 18, over the whole, 24, and reduce. $\frac{18}{24} = \frac{3}{4}$

Step 2. Change $\frac{3}{4}$ to a percent. $\frac{3}{4} \times \frac{\overset{25}{100}\%}{1} = 75\%$

In some problems, the whole you are comparing to the part may be smaller than the part.

Example 2: 42 is what percent of 30?

Step 1. Make a fraction with the part, 42, over the whole, 30, and reduce. $\frac{42}{30} = \frac{7}{5}$

Step 2. Change $\frac{7}{5}$ to a percent. $\frac{7}{5} \times \frac{\overset{20}{100}\%}{1} = 140\%$

Lesson 4   Exercise

Directions: Solve each problem.

1:   28 is what % of 70?   104 is what % of 160?   16 is what % of 20?

2.   45 is what % of 135?   24 is what % of 32?   30 is what % of 48?

3.   225 is what % of 375?   27 is what % of 72?   36 is what % of 24?

4.   15 is what % of 120?   36 is what % of 40?   150 is what % of 75?

5.   15 is what % of 90?   14 is what % of 20?   55 is what % of 220?

Answers are on page 334.

Lesson 5   Finding a Number When a Percent of It Is Given

Some percent problems seem backwards. In these problems, you have a number that is a percent of a missing number. Remember that when you found the percent of a number, you had to multiply. These problems are the opposite. To solve them, you must perform the opposite of multiplication, which is division.

To find a number when a percent of it is given, first change the percent to a decimal or to a fraction. Then divide the number you have by the decimal or the fraction.

Example 1: 35% of what number is 21?

Step 1. Change 35% to a decimal.

$$35\% = 35\% = .35$$

Step 2. Divide 21 by .35.

$$.35\overline{)21.00}$$
$$\underline{21\,0}$$
$$00$$

result 60

Example 2: $16\frac{2}{3}\%$ of what is 13?

Step 1. Change $16\frac{2}{3}\%$ to a fraction.

$$16\frac{2}{3}\% = \frac{1}{6}$$

Step 2. Divide 13 by $\frac{1}{6}$.

$$13 \div \frac{1}{6} =$$
$$\frac{13}{1} \div \frac{1}{6} =$$
$$\frac{13}{1} \times \frac{6}{1} = 78$$

It is a good idea to check these problems. When you find the missing number, multiply the percent by the number you found. The answer should be the original number in the problem.

For Example 1, find 35% of 60. $35\% = \underset{\smile}{35}\% = .35$ $\begin{array}{r} .35 \\ \times\ 60 \\ \hline 21.00 = 21 \end{array}$

For Example 2, find $16\frac{2}{3}\%$ of 78. $16\frac{2}{3}\% = \frac{1}{6}$ $\frac{1}{\underset{1}{6}} \times \frac{\overset{13}{78}}{1} = 13$

Lesson 5   Exercise

Directions: Solve each problem.

1. 60% of what number is 24?   25% of what number is 18?
2. $12\frac{1}{2}\%$ of what number is 15?   4% of what number is 9.6?
3. $66\frac{2}{3}\%$ of what number is 52? $87\frac{1}{2}\%$ of what number is 112?
4. 2.5% of what number is 8?   35% of what number is 140?
5. 90% of what number is 23.4? $62\frac{1}{2}\%$ of what number is 45?
6. 8.5% of what number is 17? $16\frac{2}{3}\%$ of what number is 36?
7. $37\frac{1}{2}\%$ of what number is 24?   40% of what number is 48?
8. $33\frac{1}{3}\%$ of what number is 112?   18% of what number is 36?
9. $83\frac{1}{3}\%$ of what number is 50?   22% of what number is 77?
10. 32% of what number is 112?   16% of what number is 32?

Answers are on page 334.

Level 1   Review

Directions: Solve each problem.

1. Change 9.6% to a decimal.
2. Change the decimal .0145 to a percent.
3. Change $8\frac{1}{3}\%$ to a fraction and reduce.
4. Change the fraction $\frac{9}{16}$ to a percent.
5. Find 4.5% of 2400.
6. What is $83\frac{1}{3}\%$ of 330?
7. 24 is what percent of 72?
8. 105 is what percent of 60?
9. 65% of what number is 52?
10. $16\frac{2}{3}\%$ of what number is 25?

Check your answers. Correct answers are on page 335. If you have at least eight answers correct, go to the Level 2 Preview. If you have fewer than eight answers correct, go back to Lesson 1 and study Level 1.

Level 2   Percent Applications

Preview

Directions: Solve each problem.

1. Out of 60 questions on a test, Joe got 80% of them right. How many questions did he get right?

2. Of the 24 students in Adrienne's class, 20 drive to school. What percent of the students in the class drive to school?

3. Out of 500 people who were interviewed, only 65 agreed with the president's foreign policy. Of the people who were interviewed, what percent agreed with the president's foreign policy?

4. So far, Sam has saved $1350 for a new motorcycle. This is 75% of what he needs. Find the total that he needs.

5. Find the interest on $1400 at 6.5% annual interest for one year.

Check your answers. Correct answers are on page 335. If you have at least four answers correct, do the Level 2 Review on page 150. If you have fewer than four answers correct, study Level 2 beginning with Lesson 1.

Lesson 1   Finding a Percent of a Number: Word Problems

By far the most common application of percents is in problems where you must find a percent of a number. Remember that this means to multiply.

Students often find percent problems difficult because there is more than one way to solve them. You can change the percent to a decimal or you can change the percent to a fraction. To build your confidence, try both methods. The solution to each example and exercise shows the method the author finds easier.

Example 1: Max bought a shirt for $19.80. The sales tax in his state is 6%. How much was the sales tax for the shirt?

Step 1. Change 6% to a decimal. $6\% = \underline{06}\% = .06$

Step 2. Multiply the price of the shirt by .06, $19.80
and round your answer off to the nearest penny. $\times$ .06
$1.1880 to the nearest
penny = $1.19

In many problems where you find the percent of a number, you will have to add or subtract your answer to another number in the problem.

Example 2: An electric drill normally selling for $35 was on sale for 20% off. Find the sale price of the drill.

Step 1. Change 20% to a fraction. $20\% = \frac{1}{5}$

Step 2. Find $\frac{1}{5}$ of $35. $\frac{1}{\overset{}{5}} \times \frac{\overset{7}{35}}{1} = 7$

Step 3. Subtract $7 from the original price. $35
                                                        $-$ 7
                                                      $28

Lesson 1   Exercise

Directions: Solve each problem.

1.   The Greens put a 6% down payment on a $48,000 house. How much was the down payment.

2.   Of the 320 employees at the County Hospital, 65% of them are women. How many women work at the hospital?

3.   A jacket that regularly sells for $69 was on sale for 15% off.
     a. How much can you save by buying the jacket on sale?
     b. What was the sale price of the jacket?

4.   Jeff sells cars on a 5% commission. One month he sold cars for a total value of $36,400. Find the total of his commissions for the month.

5.   Deborah owes $430 on her credit card. She has to pay a fee of 1.5% every month on the amount she owes. Find the monthly fee on the amount she owes.

6.   In the three years the Smiths have owned their house, the value of the house has increased $37\frac{1}{2}\%$. They paid $33,600 for their house. Find the value of the house now.

7.   Selma bought a portable cassette player for $49 and cassettes for $12. She had to pay 7% sales tax. Find the total price of the player and the cassettes, including tax.

8.   Jack pays $28 for each pair of shoes that he sells in his store. He puts a 40% markup on each pair of shoes. To the nearest dollar, how much does he charge for a pair of shoes?

9. The amount Greg owes on his credit card is $1650. The monthly interest charge is 1.8%. What is the monthly charge on the amount he owes?

10. The Korskys' house has increased in value $16\frac{2}{3}\%$ since they bought it five years ago. If they paid $90,000 for the house, what is its current value?

Answers are on page 335.

Lesson 2 Finding What Percent One Number Is of Another: Word Problems

Finding what percent one number is of another is the easiest kind of percent problem to recognize. In these problems, you are always looking for the percent. Make a fraction with the part over the whole. Then change the fraction to a percent.

Example: Of the 20 students in George's math class, 12 of them are women. Women are what percent of the class?

Step 1. Make a fraction with the part (the number of women) over the whole (the total number of students) and reduce.

$$\text{women} \atop \text{total}\quad \frac{12}{20} = \frac{3}{5}$$

Step 2. Change $\frac{3}{5}$ to a percent.

$$\frac{3}{5} \times \frac{\overset{20}{\cancel{100}}\%}{1} = 60\%$$

Lesson 2 Exercise

Directions: Solve each problem.

1. On a test with 40 questions, Sandy got 34 right. What percent of the questions did she get right?

2. In the shop where Fred works, 6 of the 48 employees were late one morning because of icy roads. What percent of the employees were late?

3. For a shirt that cost $22.50, Bill paid $1.80 in sales tax. The sales tax was what percent of the price?

4. Of the 320 people who ate at the Riverside Restaurant on Saturday, 280 paid cash. What percent of the diners paid cash?

5. In 1980, Suzanne's car was worth $6400. By 1986, the car was worth only $2560. The 1986 value of the car was what percent of the 1980 value?

6. Paul takes home $1200 a month and pays $252 a month for rent. His rent is what percent of his take-home pay?

7. The Browns bought a house for $24,000. Fifteen years later they sold it for $42,000. The sale price was what percent of the price they paid for the house?

8. John weighed 250 pounds in January. By the following December he had lost 75 pounds. What percent of his January weight did he lose?

9. Of 72 people interviewed, 60 agreed that expanding the fitness program at the community center was a good idea. What percent of the people interviewed agreed?

10. Brenda bought an automatic coffee maker for $60. If she paid $5 in sales tax, what percent of the price was the sales tax?

Answers are on page 335.

Lesson 3 Finding a Number When a Percent of It Is Given: Word Problems

Finding a number when a percent of it is given is the hardest type of percent problem to recognize. Remember that these problems are "backwards." You must divide by the percent to solve them. In these problems, you have the part. You are looking for the whole that the part is based on. Again, remember that you can change the percent to either a decimal or a fraction.

Example: Sarah got 15 problems right on a test. These problems were 75% of the test. How many problems were on the test?

Step 1. Change 75% to a decimal. $75\% = 75\% = .75$

Step 2. Divide the part (the number of problems she got right) by .75. There were 20 problems on the test.

$$.75\overline{)15.00} = 20$$

Lesson 3 Exercise

Directions: Solve each problem.

1. Phil got 16 problems right on a test. This was 80% of the total. How many problems were on the test?

2. Alfonso paid $630 as a down payment on a car. This was 9% of the total price. Find the total price.

3. So far, the Allens have driven 420 miles on their way to visit their

grandchildren. Mr. Allen figures that this is 75% of the total distance they must drive. How far do they have to drive altogether?

4.   Twenty-four women work at the Atlas Foundry. Women make up 15% of the total number of workers. How many people work at the foundry?

5.   By April, the Fifth Street Block Association had $1440 in a fund to buy trees. This was 60% of their goal. How much do they need altogether?

6.   Julio pays 22% of his gross income for taxes. Last year he paid $5280 in taxes. What was his gross income for the year?

7.   Mark weighs 180 pounds. This is $66\frac{2}{3}$% of the amount he weighed five years ago. Find his weight five years ago.

8.   Of the total number of people who were interviewed, 450 agreed to pay higher taxes for new school construction. That was $37\frac{1}{2}$% of the people who were interviewed. How many people were interviewed?

9.   Ellyn has driven 40% of the way to Lawrenceville. If she has already driven 84 miles, what is the total distance to Lawrenceville?

10. Margaret pays 18% of her gross income in taxes. If she paid $3960 in taxes, what was her gross income?

Answers are on pages 335–336.

Lesson 4   Interest

Interest is the money that someone pays for using someone else's money. A bank pays interest for the use of a customer's money in a savings account. A customer pays a bank interest for using the bank's money on a loan.

The formula for finding interest is:

$i = prt$, where p = principal (the money borrowed or loaned), r = rate (measured in percent), and t = time (the number of years)

Remember that when letters stand next to each other in a formula, you must multiply the values together.

Example 1: Find the interest on $700 at 9% annual interest for one year.

Step 1. Change 9% to a fraction. $9\% = \frac{9}{100}$

Step 2. Replace p with $700, r with $\frac{9}{100}$, and t $i = \frac{\overset{7}{\cancel{700}}}{1} \times \frac{9}{\underset{1}{\cancel{100}}} \times 1 = \$63$
with 1 in the formula $i = prt$, and evaluate the
formula.

When the time in an interest problem is not one year, first change the time to a fraction of a year.

Example 2: Find the interest on $900 at 11% annual interest for six months.

Step 1. Change 6 months to a fraction of a year. $6 \text{ mo} = \frac{6}{12} = \frac{1}{2} \text{ yr.}$

Step 2. Change 11% to a fraction. $11\% = \frac{11}{100}$

Step 3. Replace p with $900, r with $\frac{11}{100}$, and t with $\frac{1}{2}$ in the formula $i = prt$, and evaluate the formula.

$i = \frac{\overset{9}{\cancel{900}}}{1} \times \frac{11}{\cancel{100}} \times \frac{1}{2} = \frac{99}{2}$
$= \$49.50$

Remember that you can change the percent to either a fraction or a decimal. However, the fraction form fits most easily into the formula.

When the percent contains a fraction, you can put the denominator 100 under the principal. Notice carefully how the next example is set up.

Example 3: Find the interest on $400 at $3\frac{1}{2}$% annual interest for one year and three months.

Step 1. Change 1 year 3 months to a mixed number. $1 \text{ yr. 3 mo.} = 1\frac{3}{12} = 1\frac{1}{4} \text{ yr.}$

Step 2. Change $3\frac{1}{2}$% to a fraction. $3\frac{1}{2}\% = \dfrac{3\frac{1}{2}}{100}$

Step 3. Replace p with $400, r with $3\frac{1}{2}$%, and t with $1\frac{1}{4}$. Notice that the 100 from the percent fits under the 400. Then evaluate the formula.

$i = \frac{\overset{4}{\cancel{400}}}{\cancel{100}} \times \frac{7}{2} \times \frac{5}{4} = \frac{35}{2} =$
$\$17.50$

Lesson 4   Exercise

Directions: Find the interest for each of the following:

1. $600 at 4% annual interest for one year

2. $400 at 8.5% annual interest for one year

3. $4000 at 12% annual interest for one year

4. $1200 at $5\frac{1}{4}$% annual interest for one year

5. $500 at 8% annual interest for six months

6. $1000 at 9% annual interest for eight months

7. $400 at $5\frac{1}{2}$% annual interest for one year and three months

8. $800 at 4.5% annual interest for nine months

9. $2500 at 15% annual interest for three years

10. $1800 at $3\frac{3}{4}$% annual interest for ten months

Answers are on page 336.

Level 2   Review

Directions: Solve each problem.

1. Of the 1250 buckets produced each week at Plastic Products, Inc., 2% are defective. How many defective buckets are made each week?

2. Miriam made a down payment of 6% on a car that cost $6850. Find the amount of the down payment.

3. When Pete started working at his present job, he made $6.50 an hour. Now, three years later, he makes 18% more. How much does he make now?

4. The auditorium in the Greenport Town Hall has 520 seats. At a recent public meeting, 65 seats were empty. What percent of the seats were empty?

5. Louise paid $.75 sales tax on a book which cost $12.50. The tax was what percent of the price of the book?

6. The Johnsons spend $115 a week for food. Their weekly budget is $345. What percent of their budget goes for food?

7. The Salgados have paid off 60% of their mortgage. So far, they have paid $16,800. Find the total amount of the mortgage.

8. Pat paid $4.96 sales tax on a typewriter. Tax in her state is 4%. Find the price of the typewriter.

9. Find the interest on $2400 at 14.5% annual interest for one year.

10. Find the interest on $360 at $7\frac{1}{2}$% annual interest for one year and eight months.

Check your answers. Correct answers are on page 336. If you have at least eight answers correct, go to the Level 3 Preview. If you have fewer than eight answers correct, go back to Lesson 1 and study Level 2.

Level 3   Percent Problem Solving

Preview

Directions: Solve each problem.

1. Gail paid $192 for a television which was on sale. The sale price was 80% of the original price. How much did Gail save by buying the television on sale?

2. **Chris is spending his vacation hiking. So far he has walked 54 miles, which is 75% of his goal. How much farther does he have to walk?**

Items 3 to 5 are based on the following passage.

Gambol's Department Store had a three-month clearance sale when they went out of business. In July, every item in the store was on sale for 15% off. In August, every item was on sale for 25% off. For the first three weeks of September, every item was on sale for 50% off. Then, during the last week of September, every item was on sale for 80% off. Bill and Caroline have been thinking of buying a carpet originally selling at Gambol's for $700 and a stereo set originally selling for $400.

3. **What was the price of the carpet at Gambol's in August?**

4. **Find the difference in the price of the carpet between August and early September.**

5. **What was the combined price of the stereo and the carpet, including 5% sales tax, during the last week of September?**

Check your answers. Correct answers are on page 336. If you have at least four answers correct, do the Level 3 Review on pages 155–156. If you have fewer than four answers correct, study Level 3 beginning with Lesson 1.

Lesson 1   Finding What Percent One Number Is of Another: Multistep Problems

In the Applications section of this chapter, you learned to solve word problems where you had to find what percent one number is of another. The key to solving these problems is to first make a fraction with the part over the whole. Sometimes one of the pieces you need is missing.

Example 1: Joyce got 16 problems right and 4 problems wrong on a test. What percent of the problems did she get right?

Step 1. Find the whole (the total number of problems).

$$\begin{array}{r} 16 \text{ right} \\ + \ 4 \text{ wrong} \\ \hline 20 \text{ total} \end{array}$$

Step 2. Make a fraction with the part (the number right) over the whole (the total), and reduce the fraction.

$$\frac{\text{right}}{\text{total}} \ \frac{16}{20} = \frac{4}{5}$$

Step 3. Change $\frac{4}{5}$ to a percent.

$$\frac{4}{5} \times \frac{100\%}{1} = 80\%$$

In problems where an amount changes over a period of time, think of the part over the whole as the change over the original. Remember that the original is always earlier in time. These problems are called "percent of increase" or "percent of decrease" problems.

Example 2: Membership in the Southside Tenants' Union rose from 250 to 300. By what percent did the membership increase?

Step 1. Find the amount of change. Subtract the new membership from the old membership.

$$\begin{array}{r} 300 \\ -250 \\ \hline 50 \text{ new members} \end{array}$$

Step 2. Make a fraction with the change (the number of new members) over the original (250), and reduce.

$$\frac{\text{change}}{\text{original}} \quad \frac{50}{250} = \frac{1}{5}$$

Step 3. Change $\frac{1}{5}$ to a percent.

$$\frac{1}{\overset{1}{5}} \times \frac{\overset{20}{100\%}}{1} = 20\%$$

You will often see this last kind of problem in business applications. When an item is on sale, there is a percent of decrease, or discount. When a store owner determines a price for an item, he puts a percent of increase, or markup, on the item.

Lesson 1   Exercise

Directions: Solve each problem.

1. The price of gas dropped from $.96 to $.88 a gallon. Find the percent of decrease.

2. A television which originally sold for $360 was on sale for $306. Find the discount rate (percent of decrease).

3. The population of Greenport was 12,000 ten years ago. Now it is 15,000. By what percent did the population increase?

4. Frank pays $16 each for the shirts he sells for $24 in his store. Find the markup rate (percent of increase) on each shirt.

5. In the factory where Gladys works there are 84 women and 126 men. What percent of the workers are women?

6. Sally spends $270 a month for food for her family. She has $630 left each month to cover all other expenses. What percent of her monthly budget goes for food?

7. Steve bought a car for $4800 and sold it three years later for $1800. By what percent did the value of the car drop?

8. There were 720 unemployed people in Greenport in October. By November, the number increased to 792. By what percent did the number of unemployed people increase from October to November?

9. The cost of renting a car increased from $40 to $45 a day. Find the percent of increase.

10. Laura sells a coat at her store for $200. If she paid $120 for the coat, what is the percent of markup?

Answers are on page 336.

Lesson 2   Finding a Number When a Percent of It Is Given: Multistep Problems

In the Applications section of this chapter, you learned that problems where you must find a number when a percent of it is given are "backwards." You have the part, and you are looking for the whole. These problems are even more difficult when the question asks for something besides the whole. Be sure you answer what the question asks. Think carefully about the next example.

Example: Alvaro has driven 240 miles. This is 80% of his goal. How much farther does he have to drive?

Step 1. Change 80% to a decimal. $80\% = .8$

Step 2. To find out how far he has to go, you must first find his goal. Divide the distance he has driven by .8.

$$\begin{array}{r} 30\,0 \text{ mi.} \\ .8\overline{)240.0} \end{array}$$

Step 3. Subtract the distance he has driven from the total distance he wants to drive.

$$\begin{array}{r} 300 \\ -\,240 \\ \hline 60 \text{ mi.} \end{array}$$

Lesson 2   Exercise

Directions: Solve each problem.

1. Jack has saved $270, which is 60% of the down payment he wants to make on a car. How much does he still need?

2. The Farmers' Party of Greenport now have 900 signatures on a petition to get their candidate on the ballot for mayor. This is 75% of the total that they need. How many more signatures do they need?

3. One-hundred twelve workers in a factory have agreed to join a union. This is 40% of the number the organizers need. How many more workers do they need?

4. Celeste paid $102 for a coat which was on sale. This was 85% of the original price. How much did Celeste save by buying the coat on sale?

5. Four-hundred twenty students in Capital City's night-school program passed the GED Test. This is 70% of the number who took the test. How many students failed?

6. Herb weighs 216 pounds, which is 90% of his weight one year ago. How much did Herb lose in a year?

7. The Greens have driven 338 miles. This is 65% of the total distance they want to drive. How much farther do they have to drive?

8. Bill's employer deducts $3465 a year for taxes and social security. This is 21% of Bill's gross salary. Find his net salary for the year.

9. Cal drove 230 miles before stopping. This was 40% of the distance he had to travel. How much further did he have to drive?

10. Al paid $450 for a video recorder which was on sale at 75% of the original price. How much did he save over the original price?

Answers are on page 337.

Lesson 3 Using Percents in Comparisons

There is a type of problem that you may see on the GED Test which takes a while to solve. In these problems, you must compare four or five choices in order to choose the best solution. These problems usually include some percent calculations in which you must find the percent of a number. The next exercise gives you a chance to try this type of problem. Be sure to work through each choice before you make your decision.

Lesson 3 Exercise

Directions: Solve each problem.

1. Mr. and Mrs. Cooke want to rent a cottage for the month of June. They have looked at four similar cottages. Below are the rent schemes for each cottage. Which one is least expensive?
   A. $35 a day for 30 days
   B. 15% of the yearly rent of $6000
   C. $25 a day plus a nonrefundable deposit of $250
   D. $285 a week for four weeks

2. The Parents' Association of the Fourth Street School have raised $10,000 for after-school programs. They need more money for the next six months. Which of the following plans will give them the most?
   A. 50% more than they currently have
   B. $800 a month for six months
   C. Three gifts of $1500 each
   D. $250 a week for half a year

3. The town of Eastport has a population of 25,000. Four different projections have been made about the growth of the town over the next three years. Which of the following would result in the largest growth in population?
   A. An increase of 15% over the entire three years
   B. 1300 more people per year for three years
   C. A 5% increase for the first year followed by an increase of 1000 people per year for each of the next two years
   D. 80 more people per month for the next three years

4. Ovidio wants to ship 500 pounds of freight from Boston to Colombia. Four different shippers have given him the following rates. Which is the least expensive?

    A. $300 for the first 100 pounds and $12.50 for each additional 10 pounds

    B. $100 down and $65 for every 50 pounds plus a 5% tax on the total, including the down payment

    C. $150 for each 100 pounds

    D. $1.25 a pound plus a $150 handling charge

5. The Crawleys are renting a car this summer for one month. The rates of four rental agencies are shown below. If the Crawleys expect to drive approximately 2000 miles, which one of the rental schemes is the best buy?

    A. $125 a week with free mileage and a 15% discount

    B. $50 a week and $.10 a mile

    C. $295 a month and $.05 a mile

    D. $450 with no mileage charge and a 10% discount

Answers are on page 337.

Level 3   Review

Directions: Solve each problem.

1. A shirt which originally cost $24 was on sale for $20.40. Find the percent of decrease in the price of the shirt.

2. Mike has saved $390 toward a down payment on a car. He still needs $260. What percent of the down payment has he saved so far?

3. Janet weighs 136 pounds, which is 85% of her weight two years ago. How much weight did she lose in two years?

4. At the Civic Auditorium on amateur night, there were 675 occupied seats. This is 75% of the capacity of the auditorium. How many seats were empty on amateur night?

5. Arturo now makes $1200 a month. He has four choices for raises. Which of the following will give him the biggest yearly raise?

    A. $40 a week

    B. A raise of $2000

    C. A 10% raise

    D. $550 each quarter

Items 6 to 10 are based on the following information.

Danny is thinking of taking a job selling farm equipment. He has a choice of three pay structures. He can work at a straight salary of

$14,000, from which his employer will deduct 20% for taxes and social security. He can work at a combined salary and commission rate. In this plan, he will make a salary of $6000 and he will get 6% commissions on everything that he sells. The third structure is 9% commissions only.

6. If Danny chooses the commissions-only structure, how much will he make in a year in which he sells $100,000 in farm equipment?

7. If Danny chooses the combination of salary and commissions, how much will he make if he sells $100,000 worth of equipment?

8. If Danny chooses the straight salary structure, what will his net income be for the year?

9. How much would Danny make on the commissions-only plan if he sells $250,000 worth of equipment?

10. Danny chose the combination of salary and commissions. His first year, he sold $180,000 in equipment. How much did he make?

Check your answers. Correct answers are on pages 337–338. If you have at least eight answers correct, go to the Quiz. If you have fewer than eight answers correct, go back to Lesson 1 and study Level 3.

Chapter 4 Quiz

Directions: Solve each problem.

1. Change .095 to a percent.

2. Change $41\frac{2}{3}$% to a fraction and reduce.

3. Find 7.6% of 230.

4. Of the 48 employees of the post office where Louie works, $16\frac{2}{3}$% of them walk to work. How many of the employees walk to work?

5. Herb bought a suit for $128. Find the cost of the suit including 6.5% sales tax.

6. What percent of 120 is 42?

7. Out of Laverne's gross salary of $11,200 a year she pays $1680 in federal tax. This tax is what percent of her salary?

8. For a power saw that cost $40, Paul paid $2.40 in sales tax. The tax was what percent of the price of the saw?

9. Forty-five percent of what number is 72?

10. A ship has traveled 210 miles on its way from New York City to Halifax, Nova Scotia. This distance is 35% of the total distance to Halifax. How far is it by ship from New York City to Halifax?

11. Find the interest on $600 at 8.5% annual interest for one year and six months.

12. Carmela pays $20 for each pair of shoes she sells in her store. She charges her customer $29 for the shoes. Find the percent of markup which she charges.

13. In the Central High School adult-education program there are 165 women and 135 men. What percent of the students are women?

14. In a recent election in Greenport, 7200 people voted. They were 60% of the eligible voters. How many people failed to vote in that election in Greenport?

15. David spends $1020 a year on car payments. This represents 8% of his net income. Find how much David has in a year for everything besides car payments.

16. Jeff is selling an old car for which he paid $4500. Which of the following offers is the best?
    A. $50 a month for two years
    B. $500 down and $60 a month for nine months
    C. 30% of the amount Jeff paid for the car
    D. $30 a week for 44 weeks

Items 17 to 20, are based on the following information.

The Andersons want to build a cabin on a small piece of land which they bought for $3500. They had to pay the agent who sold them the land a 6% commission. Mrs. Anderson's brother Ron is a builder. He estimates the total cost of materials for building a cabin will be $16,000 and that assorted expenses will be 10% of the price of materials. Ron will use an assistant whom he will pay $8 an hour. Ron expects the assistant to work four 40-hour weeks. Ron will charge the Andersons 20% of the total of materials, expenses, and the assistant's wages.

17. What is the total cost of the land, including the agent's commission?
    (1) $5600
    (2) $3710
    (3) $3500
    (4) $3290
    (5) Insufficient data is given to solve the problem.

18. Which of the following amounts must Ron pay for the assistant's employment insurance?
    (1) $ 800
    (2) $1120
    (3) $1280

(4) $3200

(5) Insufficient data is given to solve the problem.

19. What is Ron's estimate for assorted expenses in building the cabin?

(1) $3200

(2) $1600

(3) $1440

(4) $ 960

(5) Insufficient data is given to solve the problem.

20. How much does Ron expect to make for building the cabin?

(1) $3776

(2) $3520

(3) $3020

(4) $1888

(5) Insufficient data is given to solve the problem.

Check your answers. Correct answers are on page 338. If you have at least sixteen answers correct, go to the next chapter. If you have fewer than sixteen answers correct, study the Percents chapter before you go on.

5 Graphs

Objective

In this chapter, you will read and apply the information contained in

- pictographs
- circle graphs
- bar graphs and divided bar graphs
- line graphs

Introduction to Graphs

A **graph** is a diagram that compares the relative sizes of numbers. You often see graphs in newspapers and magazines. In this chapter, you will learn to read and interpret pictographs, circle graphs, bar graphs, and line graphs.

The questions based on graphs on the GED Test range from very simple ones where you read a piece of information from a graph, to difficult questions where you must use the information shown on a graph to arrive at logical conclusions.

This chapter is organized differently from the other chapters in this book. There is no Preview at the beginning. Work through the entire chapter in order to get practice with some of the types of problems you may see on the GED Test.

Lesson 1   Pictographs

A **pictograph** uses symbols to compare statistical data. On every pictograph, a **key** or legend explains what each symbol stands for. To understand a graph, first read the title. Then read the labels for the columns and rows of information.

The title of the pictograph on the following page is "Pigs Raised in One Year." The vertical column contains state names. The key at the end of the graph shows that each symbol stands for 1,000,000 pigs.

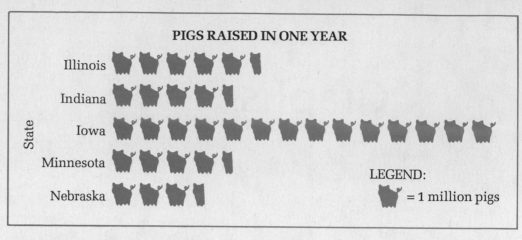

Example 1: How many pigs were produced in Nebraska in one year?

Count the number of symbols to the right of Nebraska. There are $3\frac{1}{2}$ symbols to the right of Nebraska. Since each symbol stands for 1 million pigs, multiply $3\frac{1}{2}$ by 1 million.

$3\frac{1}{2} \times 1$ million $=$

$3\frac{1}{2}$ million

Some graph questions give information which you must apply to the numbers that you get from the graphs.

Example 2: At a value of $200 a pig, find the total value of the pigs produced in Nebraska in one year.

Multiply the number of pigs produced in one year in Nebraska by the value of one pig.

$$\begin{array}{r} 3.5 \text{ million} \\ \times\, 200 \\ \hline 700.0 = \$700 \text{ million} \end{array}$$

Lesson 1 Exercise

Directions: Use the graph about pig production to answer the following questions.

1. Tell the number of pigs produced in one year in each of the following states.

    a. Illinois   b. Indiana   c. Iowa   d. Minnesota

2. Which state shown on the graph produces the most pigs in one year?

3. Together the states of Minnesota and Nebraska produce how many pigs in one year?

4. The combined production of pigs in the states of Illinois, Indiana, and Minnesota is which of the following:

    (1) about the same as the production in Nebraska
    (2) slightly more than the production in Iowa
    (3) slightly less than the production in Iowa
    (4) about one-third of the production in the United States
    (5) less than half the production in Iowa

5. The number of pigs produced in a year in Nebraska is what percent of the number produced in Iowa?

Items 6 to 10 refer to the following pictograph.

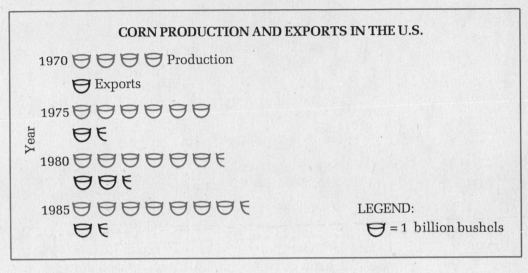

CORN PRODUCTION AND EXPORTS IN THE U.S.

LEGEND:

= 1 billion bushels

6. What does one symbol of a bushel of corn stand for?

7. Corn production in 1980 was how much greater than corn production in 1975?

8. The amount of corn produced in 1970 was what fraction of the amount produced in 1975?

9. For what year shown on the graph was the export of corn the highest percentage of the amount produced?

10. For what year shown on the graph was the export of corn the lowest percentage of the amount produced?

Answers are on page 339.

Lesson 2   Circle Graphs

A **circle graph** shows the parts of a whole. A pie-shaped piece of a circle graph usually stands for the percent of some whole or the number of cents that form a part of a whole dollar. When the pieces of a circle graph are measured in percent, the pieces add up to a total of 100%. When the pieces of a circle graph are measured in cents, the pieces add up to one dollar.

To understand a circle graph, read the title and the names of each category carefully.

The circle graph below shows how a dollar of the federal budget is spent in a year.

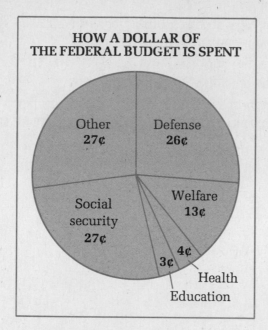

**HOW A DOLLAR OF
THE FEDERAL BUDGET IS SPENT**

Other 27¢

Defense 26¢

Social security 27¢

Welfare 13¢

4¢

3¢

Health

Education

Example: What percent of federal expenditures   4%
goes for health? Notice that all the parts of the
circle graph add up to $1.00. 4¢ of each dollar
goes to health. 4¢ is the same as $\frac{4}{100}$ or 4%.

Lesson 2   Exercise

Directions: Use the circle graph above to answer the following questions.

1. What percent of federal expenditures goes for each of the following
    categories?
    a. defense   b. welfare   c. social security   d. education

2. Which category shown on the graph represents the smallest federal
    expenditure?

3. The category of health makes up what fraction of the total yearly federal
    expenditures?

4. Together, health, education, and welfare make up what fraction of the
    total federal expenditures?

5. Together, defense and social security make up which of the following
    parts of the total federal expenditures in a year?

    (1) a little over $\frac{1}{4}$

    (2) a little over $\frac{1}{2}$

    (3) a little over 80%
    (4) approximately 95%
    (5) approximately 40%

Items 6 to 10 refer to the following graphs.

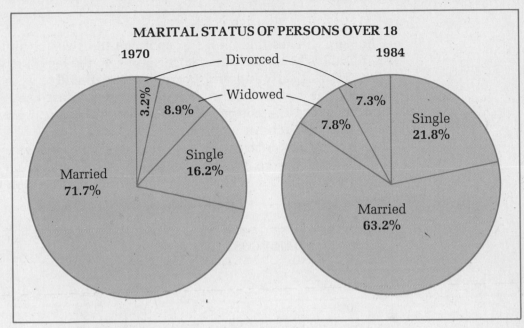

6. For both years shown on the graphs, which category represents the fewest people?

7. By how many percentage points did the number of adults who were divorced change from 1970 to 1984?

8. In 1970, widowed and single people made up about what fraction of the adults over 18?

    (1) $\frac{1}{6}$

    (2) $\frac{1}{4}$

    (3) $\frac{1}{3}$

    (4) $\frac{1}{2}$

    (5) $\frac{3}{4}$

9. By how many percentage points did the number of adults who were single change from 1970 to 1984?

10. Which of the following statements describes the change from 1970 to 1984 as shown by the two graphs?

    (1) While the percent of adults who were married or widowed decreased, the percent who were single or divorced increased.
    (2) While the percent of adults who were married or divorced increased, the percent who were single or widowed decreased.
    (3) The percent of each category increased.
    (4) While the percent of adults who were married or widowed increased, the percent who were single or divorced decreased.
    (5) While the percent of adults who were married or widowed decreased, the percent who were divorced tripled.

Answers are on page 339.

Lesson 3   Bar Graphs

A **bar graph** uses thick lines to compare the sizes of numbers. The size of the numbers being compared corresponds to the length of each bar. The bars may run vertically (up and down) or horizontally (from side to side). To estimate the number each bar stands for, you must use the **scale** along the side or across the bottom of the bar graph. Again, read the title and the categories on each bar graph carefully.

The bar graph below shows the average price of a semiprivate hospital room for six different years. The scale at the left of the graph shows the price in dollars. The scale across the bottom of the graph tells the different years.

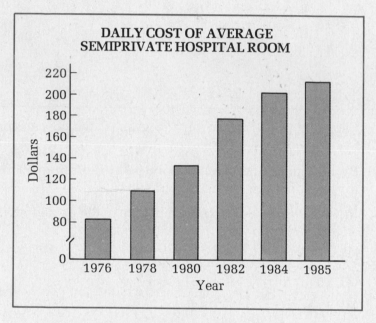

Which of the following answers is the average cost of a semiprivate hospital room in 1978?

(1) $203   **(2)** $178   **(3)** $148   **(4)** $109   **(5)** $78

Choice **(4)**, $109, is correct. Find 1978 on the scale at the bottom of the graph. Follow the heavy line above 1978 to the top and look directly left on the vertical scale. The bar above 1978 ends about half-way between $100 and $120 on the vertical scale. $109 is the closest answer.

Lesson 3   Exercise

Directions: Use the bar graph above to answer the following questions.

1.   What was the average cost of a semiprivate hospital room in 1976?

    (1) $78   **(3)** $88   **(5)** $109
    (2) $83   **(4)** $98

2. What was the average cost of a semiprivate hospital room in 1982?
    (1) $148   **(3)** $168   **(5)** $203
    (2) $158   **(4)** $178

3. During which year was the average cost of a semiprivate hospital room $203?
    (1) 1978   **(3)** 1982   **(5)** 1985
    (2) 1980   **(4)** 1984

4. Which of the following answers is closest to the difference in price for a semiprivate hospital room in 1980 and in 1985?
    (1) $50   **(3)** $70   **(5)** $90
    (2) $60   **(4)** $80

5. The average price of a semiprivate hospital room in 1978 was about what fraction of the price of a semiprivate room in 1985?
    (1) $\frac{1}{2}$ **(3)** $\frac{1}{4}$ **(5)** $\frac{3}{4}$
    (2) $\frac{1}{3}$ **(4)** $\frac{2}{3}$

6. The average price of a semiprivate hospital room in 1984 was about how many times that of the average price of a semiprivate hospital room in 1976?
    (1) $1\frac{1}{2}$ times   **(3)** 3 times   **(5)** $4\frac{1}{2}$ times
    (2) $2\frac{1}{2}$ times   **(4)** 4 times

7. During which two years shown on the graph did the average price of a semiprivate hospital room rise the most?
    (1) 1976 to 1978   **(3)** 1980 to 1982   **(5)** 1984 to 1985
    (2) 1978 to 1980   **(4)** 1982 to 1984

8. Sam was in the hospital for eight days in 1982. If the price of his semiprivate room followed the national average, which of the following answers is closest to the cost of the room for his total stay in the hospital?
    (1) $1600   **(3)** $1100   **(5)** $ 800
    (2) $1400   **(4)** $ 900

9. Letty was in the hospital in 1976 for ten days. If the price of her semiprivate room followed the national average, what was the approximate cost of her stay?
    (1) $500   **(3)** $700   **(5)** $900
    (2) $600   **(4)** $800

10. Alfredo was in the hospital for ten days in 1980 and again for ten days in 1985. If the price of his semiprivate room followed the national average, how much more did his room cost in 1985 than in 1980?
    (1) $600   **(3)** $1000   **(5)** $1400
    (2) $800   **(4)** $1200

Answers are on page 339.

Lesson 4   Divided Bar Graph

A **divided bar graph** separates the totals that are shown by each thick line into parts. The divided bar graph below shows the number of new houses sold in various parts of the United States for four different years.

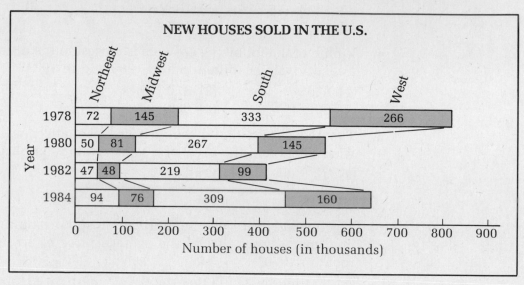

Example: Find the total number of new houses sold in the United States in 1980.

The total for 1980 lies between 500 and 600 thousand. To get the exact total, add the numbers for each region in 1980.

$$
\begin{array}{r}
50 \\
81 \\
267 \\
+\,145 \\
\hline
543 \text{ thousand}
\end{array}
$$

Lesson 4   Exercise

Directions: Use the divided graph above to answer the following questions.

1. Tell the number of new houses which were sold in each of the following categories.

   a. the Midwest in 1978   b. the South in 1980
   c. the Northeast in 1982   d. the West in 1984

2. What was the total number of new houses sold in the United States in 1982?

3. What was the total number of new houses sold in the United States in 1984?

4. For which year shown on the graph was the number of new houses sold the greatest?

5. For which year shown on the graph was the number of new houses sold in the Northeast the greatest?

6. For which year shown on the graph was the number of new houses sold in the West the smallest?

7. The total number of houses sold in the United States in 1982 was about what fraction of the total number sold in 1978?

    (1) $\frac{1}{5}$ **(3)** $\frac{1}{3}$ **(5)** $\frac{3}{4}$

    (2) $\frac{1}{4}$ **(4)** $\frac{1}{2}$

8. For which year shown on the graph was the number of new houses sold in the Northeast more than the number sold in the Midwest?

9. In 1984, the number of new houses sold in the South was about what percent of the total number of new houses sold?

    (1) 75%   **(3)** 25%   **(5)** 5%

    (2) 50%   **(4)** 10%

10. For the years shown on the graph, which region consistently sold the most new houses?

Answers are on page 339.

Lesson 5   Line Graphs

A **line graph** is a way to show changing amounts over a period of time. The vertical scale on the left can be in almost any unit of measurement—dollars, numbers of people, percent, pounds, etc. The horizontal scale across the bottom usually shows units of time. A line that rises from left to right shows an increasing or upward trend. A line that falls from left to right shows a decreasing or downward trend.

To understand a line graph, read the title and the labels on both scales carefully. The line graph below shows the unemployment rate in Capital City from 1975 to 1985. Notice that the vertical scale is measured in percent.

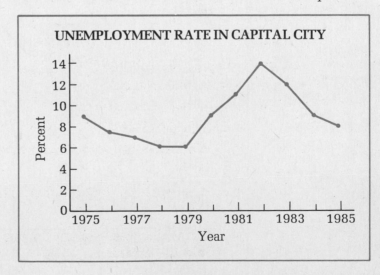

Example: What was the unemployment rate in Capital City in 1981?

Step 1. Find 1981 on the horizontal scale at the bottom of the graph. Then follow the line marked 1981 straight up until you reach the changing line.

Step 2. Look straight across to the vertical scale. You should be half-way between 10% and 12%. The unemployment rate in 1981 weas 11%.   11%

Lesson 5 Exercise

Directions: Use the line graph above to answer the following questions.

1. What was the unemployment rate in Capital City in the following year?
   a. 1975   b. 1982   c. 1985

2. For what year shown on the graph did the unemployment rate in Capital City first get as low as 6%?

3. For which year on the graph was the unemployment rate highest?

4. What is the difference in percentage points between the lowest and the highest unemployment rates shown on the graph?

5. The 1979 unemployment rate was what fraction of the 1983 rate?

Items 6 to 10 refer to the following graph. Notice that there are two lines. One shows the changing income, and the other shows the changing expenses.

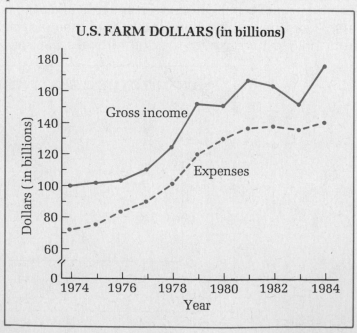

6. What were the total farm expenses in 1983?

    (1) $145 billion   **(3)** $125 billion   **(5)** $110 billion
    (2) $135 billion   **(4)** $115 billion

7. The net income for farms is the difference between the gross income and the expenses. What was the approximate net income for farms in 1975?

    (1) $40 billion   **(3)** $20 billion   **(5)** $5 billion
    (2) $25 billion   **(4)** $10 billion

8. What was the net income for farms in 1981?

    (1) $10 billion   **(3)** $20 billion   **(5)** $40 billion
    (2) $15 billion   **(4)** $30 billion

9. In what year did the gross income for farms first reach $150 billion?

    (1) 1979   **(3)** 1982   **(5)** 1984
    (2) 1981   **(4)** 1983

10. Between what two consecutive years did the expenses on farms drop?

    (1) 1976–1977   **(3)** 1980–1981   **(5)** 1983–1984
    (2) 1978–1979   **(4)** 1982–1983

Answers are on page 339.

Chapter 5 Quiz

Directions: Solve each problem.

Items 1 to 4 are based on the following pictograph.

PER CAPITA CONSUMPTION OF SOFT DRINKS

LEGEND:
= 5 gallons of soft drinks

1. What was the per capita consumption of soft drinks in gallons in 1970?

    (1) $22\frac{1}{2}$ **(3)** $32\frac{1}{2}$ **(5)** $39\frac{1}{2}$
    (2) $27\frac{1}{2}$ **(4)** $37\frac{1}{2}$

2. The per capita consumption of soft drinks in 1985 was how many times that of the per capita consumption in 1965?

    (1) 8 times   (3) 3 times   (5) $1\frac{1}{2}$ times

    (2) 4 times   (4) 2 times

3. The 1980 per capita consumption of soft drinks was how many gallons greater than the per capita consumption in 1975?

    (1) 20   (3) 12   (5) 5

    (2) 15   (4) 10

4. The per capita consumption of soft drinks in 1985 was how many gallons greater than the per capita consumption in 1965?

    (1) 10   (3) 30   (5) 50

    (2) 20   (4) 2 times

Items 5 to 10 are based on the following circle graph.

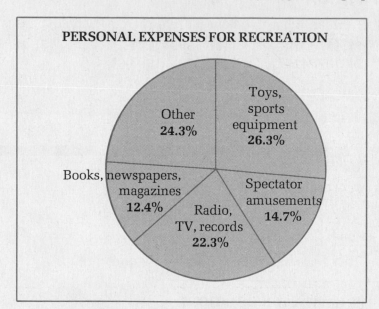

PERSONAL EXPENSES FOR RECREATION

5. What is the difference in percentage points between the amount spent on toys and sports equipment and the amount spent on books, magazines, and newspapers?

    (1) 15.9%   (3) 11.9%   (5) 7.9%

    (2) 13.9%   (4) 9.9%

6. What is the difference in percentage points between the amount spent on radio, TV, and records and the amount spent on spectator amusements?

    (1) 22.3%   (3) 11.9%   (5) 6%

    (2) 15%   (4) 7.6%

7. The amount spent on spectator amusements and the amount spent on books, magazines, and newspapers combined make up about what fraction of the total expenditures on recreation?

    (1) a little over $\frac{1}{2}$ (3) a little over $\frac{1}{4}$ (5) a little over $\frac{1}{10}$

    (2) a little over $\frac{1}{3}$ (4) a little over $\frac{1}{5}$

8. The amount spent on toys and sports equipment and the amount spent on the category called "other" combined made up about what percent of the total?

    (1) 10%   **(3)** 30%   **(5)** 50%
    (2) 20%   **(4)** 40%

9. The Joneses spend $3000 a year on recreation. If they spend at the same rate as the national average, how much do they spend each year on books, magazines, and newspapers?

    (1) $372   **(3)** $669   **(5)** $729
    (2) $441   **(4)** $744

10. If the Joneses spend at the same rate as the national average, and they spend $3000 a year on recreation, how much do they spend each year on spectator amusements?

    (1) $372   **(3)** $669   **(5)** $744
    (2) $441   **(4)** $729

Items 11 to 15 are based on the following bar graph.

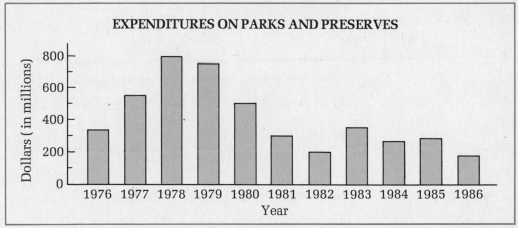

11. For the years shown on the graph, what was the largest amount spent in a single year on parks and preserves?

    (1) $400 million   **(3)** $600 million   **(5)** $800 million
    (2) $500 million   **(4)** $700 million

12. The amount spent in 1982 on parks and preserves was about what fraction of the amount spent in 1978?

    (1) $\frac{1}{8}$ **(3)** $\frac{3}{8}$ **(5)** $\frac{3}{4}$
    (2) $\frac{1}{4}$ **(4)** $\frac{5}{8}$

13. The amount spent on parks and preserves in 1982 was what fraction of the amount spent in 1981?

    (1) $\frac{1}{4}$ **(3)** $\frac{1}{2}$ **(5)** $\frac{3}{4}$
    (2) $\frac{1}{3}$ **(4)** $\frac{2}{3}$

14. Between what two years was there the largest drop in expenditures on parks and preserves?

    (1) 1978–1979   **(3)** 1980–1981   **(5)** 1983–1984
    (2) 1979–1980   **(4)** 1981–1982

15. For which period shown below did the yearly amount spent on parks and preserves rise steadily?

    (1) 1984–1986   **(3)** 1981–1982   **(5)** 1976–1978
    (2) 1982–1984   **(4)** 1978–1981

Items 16 to 20 are based on the following line graph.

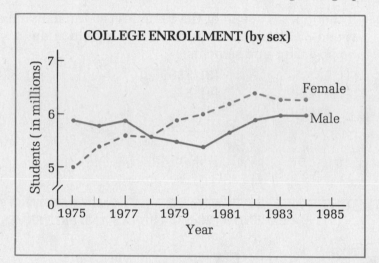

16. What is the unit of measurement of the vertical scale?

    (1) percent   **(3)** millions of students   **(5)** thousands of students

    (2) hundreds of   **(4)** years
    female students

17. In what year was the number of female college students the same as the number of male college students?

    (1) 1978   **(3)** 1981   **(5)** 1984
    (2) 1980   **(4)** 1982

18. Approximately how many more female college students than male college students were there in 1980?

    (1) 1 million   **(3)** 100,000   **(5)** 25,000
    (2) 500,000   **(4)** 50,000

19. Approximately how many more male college students than female college students were there in 1975?

    (1) 1 million   **(3)** 200,000   **(5)** 10,000
    (2) 500,000   **(4)** 100,000

20. Which of the following statements best describes college enrollment from 1975 to 1981?

    (1) Both the number of male students and the number of female students decreased.

(2) The number of male students and the number of female students increased gradually.

(3) The number of female students increased steadily, while the number of male students gradually decreased.

(4) The number of male students decreased while the number of female students decreased more slowly.

Check your answers. Correct answers are on pages 339–340. If you have at least sixteen answers correct, go to the next chapter. If you have fewer than sixteen answers correct, study the Graphs chapter before you go on.

Chapter

6 Algebra

Objective

In this chapter, you will

- Use a number line
- Add, subtract, multiply, and divide signed numbers
- Add, subtract, multiply, and divide monomials
- Use the order of operations
- Use inverse operations to solve one-step and multistep equations
- Solve inequalities
- Write algebraic expressions and equations
- Multiply binomials
- Factor binomials and quadratic equations
- Use factoring to simplify square roots
- Solve quadratic equations
- Use algebra to solve formulas
- Apply the basic skills involved in solving word problems

Level 1   Algebra Skills

Preview

Directions: Solve each problem.

1. **Which point on the number line corresponds to the value 2.25?**

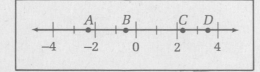

2. $(12) + (-9) + (-15) = ?$

3. $(3ab)(-4a) = ?$

4. $\dfrac{-4x^2y^2}{-8xy} = ?$

5. **Find the value of $3m - mn$, if $m = -5$ and $n = -2$.**

Check your answers. Correct answers are on page 340. If you have at least four answers correct, do the Level 1 Review on pages 188–189. If you have fewer than four answers correct, study Level 1 beginning with Lesson 1.

Lesson 1   What Algebra Is

Algebra is a branch of mathematics that builds on the basic skills you learned in arithmetic. Algebra is different from arithmetic in two important ways. In arithmetic, you used numbers that are equal to zero or more. In algebra, you will also use numbers that are less than zero, and you will use letters to stand for numbers.

The Number Line

The whole numbers, decimals, and fractions we use in arithmetic are called **positive** numbers. The value of a positive number is more than zero. The **number line** below represents all positive numbers.

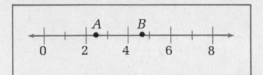

The arrow at the right end of the line means that the numbers go on and on. Point A is about half-way between 2 and 3. Point A stands for the number $2\frac{1}{2}$, or 2.5. Point B is about three-fourths of the way between 4 and 5. Point B stands for the number $4\frac{3}{4}$, or 4.75.

In algebra, we often use numbers that are less than zero. We call these **negative** numbers. The number line below represents both positive and negative numbers. In algebra, we call these **signed** numbers.

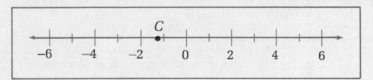

The arrows mean that the numbers go on and on in both directions. Point C is about one-fourth of the way between -1 and -2. Point C stands for the number $-1\frac{1}{4}$, or -1.25. We read this number as "minus one and one-fourth" or "negative one and one-fourth."

In arithmetic, the $+$ sign means to add. In algebra, the $+$ sign also means a positive or (plus) number. When a number has no sign, it is understood to be positive.

In arithmetic, the $-$ sign means to subtract. In algebra, the $-$ sign also means a negative (or minus) number. A negative number always has a $-$ sign in front of it.

The number zero has no sign. Zero is neither negative nor positive.

Example: Which letter on the number line below corresponds to the number $-2\frac{2}{3}$?

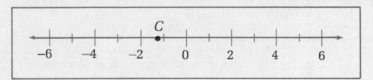

Point D is more than half-way between -2 and -3. Point D corresponds to $-2\frac{2}{3}$.

Lesson 1   Exercise

Directions: Use the number line in the preceding example and tell which letter stands for each of the following values.

1. $+4$ 2. -5 3. $+\frac{3}{4}$ 4. $-6\frac{1}{3}$ 5. $1\frac{2}{3}$
6. $-5\frac{1}{2}$ 7. $-\frac{3}{4}$ 8. $2\frac{1}{4}$ 9. $-2\frac{2}{3}$ 10. $+6\frac{1}{4}$

Answers are on page 340.

Lesson 2   Adding Signed Numbers

Adding positive numbers means moving to the right on the number line. To add $+3 + 2$, start at $+3$ on the number line and move two places to the right. The answer is $+5$, or 5.

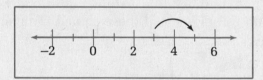

Adding negative numbers means moving to the left on the number line. To add $-1 - 5$, start at -1 on the number line and move five places to the left. The answer is -6.

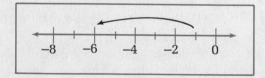

A thermometer is a practical application of the number line. If the temperature on a winter night is $-1°$ and then drops another $5°$, the temperature becomes $-6°$. To drop $5°$ means to increase the negative number.

When you add a positive number and a negative number, the answer depends on the size of the two numbers. To add $+2 - 6$, start at $+2$ on the number line and move 6 places to the left. The answer is -4.

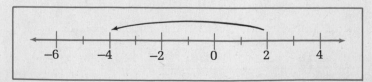

To add $-3 + 7$, start at -3 on the number line and move 7 places to the right. The answer is $+4$, or 4.

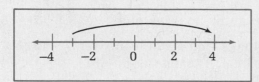

You do not have to draw the number line to solve number problems. Follow these steps to add two signed numbers.

1. If the signs are alike, add and give the answer the sign of the numbers.
2. If the signs are different, subtract and give the answer the sign of the larger number.

Example 1: $-7 - 9 = ?$

Since the signs are alike, add and make the answer negative. The answer is -16. $-7 - 9 = -16$

Example 2: $-8 + 13 = ?$

Since the signs are different, subtract and make the answer positive because $+13$ is larger. The answer is $+5$, or 5. $-8 + 13 = +5$, or 5

Example 3: $-6 + 2 = ?$

Since the signs are different, subtract and make the answer negative because -6 is larger. The answer is -4. $-6 + 2 = -4$

Follow these steps to add three or more signed numbers.

1. Add the positive numbers and make the total positive.
2. Add the negative numbers and make the total negative.
3. Subtract the two totals and give the answer the sign of the larger total.

Example 4: $+8 - 9 + 3 = ?$

Step 1. Add the positive numbers, $+8$ and $+3$, $+8 + 3 = +11$
and make the total positive.

Step 2. Since there is only one negative number, the negative total is -9.

Step 3. Subtract the two totals and make the $+11 - 9 = +2$
answer positive since $+11$ is larger. The answer
is $+2$.

Words like *larger* and *smaller* can be confusing with signed numbers. Any number on the number line is larger than the numbers to its left. For example, $+5$ is larger than -12. However, we say that -12 has a larger **absolute value** than $+5$. When a rule tells you to give an answer the sign of the larger total, give the answer the sign of the total with the larger absolute value.

Example 5: $-7 + 2 + 5 - 6 = ?$

Step 1. Find the total of the positive numbers. $+2 + 5 = +7$

Step 2. Find the total of the negative numbers. $-7 - 6 = -13$

Step 3. Subtract the totals, and give the answer $-13 + 7 = -6$
the sign of the larger total. The answer is -6.

A $+$ sign between parentheses is another way to indicate addition of signed numbers.

Example 6: $(+3) + (-4) + (+9) + (-10) = ?$

Step 1. Add the positive numbers and make the $+3 + 9 = +12$
total positive.

Step 2. Add the negative numbers and make the $-4 - 10 = -14$
total negative.

Step 3. Subtraction the totals and make the an- $-14 + 12 = -2$
swer negative because -14 is larger. The an-
swer is -2.

Example 7: $(-25) + (+25) = ?$

Since the signs are different, subtract. The an- $-25 + 25 = 0$
swer is zero, which is neither positive nor neg-
ative.

Lesson 2   Exercise

Directions: Solve each problem.

1. $+7 - 6 = ?$ $-13 + 13 = ?$ $-14 + 18 = ?$

2. $(+9) + (-15) = ?$ $(-8) + (-11) = ?$ $(-12) + (+12) = ?$

3. $(-24) + (+7) = ?$ $(-3) + (+14) = ?$ $(-19) + (-19) = ?$

4. $+3 - 9 + 7 = ?$ $-4 - 6 - 3 = ?$ $+2 + 8 - 10 = ?$

5. $(+3) + (+11) + (+5) = ?$ $(-9) + (-1) + (+4) = ?$

   $-8 - 1 - 6 + 10 = ?$

6. $(-8) + (+12) + (+16) + (-11) = ?$ $(+7) + (-12) + (+3) + (+2) = ?$

7.   Find the value of $a + b$, if $a = -6$ and $b = -4$.

8.   Find the value of $m + n + p$, if $m = 3, n = 9$, and $p = -7$.

9.   Find the value of $x + y$, if $x = 15$ and $y = -15$.

10.   Find the value of $r + s + t$, if $r = -1, s = -8$, and $t = 7$.

Answers are on page 340.

Lesson 3   Subtracting Signed Numbers

When you subtract signed numbers, you find the distance between two numbers on the number line. A − sign between parentheses means to subtract. To subtract $(-5) - (-3)$, start at -3 on the number line. Then count the number of spaces to -5. From -3 to -5 is two spaces to the left, or -2.

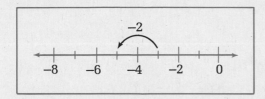

Again, you do not have to use the number line to solve every subtraction problem. The following rule is based on the fact that subtraction is the opposite of addition.

To subtract a signed number, change the sign of the number *being subtracted* and follow the rules for adding signed numbers.

Example 1: $(+5) - (-3) = ?$

Step 1. The − sign between the parentheses means that -3 is being subtracted. Change the sign of -3 to $+3$.

$$(+5) - (-3) =$$
$$+5 + 3 =$$

Step 2. Since the signs are alike, add the numbers and make the answer positive. The answer is $+8$, or 8.

$$= +8$$

Example 2: $(\ 1)\ \ (6) = ?$

Step 1. 6 is being subtracted. Change the sign of 6 to -6.

$$(-1) - (6) =$$
$$-1 - 6 =$$

Step 2. Since the signs are the same, add and make the answer negative. The answer is -7.

$$= -7$$

Some problems include both addition and subtraction. Change the sign of every number being subtracted and follow the rules for adding signed numbers.

Example 3: $(-5) + (-3) - (-2) = ?$

Step 1. Only -2 is being subtracted. Change the sign of -2 to $+2$.

$$(-5) + (-3) - (-2) =$$
$$-5 - 3 + 2 =$$

Step 2. Find the total of the negative numbers -5 and -3.

$$-8 + 2 =$$

Step 3. Subtract the two totals and give the answer the sign of the larger total. The answer is -6.

$$= -6$$

Lesson 3   Exercise

Directions: Solve each problem.

1. $(+8) - (+9) = ?$ $(+7) - (-6) = ?$ $(-9) - (+3) = ?$
2. $(-3) - (-14) = ?$ $(-10) - (+15) = ?$ $(+6) - (-11) = ?$
3. $(+20) - (-4) = ?$ $(+2) - (+18) = ?$ $(-16) - (-5) = ?$
4. $(-11) - (+15) = ?$ $(+7) - (-8) = ?$ $(-9) - (+9) = ?$
5. $(+8) - (+7) + (6) = ?$ $(-4) + (-5) - (+3) = ?$
6. $(-10) - (-7) - (-1) = ?$ $(+11) - (-14) + (-3) = ?$
7. Find the value of $a - b$, when $a = -6$ and $b = -7$.
8. Find the value of $m - n$, when $m = 14$ and $n = -4$.
9. Find the value of $w - y$, when $w = -6$ and $y = +7$.
10. Find the value of $c - d$, when $c = 15$ and $d = -9$.

Answers are on page 340.

Lesson 4   Multiplying Signed Numbers

Think about the following four situations. In these examples, a + sign refers to money somebody owes you, and a − sign refers to money that you owe somebody else. A + sign also means days in the future, and a − sign means days in the past.

If somebody pays you $3 a day, then in 5 days you will have $15 more than you have now. In algebra, this is $(+3)(+5) = +15$.

If you have to pay somebody $3 a day, then in 5 days you will have $15 less than you have now. In algebra, this is $(-3)(+5) = -15$.

If somebody has been paying you $3 a day, then 5 days ago you had $15 less than you have now. In algebra, this is $(+3)(-5) = -15$.

If you have been paying somebody $3 a day, then 5 days ago you had $15 more than you have now. In algebra, this is $(-3)(-5) = +15$.

We can write a simple rule from these examples. Follow these steps when you multiply two signed numbers.

1. Make the answer positive if the signs are alike.

2. Make the answers negative if the signs are different.

Example 1: $(-3)(20) = ?$

Multiply and make the answer negative since $(-3)(20) = -60$
the signs are different.

Example 2: $(-6)(-7) = ?$

Multiply and make the answer positive since $(-6)(-7) = +42$
the signs are alike.

Follow these steps when you multiply more than two signed numbers.

1. Make the answer positive if there is an even number of negative signs (2, 4, 6, etc.).

2. Make the answer negative if there is an odd number of negative signs (1, 3, 5, etc.).

Example 3: $\left(\frac{2}{3}\right)\left(-\frac{3}{4}\right)\left(-\frac{4}{5}\right) = ?$

Cancel and multiply across. Make the answer positive since the number of negative signs is even. $\frac{2}{3} \times -\frac{\overset{1}{\cancel{3}}}{\underset{1}{\cancel{4}}} \times -\frac{\overset{1}{\cancel{4}}}{5} = \frac{2}{5}$

Example 4: $(-1)(+8)(-3)(-2) = ?$

Multiply across and make the answer negative $(-1)(+8)(-3)(-2) = -48$
since there are three negative signs.

In algebra, multiplication can be shown in three ways:

1. A raised dot (·): $9 \cdot 2 = 18$
2. No sign between a number and a letter: $5x = 5$ times x
3. No sign before or between parentheses: $8(4) = 32$ or $(-1)(+2) = -2$

Lesson 4   Exercise

Directions: Solve each problem.

1. $(+6)(-10) = ?$ $(+12)(-1) = ?$ $(-5)(-13) = ?$

2. $(+8)(+\frac{1}{2}) = ?$ $(-\frac{1}{3})(-15) = ?$ $(-9)(+8) = ?$

3. $(-\frac{3}{4})(+24) = ?$ $(+20)(-3) = ?$ $(-15)(-2) = ?$

4. $(+4)(-10)(+8) = ?$ $(+5)(-9)(-\frac{2}{3}) = ?$

5. $(-\frac{1}{2})(-12)(-5) = ?$ $(-8)(+12)(-1) = ?$

6. $(-4)(+10)(+\frac{3}{4})(+2) = ?$ $(+5)(-1)(+9)(-6) = ?$

7. Find the value of ab, when $a = -7$ and $b = -6$.

8. Find the value of pqr, when $p = 5$, $q = -3$, and $r = -10$.

9. Find the value of mn, when $m = -9$ and $n = +9$.

10. Find the value of xyz, when $x = -1$, $y = -\frac{1}{2}$, and $z = -\frac{1}{4}$.

Answers are on page 340.

Lesson 5 Dividing Signed Numbers

The rules for dividing signed numbers are the same as the rules for multiplying signed numbers.

Follow these steps when you divide signed numbers.

1. Make the answer positive if the signs are alike.
2. Make the answer negative if the signs are different.

When you divide signed numbers, the problems usually look like fractions.

Example 1: $\frac{-36}{-9} = ?$

Divide and make the answer positive since the signs are alike. $\qquad \frac{-36}{-9} = +4$

Example 2: $\frac{56}{-8} = ?$

Divide and make the answer negative since the signs are different. $\qquad \frac{56}{-8} = -7$

In some division problems, you can only reduce the answer.

Example 3: $\frac{-12}{+16} = ?$

Reduce and make the answer negative since the signs are different. $\qquad \frac{-12}{+16} = -\frac{3}{4}$

A negative fraction can be written two ways. The minus sign can go with the numerator or in front of the entire fraction. The answer to the last example is either $\frac{-3}{4}$ or $-\frac{3}{4}$.

Lesson 5 Exercise

Directions: Solve each problem.

1. $\frac{-24}{-8} = ?$ $\qquad$ $\frac{-30}{+10} = ?$ $\qquad$ $\frac{+5}{-10} = ?$ $\qquad$ $\frac{+8}{+2} = ?$

2. $\frac{-12}{+12} = ?$ $\qquad$ $\frac{-15}{-20} = ?$ $\qquad$ $\frac{+100}{-20} = ?$ $\qquad$ $\frac{-20}{-40} = ?$

3. $\frac{63}{-9} = ?$ $\qquad$ $\frac{-48}{-12} = ?$ $\qquad$ $\frac{-80}{+100} = ?$ $\qquad$ $\frac{+18}{-6} = ?$

4. $\frac{150}{-25} = ?$ $\qquad$ $\frac{-35}{49} = ?$ $\qquad$ $\frac{-20}{-24} = ?$ $\qquad$ $\frac{-96}{+12} = ?$

5. Find the value of $\frac{a}{b}$, when $a = -14$ and $b = -2$.

6. Find the value of $\frac{x}{y}$, when $x = -12$ and $y = 15$.

7. Find the value of $\frac{m}{n}$, when $m = 23$ and $n = -1$.

8. Find the value of $\frac{c}{d}$, when $c = +18$ and $d = -24$.

9. Find the value of $\frac{e}{f}$, when $e = 22$ and $f = -11$.

10. Find the value of $\frac{u}{v}$, when $u = -8$ and $v = 32$.

Answers are on page 340.

Lesson 6   Adding Monomials

Algebra uses expressions to show mathematical relationships. An algebraic expression contains a combination of letters and numbers. $x + y$ is an expression that shows the sum of two numbers. cd is an expression that shows the product of two numbers.

Algebraic expressions are made up of terms. A term is not separated by a $+$ or $-$ sign. A **monomial** is an expression that contains only one term. $9m$ is a monomial. $-12pq$ is a monomial. And c^2 is also a monomial. A **coefficient** is the number in front of a monomial. The **variables** are the letters. In the expression $9m$, the coefficient is 9, and the variable is m. In the expression $-12pq$, -12 is the coefficient, and pq are the variables. In the expression c^2, the coefficient is 1 even though the 1 is not written, and c is the variable.

Follow these steps to add monomials.

1. Be sure the monomials have the same variables.
2. Use the rules for adding signed numbers.

Example 1: $2x + 7x = ?$

Since the signs are alike, add and make the $2x + 7x = 9x$
answer positive.

Example 2: $6y - y = ?$

Since the signs are different, subtract and make $6y - y = 5y$
the answer positive since $6y$ is larger. Notice
that $-y$ is understood to be $-1y$.

Example 3: $(8ab) + (-3ab) + (-4ab) = ?$

Step 1. Rewrite the monomials to eliminate the $(8ab) + (-3ab) + (-4ab) =$
parentheses.

$8ab - 3ab - 4ab =$

Step 2. Add the negative monomials. $8ab - 7ab =$

Step 3. Subtract and make the answer positive. $= ab$
Notice that you do not have to write the 1.

Lesson 6   Exercise

Directions: Solve each problem.

1. $5m + 3m = ?$ $2c - 6c = ?$ $8x - x = ?$

2. $-4p + 5p = ?$ $18xy - 19xy = ?$ $-6st - 5st = ?$

3. $9w - 3w + 4w = ?$ $-8y - 3y - y = ?$ $-d + 5d - 4d = ?$

4. $(-8e) + (-6e) + (+11e) = ?$ $(7n) + (-12n) + (-4n) = ?$

5. $(-a) + (5a) + (-9a) + (+3a) = ?$

$(-7w) + (-2w) + (-3w) + (11w) = ?$

Answers are on page 341.

Lesson 7   Subtracting Monomials

To subtract monomials, be sure the variables are the same. Then follow the rules for subtracting signed numbers.

Example 1: $(8x) - (-6x) = ?$

Step 1. Rewrite the monomials to eliminate the parentheses, and change the sign of $-6x$ to $+6x$.

$(8x) - (-6x) =$
$8x + 6x =$

Step 2. Add and make the answer positive.

$= 14x$

Example 2: $(-mn) + (2mn) - (7mn) = ?$

Step 1. Rewrite the monomials to eliminate the parentheses, and change the sign of $7mn$ to $-7mn$.

$(-mn) + (2mn) -$
$(7mn) =$
$-mn + 2mn - 7mn =$

Step 2. Add the two negative monomials.

$-8mn + 2mn =$

Step 3. Subtract and make the answer negative.

$= -6mn$

Lesson 7   Exercise

Directions: Solve each problem.

1. $(+5p) - (+3p) = ?$ $(+5p) - (-3p) = ?$ $(-5p) - (-3p) = ?$
2. $(-6a) - (-3a) = ?$ $(+4c) - (-4c) = ?$ $(-7mn) - (+3mn) = ?$
3. $(-9f) - (-f) + (-2f) = ?$ $(10y) - (3y) - (-7y) = ?$
4. $(-8cd) + (-2cd) - (3cd) = ?$ $(-12t) - (-3t) + (9t) = ?$
5. $(4bc) - (-8bc) - (2bc) = ?$ $(-16x) - (2x) + (-5x) = ?$

Answers are on page 341.

Lesson 8   Multiplying Monomials

When you multiply monomials, the variables change. Remember that a number multiplied by itself is the number to the second power. A letter multiplied by itself is that letter to the second power. For example, $c \cdot c = c^2$. The letter c is understood to be c^1. To multiply letters, simply add their exponents.

Follow these steps to multiply monomials.

1. Multiply the coefficients according to the rules for multiplying signed numbers.
2. Add the exponents of the same letters.

Example 1: $a^3 \cdot a^4 = ?$

Since there are no coefficients to multiply, sim- $a^3 \cdot a^4 = a^7$
ply add the exponents of the letters.

When the letters are different, write the letters beside each other in the product.

Example 2: $c \cdot d = ?$

Write the letters beside each other in the prod- $c \cdot d = cd$
uct.

When the monomials you are multiplying have more than one letter, add the exponents of each letter separately.

Example 3: $(-8m^2n)(-3m^3n^5) = ?$

Since the signs are alike, the product is positive. $(-8m^2n)(-3m^3n^5) =$
Add the exponents of the letter m, and add the $+24m^5n^6$
exponents of the letter n.

Lesson 8   Exercise

Directions: Solve each problem.

1. $c^2 \cdot c^3 = ?$	$m^4 \cdot m = ?$	$x^3 \cdot x^5 = ?$
2. $a \cdot a = ?$	$a \cdot b = ?$	$(a^2b^2)(a^3b^4) = ?$
3. $(2x)(-3x) = ?$	$(-4y^2)(-2y^5) = ?$	$(-5p)(2p^3) = ?$
4. $(5yz)(-6yz) = ?$	$(-mn^2)(-4m^2n^3) = ?$	$(-2rs)(-9r^3s) = ?$
5. $(-ef^2)(2e^2f) = ?$	$(-4kt^3)(-k^3t) = ?$	$(-5d^2h^3)(6dh) = ?$

Answers are on page 341.

Lesson 9 Dividing Monomials

Think about dividing x^5 by x^2.

$$\frac{x^5}{x^2} = \frac{\overset{1}{\cancel{x}} \cdot \overset{1}{\cancel{x}} \cdot x \cdot x \cdot x}{\underset{1}{\cancel{x}} \cdot \underset{1}{\cancel{x}}} = x^3$$

The quotient x is the result of subtracting the exponents of x.
Follow these steps to divide monomials.

1. Follow the rules for dividing signed numbers for the coefficients.
2. Subtract the exponents of the same letters.

Remember that any number with an exponent of zero equals 1.

Example 1: $\frac{-12a^4}{-3a} = ?$

Since the signs are the same, the quotient is $\frac{-12a^4}{-3a} = 4a^3$
positive. Subtract the exponents of the letter.

Sometimes there is no variable to cancel with another.

Example 2: $\frac{-3ab}{6a} = ?$

Since the signs are different, the quotient is $\frac{-3ab}{6a} = \frac{-b}{2}$
negative. The a's cancel, and the b remains.

Lesson 9 Exercise

Directions: Solve each problem.

1. $\frac{m^6}{m^2} = ?$ $\frac{st}{s} = ?$ $\frac{c^5}{c^2} = ?$

2. $\frac{a^5}{a^4} = ?$ $\frac{x^3y^2}{xy} = ?$ $\frac{x^6}{x^6} = ?$

3. $\frac{-12x^2}{3x} = ?$ $\frac{20m^3n}{-4m} = ?$ $\frac{-36a^5}{-9a^4} = ?$

4. $\frac{-4m^2n^2}{16mn} = ?$ $\frac{-18c^3d^4}{-12c^3d} = ?$ $\frac{+24x}{+30x} = ?$

5. $\frac{2x^3y^2}{-18y} = ?$ $\frac{-4a^3b}{20ab} = ?$ $\frac{-20m^2n}{-4mn} = ?$

Answers are on page 341.

Lesson 10   Signed Numbers Review

So far, you have used only one operation at a time with signed numbers. When there is more than one operation, use the following **order of operations.**

1. Parentheses and division bars
2. Powers and square roots
3. Multiplication and division
4. Addition and subtraction

In the following examples, you will see how to use this order of operations. Each of these examples involves **substitution.** Substitution means replacing letters with number values. You have already used substitution when you evaluated formulas. You also used one-step substitution in the signed-number exercises.

Example 1: Find the value of $x + xy$, if $x = -6$ and $y = -4$.

Step 1. Replace each letter with the value in the problem. Notice how the parentheses are used to show multiplication.

$x + xy =$
$-6 + (-6)(-4) =$

Step 2. This problem includes addition and multiplication. Since multiplication comes before addition in the order of operations, first multiply -6 and -4.

$-6 + (+24) =$

Step 3. Combine -6 and $+24$ according to the rules for adding signed numbers.

$-6 + 24 = +18$

Example 2: Find the value of $a(a + b)$, if $a = -9$ and $b = 4$.

Step 1. Replace a and b with the values in the problem.

$a(a + b) =$
$-9(-9 + 4)$

Step 2. The operation $-9 + 4$ is in parentheses. Since parentheses are first in the order of operations, combine -9 and $+4$ first.

$-9(-5) =$

Step 3. Multiply -9 by -5.

$= +45$

Example 3: Find the value $x(x - y)$, if $x = 2$ and $y = -6$.

Step 1. Replace x and y with the values in the problem. Notice how the extra parentheses separate -6 from the $-$ sign in front.

$x(x - y) =$
$2(2 - (-6)) =$

Step 2. Since the operation $2 - (-6)$ is in parentheses, combine these first according to the rules for subtracting signed numbers.

$2(2 + 6) =$
$2(8) =$

Step 3. Multiply $+2$ by $+8$.

$= 16$

Example 4: Find the value of $\dfrac{m + n}{2}$, if $m = -4$ and $n = -8$.

Step 1. Replace m and n with the values in the problem.

$$\dfrac{m + n}{2} =$$
$$\dfrac{-4 + (-8)}{2} =$$

Step 2. Since the operation $-4 + (-8)$ is above the division bar, combine them first. The division bar works like parentheses to group -4 and -8 together.

$$\dfrac{-4 - 8}{2} =$$
$$\dfrac{-12}{2} =$$

Step 3. Divide -12 by 2.

$$= -6$$

Lesson 10   Exercise

Directions: Solve each problem.

1. Find the value of $ab - c$, if $a = -4$, $b = -3$, and $c = -5$.

2. Find the value of $m(m - n)$, if $m = -2$ and $n = 7$.

3. Find the value of x^2y, if $x = -3$ and $y = -4$.

4. Find the value of $s(s + t) - t$, if $s = -5$ and $t = 1$.

5. Find the value of $e + ef$, if $e = 6$ and $f = -4$.

6. Find the value of ab^2, if $a = -2$ and $b = -5$.

7. Find the value of $x(x - y)$, if $x = -8$ and $y = 2$.

8. Find the value of $(j - k)^2$, if $j = -3$ and $k = 1$.

9. Find the value of $\dfrac{a + b}{2}$ if $a = -6$ and $b = -2$.

10. Find the value of $\dfrac{m - n}{n}$ if $m = 8$ and $n = -4$.

Answers are on page 341.

Level 1   Review

Directions: Solve each problem.

1. Which point on the number line corresponds to the value $-1\frac{3}{4}$?

2. $(-13) + (8) + (-7) = ?$

3. $(-9) - (-11) - (+4) = ?$

4. $(-6)(-\frac{2}{3})(-5) = ?$

5. $\frac{+24}{-36} = ?$

6. $(4m) + (-2m) + (-7m) = ?$

7. $(-n) - (-6n) + (-4n) = ?$

8. $(-9x^2y)(-2xy) = ?$

9. $\frac{12a^2c}{-3a} = ?$

10. Find the value of $ab - 4b$, if $a = -7$ and $b = -3$.

Check your answers. Correct answers are on page 341. If you have at least eight answers correct, go to the Level 2 Preview. If you have fewer than eight answers correct, go back to Lesson 1 and study Level 1.

Level 2   Algebra Applications

Preview

Directions: Solve each of the following problems.

1. Write an algebraic expression for four times the quantity of a number decreased by seven.

2. Let x represent Sam's age now. Write an expression for Sam's age 12 years ago.

3. Factor the expression $m^2 + 12m$.

4. Simplify $\sqrt{12}$.

5. Find the solutions to the equation $x^2 - 100 = 0$.

Check your answers. Correct answers are on page 341. If you have at least four answers correct, do the Level 2 Review on pages 204–205. If you have fewer than four answers correct, study Level 2 beginning with Lesson 1.

Lesson 1   One-step Equations

An **equation** is an algebraic statement that two amounts are equal. The statement $x + 15 = 42$ is an equation which means that some number called x increased by 15 is equal to 42. The letter x is called the unknown, or **variable.** You will often see the word *side* in descriptions about equations. The = sign separates an equation into two sides.

To solve an equation means to find the value of the unknown which makes the statement true. To solve equations, you must know about opposite or **inverse** operations.

Addition is the inverse of subtraction.
Subtraction is the inverse of addition.
Multiplication is the inverse of division.
Division is the inverse of multiplication.

To solve a one-step equation, perform the inverse operation on both sides of the equation in order to get a statement that says, "Unknown = solution," or "Solution = unknown."

Example 1: Solve the equation x + 15 = 42.

In this equation, 15 is added to the unknown. The inverse of addition is subtraction. Subtract 15 from both sides of the equation. The solution is 27.

$$x + 15 = 42$$
$$\underline{-15 \quad -15}$$
$$x = 27$$

To check the solution to an equation, substitute the solution for the unknown. When you evaluate the equation, you should get the same number on both sides. For the previous example, replace x with 27.

When you evaluate the equation, you get 42 on both sides of the equal sign.

$$27 + 15 = 42$$
$$42 = 42$$

In some equations, the unknown is on the right side. In these equations, you want a statement that says, "Solution = unknown."

Example 2: Solve and check the equation 18 = y − 6.

Step 1. In this equation, 6 is subtracted from the unknown. Addition is the inverse of subtraction. Add 6 to both sides. The solution is 24.

$$18 = y - 6$$
$$\underline{+6 \qquad +6}$$
$$24 = y$$

Step 2. To check the solution, replace y with 24 in the original equation. You get 18 on both sides. This proves that 24 is correct.

$$18 = 24 - 6$$
$$18 = 18$$

Example 3: Solve and check the equation 9m = 45.

Step 1. In this equation, the unknown is multiplied by 9. Division is the inverse of multiplication. Divide both sides by 9. The solution is 5.

$$\frac{9m}{9} = \frac{45}{9}$$
$$m = 5$$

Step 2. To check the solution, replace m with 5. You get 45 on both sides. This proves that 5 is correct.

$$9 \cdot 5 = 45$$
$$45 = 45$$

Example 4: Solve and check the equation $\frac{n}{8} = 6$.

Step 1. In this equation, the unknown is divided by 8. Multiplication is the inverse of division. Multiply both sides by 8. The solution is 48.

$$8 \cdot \frac{n}{8} = 6 \cdot 8$$
$$n = 48$$

Step 2. To check the solution, replace n with 48. You get 6 on both sides. This proves that 48 is correct.

$$\frac{\overset{6}{48}}{\underset{1}{8}} = 6$$
$$6 = 6$$

Example 5: Solve and check the equation $\frac{2}{3}y = 12$.

Step 1. In this equation, the unknown is multiplied by $\frac{2}{3}$. Division is the inverse of multiplication. Remember that you must invert and multiply when you divide by a fraction. Multiply both sides by $\frac{3}{2}$. The solution is 18.

$$\frac{3}{2} \cdot \frac{2}{3}y = \overset{6}{12} \cdot \frac{3}{\underset{1}{2}}$$
$$y = 18$$

Step 2. To check the solution, replace y with 18. You get 12 on both sides. This proves that 18 is the correct solution.

$$\frac{2}{3} \cdot \overset{6}{18} = 12$$
$$12 = 12$$

Lesson 1   Exercise

Directions: Solve and check each equation.

1.	$m + 11 = 30$	$8w = 56$	$c - 12 = 5$
2.	$16 = f - 4$	$\frac{c}{4} = 5$	$6n = 9$
3.	$2 = \frac{y}{9}$	$12 = 18p$	$14 = a + 3$
4.	$g - 9 = 41$	$e + 6 = 8$	$\frac{n}{12} = 1$
5.	$15 = i - 8$	$200 = 25r$	$\frac{3}{4}s = 24$
6.	$10 = \frac{1}{2}w$	$21 = d + 16$	$10 = \frac{z}{5}$
7.	$24f = 12$	$p + 14 = 4$	$\frac{3}{8}x = 15$
8.	$16y = 4$	$a + 3 = 10$	$\frac{p}{4} = 3$
9.	$20 = 2k$	$b - 3 = 17$	$40 = \frac{x}{2}$
10.	$\frac{z}{18} = 2$	$7 = a + 1$	$\frac{4c}{5} = 40$

Answers are on page 342.

Lesson 2   Multistep Equations

To solve some equations, you must use more than one inverse operation. *Take care of addition and subtraction before multiplication and division.*

Example 1: Solve and check the equation $4a - 3 = 29$.

Step 1. On the left side of the equation there is both multiplication and subtraction. First take care of subtraction by adding 3 to both sides.

$$\begin{array}{rcl} 4a - 3 &=& 29 \\ + 3 & & + 3 \\ \hline 4a &=& 32 \end{array}$$

Step 2. Since the unknown is multiplied by 4, divide both sides by 4. The solution is 8.

$$\frac{4a}{4} = \frac{32}{4}$$
$$a = 8$$

Step 3. Replace a with 8 and evaluate the equation. Both sides equal 29. This proves that the solution 8 is correct.

$$4 \cdot 8 - 3 = 29$$
$$32 - 3 = 29$$
$$29 = 29$$

Example 2: Solve and check the equation $7 = \frac{x}{5} + 1$.

Step 1. On the right side of the equation there is both division and addition. First take care of addition by subtracting 1 from both sides.

$$\begin{array}{rcl} 7 &=& \frac{x}{5} + 1 \\ -1 & & -1 \\ \hline 6 &=& \frac{x}{5} \end{array}$$

Step 2. Since the unknown is divided by 5, multiply both sides by 5. The solution is 30.

$$5 \cdot 6 = \frac{x}{5} \cdot 5$$
$$30 = x$$

Step 3. Replace x with 30 and evaluate the equation. Both sides equal 7. This proves that the solution 30 is correct.

$$7 = \frac{\overset{6}{\cancel{30}}}{\underset{1}{\cancel{5}}} + 1$$
$$7 = 6 + 1$$
$$7 = 7$$

Lesson 2   Exercise

Directions: Solve and check each equation.

1. $6m + 5 = 47$ $3x - 2 = 28$ $\frac{c}{4} + 1 = 8$

2. $17 = 7a + 3$ $50 = 9d - 4$ $8 = \frac{x}{7} - 2$

3. $2n - 11 = 3$ $\frac{3}{4}a + 5 = 17$ $\frac{s}{10} - 6 = 3$

4. $5y + 7 = -3$ $2 = \frac{w}{9} + 11$ $\frac{4}{5}f - 7 = 1$

5. $8z + 3 = 9$ $2 = 9p - 4$ $-4 = 7t + 3$

Answers are on page 342–343.

Lesson 3   Equations with Separated Unknowns

In some equations, the unknown appears more than once. When the unknowns appear on the same side of the equal sign, combine those terms according to the rules for adding monomials.

Example 1: Solve the equation $7m - 3 - 2m = 27$.

Step 1. Combine the terms containing the unknown m.

$$\begin{array}{rcl} 7m - 3 - 2m &=& 27 \\ 5m - 3 &=& 27 \end{array}$$

Step 2. Add 3 to both sides of the equation.

$$\begin{array}{rcl} 5m - 3 &=& 27 \\ + 3 & & +3 \end{array}$$

Step 3. Divide both sides of the equation by 5. The solution is 6.

$$\frac{5m}{5} = \frac{30}{5}$$
$$m = 6$$

When the unknown appears on both sides of the = sign, use inverse operations to combine the unknowns. Remember that you want the unknowns on one side and numbers on the other side.

Example 2: Solve the equation 5x + 30 = 8x.

Step 1. To get a statement that says, "Value = unknown," subtract 5x from both sides.

$$5x + 30 = 8x$$
$$\underline{-5x \qquad -5x}$$
$$30 = 3x$$

Step 2. Divide both sides by 3. The solution is 10.

$$\frac{30}{3} = \frac{3x}{3}$$
$$10 = x$$

To check an equation with separated unknowns, replace each unknown with the solution. When you evaluate the equation, you should get the same number on both sides. For the previous example, replace both x's with 10.

$$5 \cdot 10 + 30 = 8 \times 10$$
$$50 + 30 = 80$$
$$80 = 80$$

Lesson 3   Exercise

Directions: Solve and check each equation.

1. $9a - 2a = 21$ $5m = 18 + 2m$ $8r = 15 + 3r$

2. $12 - 5x = 7x$ $16 = 13c + 7c$ $3p + 7 = 10p$

3. $6y - y = 10$ $3t = 9 + 2t$ $4 - 2n = 6n$

4. $5x + 4 = 3x + 20$ $8w - 5 = 7w + 13$ $3p + 12 = 8p - 23$

5. $7c - 3c = c + 27$ $9m - 12 = m + 20$ $2d - 8 = 7d + 12$

Answers are on pages 343–344.

Lesson 4   Equations with Parentheses

To solve an equation with parentheses, first multiply each term inside the parentheses by the number outside the parentheses.

Example: Solve the equation 6(x − 2) = 18.

Step 1. Multiply x by 6 and −2 by 6.

$$6(x - 2) = 18$$

Step 2. Add 12 to both sides.

$$6x - 12 = 18$$
$$\underline{+ 12 \qquad +12}$$

Step 3. Divide both sides by 6.
The solution is 5.

$$\frac{6x}{6} = \frac{30}{6}$$
$$x = 5$$

To check an equation with parentheses, replace the unknown with the solution. For the example, replace x with 5. To evaluate the equation, you can first subtract 2 from 5, or you can first multiply 6 by both 5 and 2.

$6(5 - 2) = 18$	or	$6(5 - 2) = 18$
$6(3) = 18$		$30 - 12 = 18$
$18 = 18$		$18 = 18$

Lesson 4   Exercise

Directions: Solve and check each equation.

1. $4(m - 3) = 20$ $5(a + 2) = 15$ $9 = 2(x - 3)$

2. $3(c + 4) = 2c + 17$ $8n - 7 = 6(n - 1)$ $9(p + 2) = p + 20$

3. $4(a - 5) = 3(a + 2)$ $6(d - 1) = 3(d + 2)$ $5(y + 2) = 3(y - 8)$

4. $20 = 4(15 - x)$ $4a - 8 = 2(a + 4)$ $4(y - 8) = 2(y + 8)$

5. $4(c - 5) = 2(c + 5)$ $12(p + 6) = 8(p + 10)$ $6(d - 3) = 3(d + 4)$

Answers are on pages 344–345.

Lesson 5   Inequalities

An **inequality** is a statement that two amounts are *not* equal. Below are the four symbols which express inequalities.

Symbol   **Example**
< means *less than* $2 < 5$
> means *more than* $9 > 1$
≤ means *less than or equal to* $7 \le 10$ or $4 \le 4$
≥ means *more than or equal to* $8 \ge 3$ or $6 \ge 6$

To solve an inequality, follow the rules you learned for solving equations.

Example: Solve the inequality $m - 9 > 6$.

Add 9 to both sides of the inequality. The solution is 15. This means that any number greater than 15 is a solution to the inequality. For example, both $15\frac{1}{4}$ and 99 are solutions.

$$
\begin{array}{rcr}
m - 9 > & & 6 \\
+ 9 & & +9 \\
\hline
m \quad > & & 15
\end{array}
$$

Lesson 5   Exercise

Directions: Solve each inequality.

1. $a + 6 > 9$ $c - 12 \leq 3$ $\frac{n}{2} < 7$

2. $16r \geq 20$ $6m - 2 < 22$ $\frac{3}{5}x \leq 18$

3. $3p - 4 > p + 6$ $9w + 2 \geq w + 10$ $3(m - 2) < 9$

4. $\frac{1}{2}y - 4 \leq 1$ $4 > 2(n - 9)$ $8t - 5 \leq 2t + 1$

5. $6(x - 12) \leq 18$ $5a - 3 \geq 3a + 3$ $c - 1 \leq \frac{3}{4}c + 2$

Answers are on page 345.

Lesson 6   Writing Algebraic Expressions

One of the most useful applications of algebra is writing verbal information in algebraic form. Below are some key words and phrases for each of the basic operations. After each key word or phrase is a verbal example, and after each example there is an algebraic expression that represents the example. In each expression, a letter stands for the unknown number.

Addition

sum   "the sum of a number and 8": $x + 8$

total   "the total of 12 and a number": $12 + c$

increased by   "a number increased by 5": $y + 5$

plus   "one plus a number": $1 + t$

more than   "three more than a number": $m + 3$

The order in the previous examples is not important. For example, $m + 3$ is the same as $3 + m$. However, in the subtraction examples that follow, the order is important.

Subtraction

less than   "ten less than a number": $a - 10$

decreased by   "seven decreased by a number": $7 - b$

subtracted from   "a number subtracted from 12": $12 - f$

minus   "nine minus a number": $9 - g$

Notice that in the following multiplication examples, the coefficient comes before the unknown.

Multiplication

product   "the product of six and a number": $6k$

times "two times a number": $2m$
a fraction of "one-fourth of a number": $\frac{1}{4}a$

Division

divided by "a number divided by 20": $\frac{d}{20}$

"30 divided by a number": $\frac{30}{n}$

The previous examples all show one operation. Below are examples of multistep operations. Notice the use of parentheses in some of these examples. A set of parentheses groups numbers together. The words *quantity* and *all* often suggest that some amount should be enclosed in parentheses.

Verbal Expression	Algebraic Expression
A number decreased by one-half the number	$s - \frac{1}{2}s$
The sum of 11 and a number, all multiplied by four	$4(n + 11)$
One-third of the quantity of a number increased by seven	$\frac{1}{3}(e + 7)$
Eight divided into the quantity of a number decreased by two	$\frac{i-2}{8}$, or $(i-2)/8$
Two less than a number divided by eight	$\frac{i}{8} - 2$, or $i/8 - 2$

You can use algebraic expressions to show the mathematical relationships in practical situations. Watch for the words and phrases which tell you what operations to use.

Example 1: Alberto makes x dollars an hour. Write an expression for his hourly wage if he gets a raise of $2 an hour.

A raise of $2 an hour means to add. $x + 2$

Example 2: Beef costs b dollars a pound. How much did Celia pay for 4 pounds of beef?

Multiply the cost of one pound by 4. $4b$

Lesson 6   Exercise

Directions: *For items 1 to 5, write an algebraic expression. Use x to stand for the unknown.*

1.   Twelve more than a number

2.   A number divided by four

3.   A number increased by five times the same number

4.   The sum of eight and a number, all multiplied by six

5.   Three divided into the quantity of a number increased by seven

For items 6 to 10, write an algebraic expression.

6. Celeste's take-home pay is p dollars a month. She spends one-fourth of her pay for rent. What is her monthly rent?

7. Jack is x years old.
   a. What was his age five years ago?
   b. What will his age be in three years?

8. If Jim drives at an average speed of r miles per hour, how far will he drive in four hours?

9. Let g stand for Larry's gross salary. Larry's employer deducts 22% of Larry's salary for taxes. Write an expression for the amount of the deductions. (Hint: Change the percent to a decimal.)

10. Colin is x years old. His son is 32 years younger. Write an expression for Colin's son's age.

Answers are on page 346.

Lesson 7   Multiplying Binomials

A **binomial** is an algebraic expression with two terms. $x + 8$ is an example of a binomial. Multiplying binomials is an important building block in an area of algebra which you will learn more about later. You already know the rules for multiplying monomials.

Follow these steps to multiply binomials.

1. Put one binomial under the other.
2. Multiply both terms in the top binomial by the term at the right in the bottom binomial.
3. Multiply both terms in the top binomial by the term at the left in the bottom binomial.
4. Add the results of steps 2 and 3.

Study the next examples carefully to see how these problems are set up.

Example 1: $(x + 3)(x + 2) = ?$

Step 1. Put one binomial under the other.

Step 2. Multiply the top terms by $+2$.
$+2 \cdot +3 = +6$ and $+2 \cdot x = +2x$.

$$\begin{array}{r} x + 3 \\ x + 2 \\ \hline + 2x + 6 \end{array}$$

Step 3. Multiply the top terms by x.
$x \cdot +3 = +3x$ and $x \cdot x = x^2$.

$$\begin{array}{r} x^2 + 3x \\ \hline x^2 + 5x + 6 \end{array}$$

Step 4. Add the results. Notice how $+2x$ and $+3x$ are written under each other.

Example 2: $(x - 5)(x + 4) = ?$

Step 1. Put one binomial under the other.

Step 2. Multiply -5 by $+4$ and x by $+4$.

$$\begin{array}{r} x - 5 \\ \underline{x + 4} \\ + 4x - 20 \end{array}$$

Step 3. Multiply -5 by x and x by x.

Step 4. Add the results.

$$\begin{array}{r} x^2 - 5x \\ \hline x^2 - x - 20 \end{array}$$

Example 3: $(x - 6)(x - 1) = ?$

Step 1. Put one binomial under the other.

Step 2. Multiply -6 by -1 and x by -1.

$$\begin{array}{r} x - 6 \\ \underline{x - 1} \\ - x + 6 \end{array}$$

Step 3. Multiply -6 by x and x by x.

Step 4. Add the results.

$$\begin{array}{r} x^2 - 6x \\ \hline x^2 - 7x + 6 \end{array}$$

Example 4: $(x - 2)(x + 2) = ?$

Step 1. Put one binomial under the other.

Step 2. Multiply -2 by $+2$ and x by $+2$.

$$\begin{array}{r} x - 2 \\ \underline{x + 2} \\ + 2x - 4 \end{array}$$

Step 3. Multiply -2 by x and x by x.

Step 4. Add the results. Notice that the answer has no middle term.

$$\begin{array}{r} x^2 - 2x \\ \hline x^2 \qquad - 4 \end{array}$$

Be sure that you understand where we got each term in the above examples before you try the next exercise.

Lesson 7   Exercise

Directions: Multiply each pair of binomials.

1. $(x + 5)(x + 2) = ?$ $(x + 3)(x + 1) = ?$ $(x + 2)(x + 6) = ?$

2. $(x - 4)(x - 3) = ?$ $(x - 1)(x - 8) = ?$ $(x - 4)(x - 5) = ?$

3. $(x + 5)(x - 2) = ?$ $(x - 6)(x + 7) = ?$ $(x + 12)(x - 10) = ?$

4. $(x + 8)(x - 8) = ?$ $(x + 10)(x - 10) = ?$ $(x - 3)(x + 3) = ?$

5. $(x - 7)(x + 3) = ?$ $(x + 6)(x - 9) = ?$ $(x - 11)(x + 6) = ?$

Answers are on page 346.

Lesson 8   Factoring

Factors are numbers that multiply together to give another number. For example, 3 and 2 are factors of 6. 6 and 1 are also factors of 6. In this book, you will learn some of the common types of factoring problems which you may see on the GED Test.

Factoring Binomials

Think about the binomial $6x + 15$. Both terms can be divided evenly by the whole number 3. Therefore, 3 is one of the factors of $6x + 15$.

Follow these steps to factor a binomial.

1. Look for a number or letter that divides evenly into both terms.

2. Divide the binomial by the number or letter to find the other factor.

Example 1: Factor the expression $6x + 15$.

Three divides evenly into both $6x$ and 15. Di- $6x + 15 = 3(2x + 5)$
vide 3 into each term. Write 3 outside the pa-
rentheses, and write the quotient of 3 divided
into each term inside a set of parentheses.

To check the factors, multiply 3 by each term inside the parentheses. 3 times $2x = 6x$, and 3 times $+5 = +15$.

In some binomials, a letter divides evenly into both terms.

Example 2: Factor the expression $s^2 + 7s$.

The letter s divides evenly into each term. Di- $s^2 - 7s = s(s + 7)$
vide s into both s and $7s$. Write s outside the
parentheses, and write the quotient from dividing each term by s inside the parentheses.

To check the previous example, multiply s by each term inside the parentheses, s times s is s^2, and s times $+7$ is $+7s$.

Factoring Quadratic Expression

One of the most common factoring problems in algebra is factoring a **quadratic** expression. In a quadratic expression, the unknown is raised to the second power. The answers to the problems you worked out in Exercise 7 were all quadratic expressions. The factors for these expressions are two binomials.

To factor a quadratic expression, you must think about the way we get each term in the expression. Look carefully at the problem $(x + 9) (x + 5)$.

$$
\begin{array}{r}
x + 9 \\
x + 5 \\
\hline
+ 5x + 45 \\
x^2 + 9x \\
\hline
x^2 + 14x + 45
\end{array}
$$

To factor the expression $x^2 + 14x + 45$, you must understand where each term comes from. The first term, x^2, is easy. It comes from multiplying x by x. The third term, 45, is also easy. It comes from multiplying $+5$ by $+9$. The middle term, $14x$, is more difficult. It comes from adding $5x$ and $9x$. The most important questions to consider when you factor a quadratic expression are, "What terms multiply to give the third term?" and "What terms add to give the middle term?'' Think about these two questions as you study the next examples.

Example 3: Factor the expression $x^2 + 10x + 24$.

Think first about $+24$. Several combinations $\qquad x^2 + 10x + 24 = ?$
multiply together to make 24, for example $+12$
by $+2$ or $+8$ by $+3$ or $+6$ by $+4$. However,
only one of these gives the middle term, 10x:
$6x + 4x = +10x$. The factors for $x^2 + 10x +$
24 are $(x + 6)$ and $(x + 4)$. $\qquad\qquad\qquad\qquad (x + 6)\,(x + 4)$

    When you factor a quadratic expression, *always* check by multiplying the factors together.

$$
\begin{array}{r}
x +\ \ 6 \\
x +\ \ 4 \\
\hline
+\ \ \ \ 4x + 24 \\
x^2 +\ \ 6x \\
\hline
x^2 + 10x + 24
\end{array}
$$

Example 4: Factor the expression $x^2 - 7x + 10$.

First think about $+10$. Both 10 times 1 and 5 $\qquad x^2 - 7x + 10 = ?$
times 2 equal 10. But look at the middle term.
When the middle term is negative and the third
term is positive, you are looking for two negative numbers. $-7x$ comes from adding $-5x$ and
$-2x$. The $+10$ comes from multiplying -5 by
-2. The factors for $x^2 - 7x + 10$ are $(x - 5)$
and $(x - 2)$. $\qquad\qquad\qquad\qquad\qquad (x - 5)\,(x - 2)$

    Check the previous problem by multiplying $x - 5$ by $x - 2$. Think carefully about the signs in every factoring problem.

Example 5: Factor the expression $x^2 + 2x - 15$.

First think about -15. Both 15 times 1 and 5 $\qquad x^2 + 2x - 15 = ?$
times 3 equal 15. Since -15 is negative, 'there
must be one positive number and one negative
number. Since $+2x$ is positive, the larger factor
must be positive. $+5x$ and $-3x$ equal $+2x$.
$+5$ times $-3 = -15$. The factors for $x^2 + 2x$
-15 are $(x + 5)$ and $(x - 3)$. $\qquad\qquad\qquad (x + 5)\,(x - 3)$
    Multiply the factors to check.

Example 6: Factor the expression $x^2 - 3x - 28$.

First think about -28. Both 14 times 2 and 7 $\qquad x^2 - 3x - 28 = ?$
times 4 equal 28. Since -28 is negative, there
must be one positive number and one negative
number. Since $-3x$ is negative, the larger factor
must be negative. $-7x$ and $+4x$ equal $-3x$. -7

times $+4 = -28$. The factors for $x^2 - 3x - 28$
are $(x - 7)$ and $(x + 4)$. $(x - 7)(x + 4)$
     Multiply the factors to check.

Example 7: Factor the expression $x^2 - 25$.

First think about -25. Both 25 times 1 and 5 $x^2 - 25 = ?$
times 5 equal 25. Since -25 is negative, there
must be one negative number and one positive
number. Since there is no middle term, the factors must be the same. $+5x$ and $-5x$ equal 0.
$+5$ times $-5 = -25$. The factors for $x^2 - 25$
are $(x + 5)$ and $(x - 5)$.
     Multiply the factors to check. Note that the
order for writing the factors can also be
$(x - 5)(x + 5)$. $(x + 5)(x - 5)$

Lesson 8   Exercise

Directions: Factor each expression. Check by multiplying the factors together.

1. $3x + 9 = ?$	$8w - 12 = ?$	$10c - 5d = ?$
2. $4x - 16 = ?$	$7a + 21 = ?$	$5y - 20z = ?$
3. $36f - 12 = ?$	$9m + 21 = ?$	$6w - 9 = ?$
4. $c^2 + 8c = ?$	$m^2 + 6m = ?$	$x^2 - 5x = ?$
5. $p^2 + 10p = ?$	$a^2 - a = ?$	$n^2 - 2n = ?$
6. $x^2 + 6x + 8 = ?$	$x^2 + 8x + 7 = ?$	$x^2 + 13x + 40 = ?$
7. $x^2 - 13x + 36 = ?$	$x^2 - 12x + 36 = ?$	$x^2 - 10x + 9 = ?$
8. $x^2 + 5x - 14 = ?$	$x^2 + 2x - 8 = ?$	$x^2 + x - 56 = ?$
9. $x^2 - 36 = ?$	$x^2 - 4 = ?$	$x^2 - 81 = ?$
10. $x^2 - 3x - 10 = ?$	$x^2 - 4x - 96 = ?$	$x^2 - x - 12 = ?$

Answers are on pages 346–347. (For more practice in factoring quadratic expressions,
try factoring the solutions to Exercise 7.)

Lesson 9   Factoring and Square Roots

A useful application of factoring is simplifying square roots. So far in this
book, you have found the square root of numbers with exact square roots.
Most numbers, however, do not have an exact square root.
     Think about the number 20. The square root of 20 is larger than 4, since

4 × 4 = 16, and smaller than 5, since 5 × 5 = 25.
    Follow these steps to simplify a square root.

1. Find factors of the number which include an exact square root.
2. Put the square root of one factor outside the $\sqrt{}$ sign, and leave the other factor inside the $\sqrt{}$ sign.

Example 1: Simplify $\sqrt{20}$.

Step 1. The whole-number factors of 20 are 1 and 20, 2 and 10, and 4 and 5. The combination that includes an exact square root is 4 and 5. Put each factor inside a $\sqrt{}$ sign.

$\sqrt{20} =$

$\sqrt{4} \cdot \sqrt{5} =$

Step 2. Put the square root of 4 next to $\sqrt{5}$. $2\sqrt{5}$

Example 2: Simplify $\sqrt{125}$.

Step 1. The factors of 125 that include an exact square root are 25 and 5. Put each factor inside a sign.

$\sqrt{125} =$

$\sqrt{25} \cdot \sqrt{5} =$

Step 2. Put the square root of 25 next to $\sqrt{5}$. $5\sqrt{5}$

    To the nearest hundredth, the square root of 5 is 2.24. To show that $2\sqrt{5}$ is a sensible answer to $\sqrt{20}$, find 2 × 2.24 to the second power.

$$
\begin{array}{r}
2.24 \\
\times\ \ \ 2 \\
\hline
4.48
\end{array}
\qquad
\begin{array}{r}
(4.48)^2 = \quad 4.48 \\
\times 4.48 \\
\hline
35\ 84 \\
1\ 79\ 2 \\
17\ 92 \\
\hline
20.07\ 04,\ \text{which is about 20}
\end{array}
$$

    To show that $5\sqrt{5}$ is a sensible answer to $\sqrt{125}$, find 5 × 2.24 to the second power.

$$
\begin{array}{r}
2.24 \\
\times\ \ \ 5 \\
\hline
11.20
\end{array}
\qquad
\begin{array}{r}
(11.2)^2 = \quad 11.2 \\
\times 11.2 \\
\hline
2\ 24 \\
11\ 2 \\
112 \\
\hline
125.44,\ \text{which is about 125}
\end{array}
$$

Lesson 9   Exercise

Directions: Simplify each square root.

1. $\sqrt{8}$ 2. $\sqrt{75}$ 3. $\sqrt{18}$ 4. $\sqrt{24}$ 5. $\sqrt{72}$

6. $\sqrt{54}$ 7. $\sqrt{500}$ 8. $\sqrt{45}$ 9. $\sqrt{96}$ 10. $\sqrt{128}$

Answers are on page 347.

Lesson 10   Quadratic Equations

In a quadratic equation, the unknown is raised to the second power. Most quadratic equations have two correct solutions. Think about the equation $x^2 = 16$. You know that the square root of 16 is 4. In algebra, however, the square root of 16 can also be -4. When you multiply -4 by itself, you get $+16$, since the signs are alike. The solutions to the equation $x^2 = 16$ are $x = +4$ and $x = -4$.

Not every quadratic equation is this easy to solve. Most quadratic equations are written in the form $x^2 + Ax + B = 0$, where A and B are numbers. This is called the standard form of a quadratic equation.

Follow these steps to solve a quadratic equation.

1. Be sure the equation is written in standard form.
2. Factor the expression on the left side of the = sign.
3. Set each factor equal to zero, and solve each new equation.

Example 1: Solve the equation $x^2 + 5x + 6 = 0$.

Step 1. Factor the expression $x^2 + 5x + 6$. The factors that give $+6$ at the right and the term $5x$ in the middle are $(x + 3)$ and $(x + 2)$.

$$x^2 + 5x + 6 = 0$$
$$(x + 3)(x + 2) = 0$$

Step 2. Set each factor equal to zero. The solution from one factor is $x = -3$. The solution from the other factor is $x = -2$.

$$
\begin{array}{rr}
x + 3 = & 0 \\
-3 & -3 \\
\hline
x = & -3 \text{ and}
\end{array}
$$

$$
\begin{array}{rr}
x + 2 = & 0 \\
-2 & -2 \\
\hline
x = & -2
\end{array}
$$

To check the solutions to a quadratic equation, substitute both values into the original equation.

When you substitute $x = -3$ into the original equation, you get:

$$
\begin{aligned}
(-3)^2 + 5(-3) + 6 &= 0 \\
+9 \quad -15 \quad\;\; +6 &= 0 \\
+15 - 15 \qquad\quad &= 0
\end{aligned}
$$

When you substitute $x = -2$ into the original equation, you get:

$$
\begin{aligned}
(-2)^2 + 5(-2) + 6 &= 0 \\
+4 \quad - \quad 10 \quad +6 &= 0 \\
+10 - \quad 10 \qquad &= 0
\end{aligned}
$$

Example 2: Solve the equation $x^2 - 3x - 28 = 0$.

Step 1. Factor the expression $x^2 - 3x - 28$. The factors that give -28 at the right and the term $-3x$ in the middle are $(x - 7)$ and $(x + 4)$.

$$x^2 - 3x - 28 = 0$$
$$(x - 7)(x + 4) = 0$$

Step 2. Set each factor equal to zero. The solution from one factor is x = 7. The solution from the other factor is x = −4.

$$x - 7 = 0$$
$$\underline{+7 \quad +7}$$
$$x \quad = \quad 7 \text{ and}$$

$$x + 4 = 0$$
$$\underline{-4 \quad -4}$$
$$x \quad = -4$$

To check the solutions, substitute each value into the original equation.

$$(7)^2 - 3(7) - 28 = 0 \qquad \text{and} \qquad (-4)^2 - 3(-4) - 28 = 0$$
$$+49 - 21 - 28 = 0 \qquad\qquad\qquad +16 + 12 - 28 = 0$$
$$+49 - 49 \quad\quad = 0 \qquad\qquad\qquad\quad\; + 28 - 28 = 0$$

Quadratic equations are difficult. It is easy to make mistakes with the signs. Be sure to check the solutions by substituting your answers into the original equations.

Lesson 10   Exercise

Directions: Solve and check each quadratic equation.

1. $x^2 + 7x + 10 = 0$ $x^2 + 10x + 9 = 0$ $x^2 + 15x + 56 = 0$
2. $x^2 - 7x + 12 = 0$ $x^2 - 17x + 60 = 0$ $x^2 - 12x + 27 = 0$
3. $x^2 + 2x - 24 = 0$ $x^2 + x - 72 = 0$ $x^2 + 7x - 30 = 0$
4. $x^2 - 4x - 21 = 0$ $x^2 - 5x - 6 = 0$ $x^2 - 10x - 24 = 0$
5. $x^2 - 49 = 0$ $x^2 - 1 = 0$ $x^2 - 144 = 0$

Answers are on pages 347–348.

Level 2   Review

Directions: For items 1 to 4, solve for each unknown.

1. $15 = 9x - 3$
2. $4a + 2 + 3a = a + 20$
3. $7x - 3 = 5(x + 5)$
4. $6n - 5 > 4n + 21$

For items 5 to 10, solve each problem.

5. Write an algebraic expression for the quantity of three less than twice a number, all divided by five.

6. Let t represent the total number of students in Ann's math class. There

are eight women in the class. Write an expression for the number of men in the class.

7.   Factor the expression $c^2 - 6c$.

8.   Factor the expression $x^2 + 6x - 16$.

9.   Simplify $\sqrt{28}$.

10.   Find the solutions to the equation $x^2 - 6x - 27 = 0$.

Check your answers. Correct answers are on page 349. If you have at least eight answers correct, go to the Level 3 Preview. If you have fewer than eight answers correct, go back to Lesson 1 and study Level 2.

Level 3   Algebra Problem Solving

Preview

Directions: Solve each problem.

1.   Five more than six times a number is 17. Find the number.

2.   Forty-three equals two less than nine times a number. Find the number.

3.   Four times the quantity of a number decreased by one is the same as two more than the same number. Find the number.

4.   At the Spring Street Day Care Center there are 65 children. There are nine more girls than boys. Find the number of girls at the center.

5.   Fred, along with his sons Tom and Bill, built a recreation room in their basement. Fred worked twice as many hours as Tom, and Bill worked ten hours more than Tom. Altogether, the three of them worked 94 hours. How many hours did Fred work?

Check your answers. Correct answers are on page 349. If you have at least four answers correct, do the Level 3 Review on pages 210–211. If you have fewer than four answers correct, study Level 3 beginning with Lesson 1.

Lesson 1   Writing Equations

Earlier you learned to write algebraic expressions from verbal descriptions. To write equations from verbal descriptions, watch for the verb in each description. Words like *is* and *equals* tell you where to put the = sign.

Example 1: Eight less than three times a number is 19. Write an equation and find the number.

Step 1. Let x stand for the number. Put the = sign where the verb *is* appears in the sentence.

$$3x - 8 = 19$$

Step 2. Add 8 to both sides.

Step 3. Divide both sides by 3. The solution is 9.

$$3x - 8 = 19$$
$$\underline{+\ 8 \qquad +\ 8}$$
$$\frac{3x}{3} = \frac{27}{3}$$
$$x = 9$$

Example 2: Twice the quantity of a number decreased by five equals 20. Find the number.

Step 1. Let x stand for the number. Put the = sign where the verb *equals* appears.

$$2(x - 5) = 20$$

Step 2. Multiply x and −5 by 2.

$$2x - 10 = 20$$

Step 3. Add 10 to both sides.

Step 4. Divide both sides by 2. The solution is 15.

$$\underline{+\ 10 \qquad +\ 10}$$
$$\frac{2x}{2} = \frac{30}{2}$$
$$x = 15$$

Lesson 1   Exercise

Directions: Write and solve an equation for each of the following verbal descriptions.

1. A number increased by fifteen is 21.

2. Six times a number is 72.

3. A number divided by 25 is six.

4. Thirty equals a number decreased by 27.

5. Seven times a number decreased by the same number is 78.

6. Nine times a number decreased by three is the same as twice the same number increased by 25.

7. Five less than ten times a number is 19 more than twice the same number.

8. Twelve times a number equals 45 more than three times the same number.

9. Five times the quantity of a number decreased by two is 30.

10. Seven times the quantity of a number decreased by one equals four times the quantity of the same number increased by two.

Answers are on page 349.

Lesson 2   Using Algebra to Solve Formulas

You can use your knowledge about solving equations to use formulas when the information you have is "backwards." For example, you have used the formula $d = rt$ to find distance. If the information you have includes the distance, and either the rate or the time, you can use the formula to find either the time or the rate.

Example 1: Jack walked 10 miles at an average speed of 4 mph. Find the total time Jack walked.

Step 1. Replace d with 10 and r with 4 in the formula $d = rt$.

$d = rt$

$10 = 4t$

Step 2. Solve for the time by dividing both sides by 4. The time is $2\frac{1}{2}$ hours.

$\frac{10}{4} = \frac{4t}{4}$

$2\frac{2}{4} = 2\frac{1}{2}$ hr. $= t$

You can use the formula $c = nr$ for cost in the same way.

Example 2: Maria spent $7.95 for 3 pounds of meat. What was the price per pound?

Step 1. Replace c with $7.95 and n with 3 in the formula $c = nr$.

$c = nr$

$7.95 = 3r$

Step 2. Solve for the rate (the price of one pound) by dividing both sides by 3. The price of one pound is $2.65.

$\frac{7.95}{3} = \frac{3r}{3}$

$\$2.65 = r$

You can use the formula $i = prt$ for interest in the same way.

Example 3: Bill earned $21 in interest on a loan of $350 in one year. Find the interest rate.

Step 1. Replace i with $21, p with $350, and t with 1 in the formula $i = prt$.

$i = prt$

$21 = 350 \cdot r \cdot 1$

Step 2. To find r (the interest rate) divide both sides by 350. The interest rate is $\frac{3}{50}$.

$\frac{21}{350} = \frac{350r}{350}$

$\frac{3}{50} = r$

Step 3. Change the fraction $\frac{3}{50}$ to a percent by multiplying by 100%. The interest rate is 6%.

$\frac{3}{50} \times \frac{\overset{2}{100\%}}{1} = 6\%$

Lesson 2   Exercise

Directions: Use the formulas $d = rt$, $c = nr$, or $i = prt$ to solve each of the following problems.

1. Alfredo drove 210 miles at an average rate of 42 mph. For how many hours did he drive?

2. A plane flew 1194 miles in three hours. Find the average speed of the plane.

3. Isabella hiked 27 miles in six hours. Find her average hiking speed.

4. Tom spent $22 each for the sweaters he sells in his store. Altogether, he spent $528 for sweaters. How many sweaters did he buy?

5. For a dozen baseball gloves, Mr. Huston spent $420. Find the average price for each glove.

6. The Rigbys spent $5625 for 4.5 acres of land. Find the price per acre.

7. One ticket for a concert at the Municipal Theater is $6.50. If the total ticket sales were $3380, how many tickets were sold?

8. Kate earned $51 interest on $850 for one year. Find the rate of interest.

9. On a loan of $350, Carmen had to pay $14 interest at 8% annual interest. Find the length of time of the loan.

10. In one year, the Santiagos pay $4800 in interest on a mortgage at an annual interest rate of 12%. Find the total amount of the mortgage.

Answers are on pages 349–350.

Lesson 3   Algebra Word Problems

Algebra is a tool for solving long, complicated word problems. In this book, you will learn the basic skills involved in solving word problems. For every algebra word problem, choose a letter to stand for the unknown. Then use this letter to write expressions for all the parts of the problem.

Example 1: One number is four times another number. The larger number decreased by two is the same as the smaller number increased by 16. Find the two numbers.

Step 1. Let x stand for the smaller number. Let 4x stand for the larger number.

smaller number = x
larger number = 4x

Step 2. Write an equation from the second sentence in the problem. Notice that the phrase *is the same as* tells you where to put the = sign.

$4x - 2 = x + 16$

Step 3. Subtract x from both sides. Then add 2 to both sides.

$$\begin{array}{rcr} 4x - 2 = & x + 16 \\ -x & -x \\ \hline 3x - 2 = & 16 \\ + 2 & +2 \\ \hline \end{array}$$

Step 4. Divide both sides by 3.

$$\dfrac{3x}{3} = \dfrac{18}{3}$$

Step 5. The solution is 6, which is the smaller number. Find the value of 4x, which is the larger number.

$$x = 6$$
$$4x = 4 \cdot 6 = 24$$

For some problems, it helps to make a chart of the information in the problem.

Example 2: Joaquin is 25 years older than his son. In ten years, the sum of their ages will be 99. Find their ages now.

Step 1. Make a chart that expresses Joaquin's age and his son's age now and in 10 years.

	age now	age in 10 years
son	x	x + 10
Joaquin	x + 25	x + 25 + 10 = x + 35

Step 2. Write an equation that shows the sum of their ages in 10 years.

$$x + 10 + x + 35 = 99$$

Step 3. Combine the unknowns and solve the equation.

$$2x + 45 = 99$$
$$\quad\ \ -45 \quad -45$$
$$\frac{2x}{2} = \frac{54}{2}$$

son's age $x = 27$

Step 4. The unknown x is the son's age. To find Joaquin's age, find the value of x + 25.

Joaquin's age x + 25 =
27 + 25 = 52

Lesson 3   Exercise

Directions: Write and solve an equation for each problem.

1. One number is three times another number. When the larger number is decreased by ten, the result is the same as when the smaller number is increased by 18. Find both numbers.

2. One number is seven more than another number. Four less than three times the smaller number is the same as twice the larger number. Find the two numbers.

3. Louise is 26 years older than her daughter. Louise's age is two more than four times her daughter's age. Find both their ages.

4. Douglas is 46 years younger than his grandfather. Three times Douglas's age is eight less than the grandfather's age. Find both their ages.

5. Juan's net pay is five times the amount his employer deducts. Juan's gross weekly pay is $324. Find his weekly take-home pay.

6. There are 15 more women than men at the Wednesday night exercise class at the Greenport Community Center. Altogether, there are 47 people in the class. Find the number of women.

7. Joe is a plumber. He makes $3 an hour more than his assistant Phil. On a job which they each spent ten hours doing, they made $350. Find the hourly wage for each of them.

8. Altogether, Jim, Carmen, and George spent 95 hours campaigning for their friend Ed, who was running for state assemblyman. Carmen worked five hours more than Jim, and George worked twice as long as Carmen. How many hours did each of them work on the campaign?

9. Chris is three years older than his brother Andy. In five years, three times Andy's age will be six more than twice Chris's age. Find their ages now.

10. For every $3 the Martins spend for food, they spend $2 on car expenses and $4 on rent. They spend a total of $648 a month on these items. How much do they spend each month on rent?

Answers are on pages 350–351.

Level 3   Review

Directions: Solve each problem.

1. Nine less than four times a number is 23. Find the number.

2. Eight times a number is equal to five times the same number increased by 12. Find the number.

3. Two times the quantity of a number decreased by three equals the same number increased by seven. Find the number.

For items 4 to 6, use the formulas $d = rt$ for distance, $c = nr$ for cost, and $i = prt$ for interest.

4. Jerry drove 234 miles in $4\frac{1}{2}$ hours. Find his average speed.

5. How many gallons of gasoine at $.95 a gallon can Carl buy for $5.70?

6. On a loan of $1200 at 15% annual interest, Shirley had to pay $135 interest. For how many months did she have the loan?

---

7. A number is one less than twice another number. Twelve less than seven times the smaller number is the same as three times the larger number. Find both numbers.

8. Dorothy, who makes curtains for a living, makes $4 an hour more than her helper Ann. Dorothy and Ann each worked 25 hours on a job for which they earned $500 altogether. How much does Dorothy make in an hour?

9. Mr. and Mrs. Nash and their daughter Sally shared the driving on a trip to see relatives. Sally drove three hours more than Mrs. Nash, and Mr. Nash drove twice as many hours as Sally. They drove 33 hours altogether. How many hours did Mr. Nash drive?

10. Alvaro is three years younger than his brother Carlos. Ten years ago, five less than three times Alvaro's age was the same as twice Carlos's age. Find Alvaro's age now.

Check your answers. Correct answers are on page 351. If you have at least eight answers correct, go to the Quiz. If you have fewer than eight answers correct, go back to Lesson 1 and study Level 3.

Chapter 6 Quiz

Directions: Solve each problem.

1. Which point below corresponds to the value $-3\frac{1}{3}$?

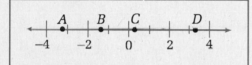

2. $(-16) + (-19) + (12) = ?$

3. $(-40)\left(-\frac{3}{4}\right)(+2) = ?$

4. $(-8m) + (-7m) + (-m) = ?$

5. $(-6a)(+3ac)(-4c) = ?$

6. $\frac{-20m^2n^3}{-10mn} = ?$

7. Find the value of $-3xy + 4x$, if $x = -3$ and $y = 7$.

8. Solve for m in $\frac{3}{4}m - 2 = 34$.

9. Solve for a in $7a - 13 = 2a + 27$.

10. Solve for s in $6(s - 3) = 5(s + 2)$.

11. Solve for p in $8p + 1 \leq p + 15$.

12. Write an algebraic expression for two-thirds times the quantity of a number decreased by six.

13. $(x + 7)(x - 3) = ?$

14. Simplify $\sqrt{48}$.

15. Solve for x in $x^2 + 13x + 40 = 0$.

16. Five times the quantity of a number decreased by two is the same as four times the quantity of the same number increased by one. Find the number.

17. Bob paid $132 interest on a $1600 loan which he had for nine months. Find the annual interest rate.

18. One number is two more than three times another number. Four less than seven times the smaller number is the same as ten more than the larger number. Find both numbers.

19. Al and his daughters Maria and Lucy are house painters. Al makes $5 an hour more than Maria. Lucy makes $2 an hour more than Maria. Each of them worked 30 hours on a job for which they earned $750 altogether. How much does Al make in an hour?

20. David is five years more than twice as old as his daughter Katherine. In 20 years, twice David's age will be the same as three times Katherine's age increased by 11. Find David's age now.

Check your answers. Correct answers are on pages 351–352. If you have at least sixteen answers correct, go to the next chapter. If you have fewer than sixteen answers correct, study the Algebra chapter before you go on.

Chapter

7 | Geometry

Objective

In this chapter, you will

- name angles according to the number of degrees they contain
- identify pairs of angles
- work with parallel lines and transversals
- identify the various kinds of triangles and their properties
- work with similar and congruent figures
- use the Pythagorean relationship
- locate points and find the distance between points on the rectangular coordinate system
- graph linear and quadratic equations
- find slope and intercept
- use algebra to solve problems in geometry

Level 1   Geometry Skills

Preview

Directions: Solve each problem.

1. Angle x pictured at the right contains 75°. What kind of angle is it?

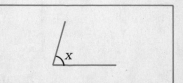

2. How many degrees are there in the complement of a 67° angle?

Items 3 and 4 are based on the following figure.

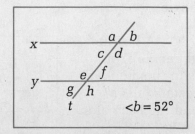

3. Find the measurement of $\angle f$.

4. Which angle is an alternate interior angle with $\angle c$?

5. In triangle WXY, $\angle W = 46°$ and $\angle X = 54°$. Which side of $\triangle WXY$ is longest?

Check your answers. Correct answers are on page 352. If you have at least four answers correct, do the Level 1 Review on pages 222–223. If you have fewer than four answers correct, study Level 1 beginning with Lesson 1.

Lesson 1   Geometry

Geometry is a branch of mathematics that involves the measuring of lines, angles, surfaces, and three-dimensional figures. You have already learned several geometric concepts in this book. You have found the perimeter and area of squares, rectangles, triangles, and circles. You have also found the volume of cubes, rectangular solids, and cylinders.

In this chapter, you will become familiar with other geometric ideas which you need for the GED Test. You will learn many new terms in this chapter. Take the time to memorize the terms you do not already know.

Angles

An **angle** is formed by two lines extending from the same point. The size of an angle depends on how open or closed the lines are.

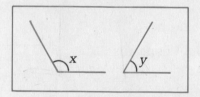

The angle called x above is larger than the angle called y because the lines that form x are open more than the lines that form y. The point that the lines of an angle extend from is called the **vertex.** The unit of measurement of an angle is a **degree.** (°).

Angles are named according to the number of degrees they contain. Below is a list of angle names. Notice that a small curve (or sometimes a small square) indicates each angle.

A **right angle** has exactly 90°. A box at the vertex always means a right angle. You have already seen right angles in this book. Each corner of a square or a rectangle is a right angle. Also, the height and the base of a triangle meet to form a right angle.

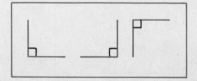

An **acute angle** has less than 90°. The sides of an acute angle are more closed than the sides of a right angle.

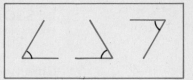

An **obtuse angle** has more than 90° and less than 180°. The sides of an obtuse angle are open more than the sides of a right angle.

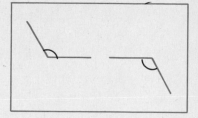

A **straight angle** has exactly 180°. A straight angle looks like a straight line.

A **reflex angle** has more than 180° and less than 360°. You can think of a reflex angle as being bent back upon itself.

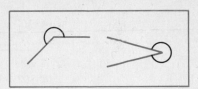

An angle with 360° does not look like an angle. 360° is the angular measurement of a complete circle.

The symbol for an angle is ∠.

There are three ways to refer to an angle. In one method, a capital letter near the vertex refers to the angle.

Example 1: What kind of angle is ∠X?

Since there is a small square at ∠X, ∠X is a right angle.

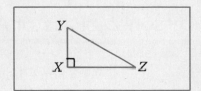

Sometimes three letters are used to refer to an angle. The middle letter is always the vertex.

Example 2: What kind of angle is ∠ABC?

Since the sides that form ∠ABC are closed more than a right angle, ∠ABC is acute.

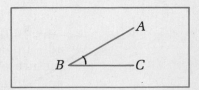

Sometimes a small letter inside the opening of an angle is used to refer to the angle.

Example 3: What kind of angle is ∠m?

Since the sides that form ∠m are open more than a right angle and are closed more than a straight angle, ∠m is an obtuse angle.

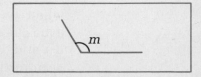

Lesson 1   Exercise

Directions: *For items 1 and 2 identify each angle pictured.*

1.

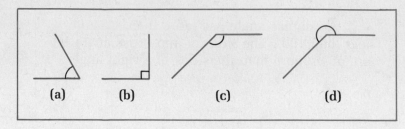

(a)   (b)   (c)   (d)

2.

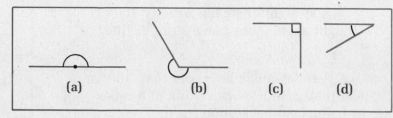

(a)   (b)   (c)   (d)

For items 3 to 5 identify each angle.

3.   a. 35°   b. 60°   c. 130°   d. 90°

4.   a. 250°   b. 162°   c. 180°   d. 85°

5.   a. 40°   b. 125°   c. 320°   d. 95°

Answers are on page 352.

Lesson 2   Pairs of Angles

Sometimes angles add together to form different types of angles.

Complementary angles are two angles that add up to 90°. Together, $\angle a$ and $\angle b$ in the picture at the right make a right angle. $\angle a$ and $\angle b$ are complementary angles. If $\angle a = 60°$, then $\angle b = 90° - 60° = 30°$.

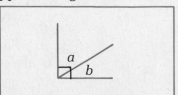

Supplementary angles are two angles that add up to 180°. Together, $\angle c$ and $\angle d$ in the picture at the right make a straight angle. $\angle c$ and $\angle d$ are supplementary angles. If $\angle c = 70°$, then $\angle d = 180° - 70° = 110°$.

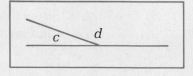

When two straight lines intersect, four angles are formed.

Vertical angles are across from each other when two lines intersect. Vertical angles are equal to each other. In the picture at the right, $\angle e$ and $\angle g$ are vertical angles. $\angle f$ and $\angle h$ are also vertical angles. If $\angle e = 50°$, then $\angle g =$

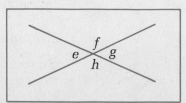

50°. Notice that the angles next to each other are supplementary. If $\angle e = 50°$, then $\angle f = 180° - 50° = 130°$. $\angle h$ is also 130°.

When two straight lines intersect, we say that the angles that are across from each other are **opposite** each other. For example, in the picture above, $\angle f$ and $\angle h$ are opposite each other and vertical. When two angles are next to each other we say that they are **adjacent.** $\angle e$ and $\angle f$ are adjacent and supplementary.

Example 1: Find the complement of a 15° angle.

Subtract 15° from 90°.

$$\begin{array}{r} 90° \\ -\ 15° \\ \hline 75° \end{array}$$

Example 2: Find the supplement of a 15° angle.

Subtract 15° from 180°.

$$\begin{array}{r} 180° \\ -\ \ 15° \\ \hline 165° \end{array}$$

Lesson 2   Exercise

Directions: Solve each problem.

1. What is the complement of an angle that measures 22°?

2. What is the supplement of an angle that measures 22°?

Items 3 to 7 are based on the figure at the right.

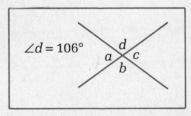

3. Find the measurement of $\angle a$.

4. Which angle is vertical to $\angle d$?

5. Which angle is vertical to $\angle a$?

6. Find the measurement of $\angle b$.

7. Find the measurement of $\angle c$.

8. Solve for m in the picture at the right.

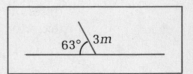

9. One angle is 20° more than another, and the two angles are complementary. Find the measurement of each angle.

10. One angle is four times another, and the two angles are supplementary. Find the measurement of each angle.

Answers are on pages 352–353.

Lesson 3   Parallel Lines and Transversals

Two lines that run in the same direction and never cross are called **parallel lines.** *AB//CD* means that the two lines at the right are parallel.

When another line, called a **transversal,** crosses two parallel lines, eight angles are formed.

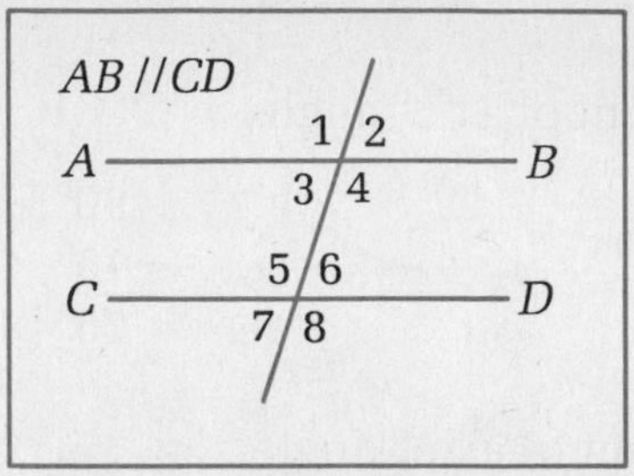

```
AB //CD
                1/2
A───────────────/───── B
                3/4
                5/6
C───────────────/───── D
                7/8
```

Following are the names of the angles formed when two parallel lines are crossed by a transversal.

There are four pairs of **corresponding angles.** In the picture above, the corresponding angles are:

∠1 and ∠5   ∠2 and ∠6   ∠3 and ∠7   ∠4 and ∠8

Corresponding angles are equal to each other.

Think about each pair of corresponding angles. For example, ∠1 is in the upper left of the top intersection, and ∠5 is in the upper left of the bottom intersection. Notice that each location "corresponds" to the other.

There are two pairs of **alternate interior angles.** Here the word *interior* refers to being inside the parallel lines. In the picture above, the alternate interior angles are:

∠3 and ∠6   ∠4 and ∠5

Alternate interior angles are equal to each other. In each pair, one angle is at the upper intersection, and the other is at the lower intersection.

There are two pairs of **alternate exterior angles.** Here the word *exterior* refers to being outside the parallel lines. In the picture above, the alternate exterior angles are:

∠1 and ∠8   ∠2 and ∠7

Alternate exterior angles are equal to each other. In each pair, one angle is at the upper intersection, and the other is at the lower intersection.

When two parallel lines are crossed by a transversal, two angles are either equal to each other or they are supplementary. For example, ∠1 and ∠4 are equal, but ∠1 and ∠2 are supplementary.

Lesson 3   Exercise

Directions: *Items 1 to 10 are based on the following picture.*

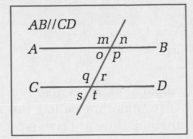

1. Which angle corresponds to ∠p?

2. Which angle is alternate exterior with ∠n?

3. Which angle is vertical to ∠t?

4. Which angle is alternate interior with ∠q?

5. If ∠m = 118°, what is the measurement of ∠t?

6. If ∠n = 49° what is the measurement of ∠q?

For items 7 to 10, fill in the blanks.

7. ∠m and ∠p are ——————— angles.

8. ∠o and ∠r are ——————— angles.

9. ∠n and ∠s are ——————— angles.

10. ∠p and ∠t are ——————— angles.

Answers are on page 353.

Lesson 4   Triangles

You have already found the perimeter and the area of triangles. In this section, you will learn the names of various triangles. You will also learn the relationships among the sides and angles of triangles.

A triangle is a flat figure with three sides. The three angles of a triangle add up to 180°. Each point where two sides meet is called a **vertex**. Below is a description of the most common triangles. Notice how the names depend on the relationships between the sides or the angles.

An **equilateral** triangle has three equal sides. An equilateral triangle also has three equal angles. Each angle measures 60°. An equilateral triangle can also be called **equiangular.** The triangle on the left is equilateral. The triangle on the right is equiangular.

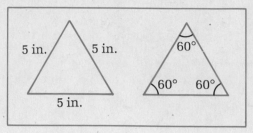

An **isosceles** triangle has two equal sides. It also has two equal angles. The equal angles of an isosceles triangle are called the **base angles.** The third angle is called the **vertex angle.** Below are two isosceles triangles.

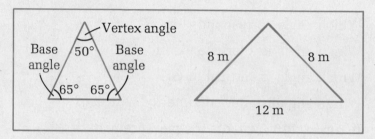

A **scalene** triangle has no equal sides and no equal angles. Below are two scalene triangles.

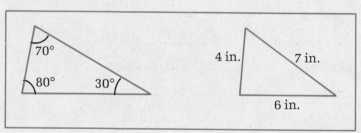

A **right triangle** has one right angle. The side opposite the right angle is called the **hypotenuse.** The other two sides are sometimes called the **legs.** Below are two right triangles.

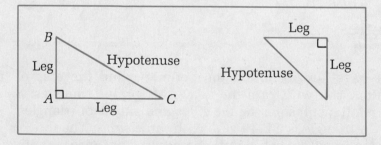

The symbol △ stands for the word triangle. Each side of a triangle is usually referred to by the two letters which mark the end points of the side. For example, the hypotenuse in △ABC above is BC. The legs are AB and AC. An angle of a triangle can be identified by one letter. The right angle in

$\triangle ABC$ is $\angle A$. Sometimes three letters are used to identify an angle of a triangle. With three letters, the middle letter is always the vertex of the angle. The right angle in $\triangle ABC$ can be called $\angle BAC$ or $\angle CAB$.

Example 1: In $\triangle XYX$, $\angle X = 70°$ and $\angle Y = 40°$. Find the measurement of $\angle Z$.

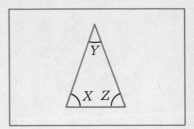

Step 1. Add the measurements of $\angle X$ and $\angle Y$.

$$\begin{array}{r} 70° \\ +40° \\ \hline 110° \end{array}$$

Step 2. Subtract the total of $\angle X$ and $\angle Y$ from 180°. $\angle Z = 70°$.

$$\begin{array}{r} 180° \\ -110° \\ \hline 70° \end{array}$$

Example 2: What kind of triangle is $\triangle XYZ$ in Example 1?

Since two angles are the same, $\triangle XYZ$ is isosceles.

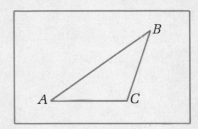

In any triangle, the longest side is opposite the largest angle. In addition, the shortest side is opposite the smallest angle. In triangle ABC at the right, notice that side AB is the longest and that it is opposite $\angle C$, which is the largest angle.

Example 3: In $\triangle PQR$, $\angle P = 45°$ and $\angle Q = 65°$. Which side of $\triangle PQR$ is longest?

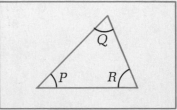

Step 1. Add the measurements of $\angle P$ and $\angle Q$.

$$\begin{array}{r} 45° \\ +65° \\ \hline 110° \end{array}$$

Step 2. Find the measurement of $\angle R$ by subtracting the total of $\angle P$ and $\angle Q$ from 180°.

$$\begin{array}{r} 180° \\ -110° \\ \hline \angle R = 70° \end{array}$$

Step 3. The largest angle is $\angle R$ with 70°. The side opposite $\angle R$ is PQ. PQ is the longest side.

PQ is longest.

Lesson 4   Exercise

Directions: Solve each problem.

1. In △KLM, ∠K = 32° and ∠L = 74°. What kind of triangle is △KLM?

2. In △DEF, ∠D = 35° and ∠E = 55°. What kind of triangle is △DEF?

3. In an isosceles triangle, each base angle measures 42°. Find the measurement of the vertex angle.

4. In another isosceles triangle, the vertex angle measures 78°. Find the measurement of each base angle.

5. One acute angle of a right triangle is 33°. Find the measurement of the other acute angle.

6. In △ABC, ∠A = 30° and ∠B = 60°. Which side of △ABC is longest?

7. DE = 6 inches and EF = 8 inches. The perimeter of the triangle is 24 inches. Which angle is largest?

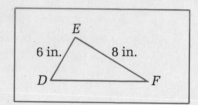

8. ∠X = 48° and ∠Y = 66°. Which side of the triangle is the shortest?

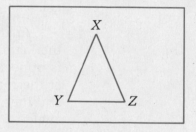

9. In triangle MNO, ∠M = 90°. Which side is the hypotenuse?

10. What kind of triangle is pictured at the right?

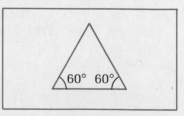

Answers are on page 353.

Level 1   Review

Directions: Solve each problem.

1. Angle m is what type of angle?

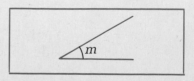

2. Angle *n* is what type of angle?

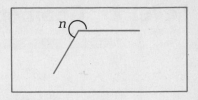

3. Find the measurement of the supplement to an angle of 84°.

4. ∠AOC = 90° and ∠AOB = 36°. Find the measurement of ∠BOC.

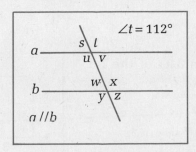

Items 5 to 7 are based on the following figure.

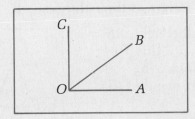

5. What is the measurement of ∠u?

6. Which angle is an alternate exterior angle with ∠s?

7. Find the measurement of ∠z.

8. In triangle *ABC*, ∠A = 33° and ∠B = 57°. Side *AB* is called the _____.

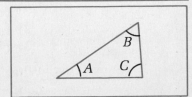

9. In triangle *XYZ*, ∠X = 49° and ∠Y = 82°. What kind of triangle is XYZ?

10. In △PQR, ∠P = 55° and ∠Q = 47°. Which side of the triangle is longest?

Check your answers. Correct answers are on page 353. If you have at least eight answers correct, go to the Level 2 Preview. If you have fewer than eight answers correct, go back to Lesson 1 and study Level 1.

Level 2 Geometry Applications

Preview

Directions: Solve each problem. Use the formulas on page 372.

1. A photograph which is 3 inches wide and 5 inches long was enlarged to be 15 inches wide. How long was the enlargement?

2. For the triangles shown at the right, $RS = UV$ and $\angle R = \angle U$. Along with this information, which of the following conditions is sufficient to insure that the triangles are congruent?

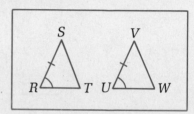

   (1) $RS = UW$ (4) $RT = UW$
   (2) $RT = UV$ (5) $UV = VW$
   (3) $UV = UW$

3. Find the diagonal distance MO of the rectangle pictured at the right.

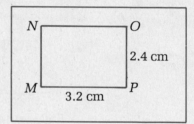

4. Which of the following points lies on the graph of the equation $y = x^2 - 3x$?

   (1) $(-3, 18)$ (4) $(2, -6)$
   (2) $(-2, 9)$ (5) $(4, 8)$
   (3) $(1, -1)$

5. What is the slope of the straight line that connects points A and B on the graph at the right?

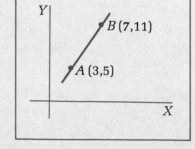

   (1) $\frac{-2}{3}$ (4) $\frac{+3}{2}$
   (2) $\frac{+2}{3}$ (5) $\frac{+1}{2}$
   (3) $\frac{-3}{2}$

Check your answers. Correct answers are on page 353. If you have at least four answers correct, do the Level 2 Review on pages 254–255. If you have fewer than four answers correct, study Level 2 beginning with Lesson 1.

Lesson 1   Similar Figures

Two geometric figures are **similar** when they have the same shape. Two figures are similar if their angles are equal. The triangles at the right are similar because they each have angles of 30°, 60°, and 90°.

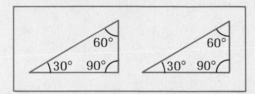

Two triangles are also similar if their sides are proportional. The triangles at the right are similar because their sides are proportional. Each side of the triangle in the figure at the right is twice as long as the corresponding side of the triangle at the left.

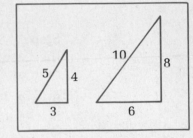

Example 1: In triangle ABC, $\angle A = 77°$ and $\angle B = 58°$. In triangle DEF, $\angle E = 58°$ and $\angle F = 45°$. Are the triangles similar?

Step 1. Find the measurement of $\angle C$. Subtract the total of $\angle A$ and $\angle B$ from 180°.

$$\begin{array}{r} 77° \\ +\,58° \\ \hline 135° \end{array} \qquad \begin{array}{r} 180° \\ -\,135° \\ \hline 45° = \angle C \end{array}$$

Step 2. Find the measurement of $\angle D$. Subtract the total of $\angle E$ and $\angle F$ from 180°. Since the triangles have the same angles, they are similar.

$$\begin{array}{r} 58° \\ +\,45° \\ \hline 103° \end{array} \qquad \begin{array}{r} 180° \\ -\,103° \\ \hline 77° = \angle D \end{array}$$

Yes

Example 2: Are these rectangles similar?

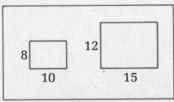

Step 1. Find the ratio of the width to the length of the first rectangle.

$\dfrac{w}{l}$ $\qquad \dfrac{8}{10} = \dfrac{4}{5}$

Step 2. Find the ratio of the width to the length of the second rectangle. Since the sides of the rectangles are proportional, they are similar.

$\dfrac{w}{l}$ $\qquad \dfrac{12}{15} = \dfrac{4}{5}$

Corresponding sides of similar figures are proportional. You can use proportion to find the measurement of the side of a figure. In the diagram for the next example, notice how small curves are used to indicate equal angles.

Example 3: In the triangles below, $\angle A = \angle D$, $\angle B = \angle E$, and $\angle C = \angle F$. $AB = 4$ inches, $AC = 6$ inches, and $DE = 18$ inches. Find the measurement of side DF.

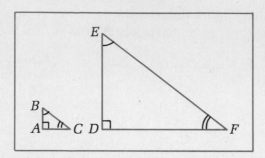

Step 1. Set up a proportion with the height to the base for each triangle. Let x stand for the base of $\triangle DEF$.

$$\text{height} \quad \frac{4}{6} = \frac{18}{x}$$
$$\text{base}$$

Step 2. Solve the proportion. The length of DF is 27 inches.

$$\frac{4x}{4} = \frac{108}{4}$$
$$x = 27 \text{ in.}$$

With some pairs of triangles it is hard to know which sides are corresponding. In the triangles below. $\angle M$ and $\angle P$ each equal 90°. The sides opposite them are corresponding. This means that side NO corresponds to side QO. $\angle N = \angle Q$. The sides opposite these angles are corresponding. This means that side MO corresponds to side OP. $\angle MON = \angle POQ$. The sides opposite these angles are corresponding. This means that side MN corresponds to side PQ.

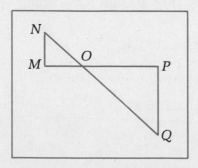

Example 4: In the picture above, $MN = 10$, $MO = 16$, and $PQ = 25$. Find the measurement of side OP.

Step 1. Set up a proportion of the short leg to the long left of each right triangle. Let x stand for the long leg of $\triangle POQ$.

$$\text{short} \quad \frac{10}{16} = \frac{25}{x}$$
$$\text{long}$$

Step 2. Solve the proportion. The length of side OP is 40.

$$\frac{10x}{10} = \frac{400}{10}$$
$$x = 40$$

Lesson 1   Exercise

Directions: Solve each problem.

1.   Is △ABC similar to △DEF?

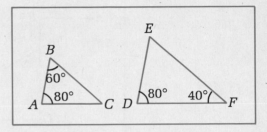

2.   In △GHI, ∠G = 65° and ∠H = 55°. In △JKL, ∠J = 60° and ∠K = 50°. Is △GHI similar to △JKL?

3.   Is rectangle MNOP similar to rectangle QRST?

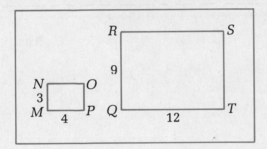

4.   Are the two rectangles at the right similar?

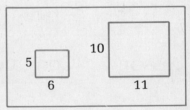

5.   A snapshot that is 4 inches wide and 5 inches long was enlarged to be 20 inches wide. How long is the enlargement?

6.   ∠M = ∠P and ∠O = ∠R. MO = 30, NO = 18, and PR = 25. Find QR.

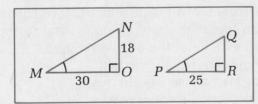

7.   A 5-foot vertical pole casts a shadow 3 feet long at the same time that a building casts a shadow 72 feet long. How tall is the building? (To solve this problem, draw a picture to show that the ground, the height of each object, and an imaginary line from the top of each object to the ground form similar triangles.)

8. $\angle B$ and $\angle D$ are each 90°, $\angle A = \angle E$, $AB = 10$ feet, $BC = 3$ feet, and $CD = 24$ feet. DE is the distance across a river. Find the measurement of DE.

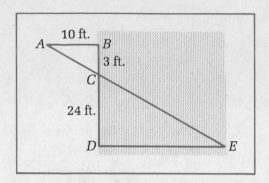

9. Every window in a new building is 6 feet high and 3 feet wide. The shape of the doors is proportional to the shape of the windows. Every door in the new building is 7 feet high. Find the width of each door.

10. Find the measurement of PQ in the figure at the right.

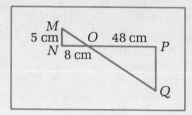

Answers are on pages 353–354.

Lesson 2   Congruent Figures

Two geometric figures are **congruent** when they have the same shape and the same size. In congruent figures, all corresponding parts are the same. Corresponding angles are equal, and corresponding sides are equal.

For triangles there are three sets of conditions that guarantee that two triangles are congruent.

1. **The *SAS* (or *side, angle, side*) requirement.** Two triangles are congruent if two sides and an included angle of one triangle are equal to two sides and a corresponding angle of the other triangle.

   The two triangles shown at the right are congruent because they satisfy the SAS requirement. Notice how the small marks on the sides indicate equal sides.

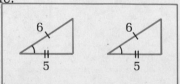

2. **The *ASA* (or *angle, side, angle*) requirement.** Two triangles are congruent if two angles and an included side of one triangle are equal to two angles and a corresponding side of the other triangle.

The two triangles shown below are congruent because they satisfy the ASA requirement. Notice that *all* corresponding angles are equal. The small curves indicate equal angles.

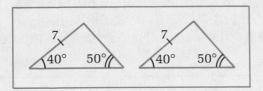

3. **The *SSS* (or *side, side, side*) requirement.** Two triangles are congruent if the three sides of one triangle are equal to the three sides of the other triangle.

The two triangles shown at the right are congruent because they satisfy the SSS requirement. Notice the small marks that indicate equal sides.

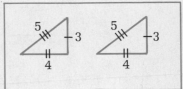

Example 1: Are these two triangles congruent?

Yes. Two sides and an included angle of one triangle are equal to two sides and a corresponding angle of the other triangle. The triangles satisfy the SAS requirement.

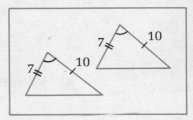

Example 2: Are these two triangles congruent?

No. Two angles of one triangle are equal to two angles of the other triangle, but the equal sides do not correspond. One 9-inch side is opposite the 45° angle. The other 9-inch side is not. These triangles do not satisfy any of the requirements for congruence.

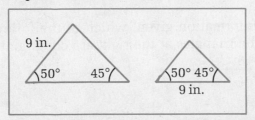

Example 3: Along with the information shown, which of the following conditions is sufficient to guarantee that these triangles are congruent?

(1) $MN = MO$ (4) $MO = PQ$
(2) $MO = PR$ (5) $PQ = PR$
(3) $MN = PR$

Choice **(2)** is enough to satisfy the SAS requirement. Choices **(1)** and **(5)** are about sides in the same triangle. Choices **(3)** and **(4)** are about sides which do not correspond.

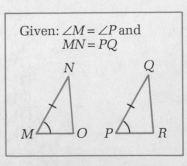

Lesson 2   Exercise

Directions: *For items 1 to 6,* decide whether the two triangles are congruent. Tell which requirement (SAS, ASA, or SSS) the triangles satisfy if they are congruent. Tell which requirement the triangles fail to satisfy if they are not congruent.

1.

2.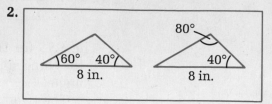

3.

4.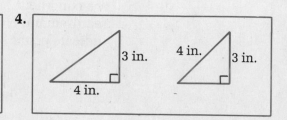

5.

6.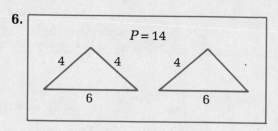

7. Along with the information given, which of the following conditions guarantees that the triangles at the right are congruent?

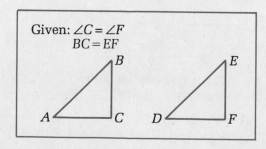

Given: ∠C = ∠F
BC = EF

(1) ∠C = ∠E
(2) ∠A = ∠E
(3) ∠B = ∠D
(4) ∠B = ∠E
(5) AB = DF

8. Along with the information given, which of the following conditions guarantees that these triangles are congruent?

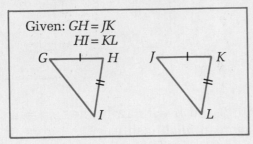

Given: $GH = JK$
$HI = KL$

(1) $GI = JL$ (4) $GI = JK$
(2) $GH = JL$ (5) $GI = KL$
(3) $HI = JL$

9. Along with the information given, which of the following conditions makes the triangles at the right congruent?

A. $JL = KL$
B. $JL = MO$
C. $KL = MN$

(1) A only   (4) A or B
(2) B only   (5) B or C
(3) C only

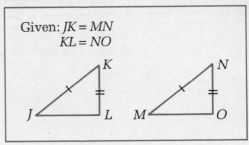

Given: $JK = MN$
$KL = NO$

10. Along with the information given, which of the conditions below insures that the triangles at the right are congruent?

A. $PR = SU$
B. $\angle P = \angle U$
C. $\angle Q = \angle T$

(1) A only   (4) A or B
(2) B only   (5) A or C
(3) C only

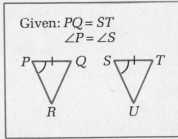

Given: $PQ = ST$
$\angle P = \angle S$

Answers are on page 354.

Lesson 3   Pythagorean Relationship

In the sixth century B.C., a mathematician named Pythagoras found that the sides of a right triangle have a special relationship. He found that the square of the hypotenuse equals the sum of the squares of the other two sides. We call this relationship the **Pythagorean relationship** or theorem.

The formula for the Pythagorean relationship is:

$$c^2 = a^2 + b^2,$$

where c = the hypotenuse, and a and b are the legs of a right triangle.

Example 1: What is the length of the hypotenuse of the triangle pictured at the right?

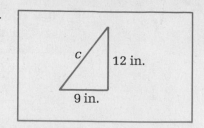

Step 1. Replace a with 9 and b with 12 in the formula for the Pythagorean relationship.

$c^2 = a^2 + b^2$
$c^2 = 9^2 + 12^2$

Step 2. Solve the formula for c. Notice that the formula gives c to the second power. To solve for c, find the square root of 225. The length of the hypotenuse is 15 inches.

$c^2 = 81 + 144$
$c^2 = 225$
$c = \sqrt{225}$
$c = 15$ in.

Remember that c always stands for the hypotenuse. In some problems, you may have to find the length of one of the legs.

Example 2: Find the length of the side labeled a in the picture at the right.

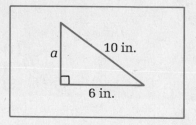

Step 1. Replace c with 10 and b with 6 in the formula for the Pythagorean relationship.

$c^2 = a^2 + b^2$
$10^2 = a^2 + 6^2$

Step 2. Solve the formula for a. First subtract 36 from both sides. Then find the square root of 64. Side a is 8 inches long.

$100 = a^2 + 36$
$ - 36$
$64 = a^2$

$\sqrt{64} = a$
$8\,\text{in.} = a$

Before you try the next exercise, review the lesson on finding square roots on page 30.

Lesson 3   Exercise

Directions: Solve each problem.

1. Find the length of the hypotenuse of the triangle shown at the right.

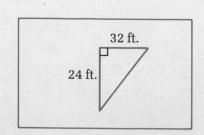

2. The legs of a right triangle measure 36 inches and 48 inches. Find the length of the hypotenuse.

3. In the right triangle shown at the right, what is the length of side *AB*?

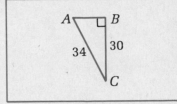

4. The hypotenuse of a right triangle measures 13 feet, and one leg measures 5 feet. Find the length of the other leg.

5. The diagram at the right shows the plan of a rectangular garden. What is the diagonal distance shown by the dotted line?

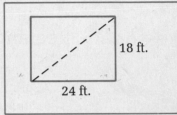

6. Find the length of the line *AC* in the rectangle pictured at the right.

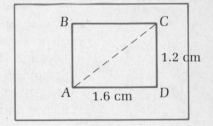

7. The picture at the right shows an isosceles triangle. The height is 24 inches. Sides *XY* and *YZ* each measure 26 inches. What is the measurement of the base?

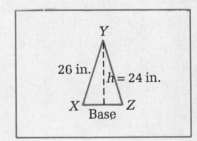

8. Find the length of the side labeled *b* in the drawing at the right.

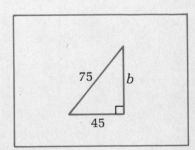

9. Figure *MNOP* at the right is a rectangle. Side *MP* is 15 feet long, and diagonal distance *MO* is 17 feet long. Find the measurement of *OP*.

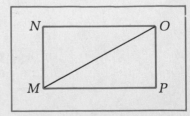

10. Geraldine drove 20 miles east and then 15 miles south. What was the shortest distance from her starting point to her ending point?

Answers are on page 354.

Lesson 4   Rectangular Coordinates

A graph called the **rectangular coordinate system** is a tool for picturing algebraic relationships. To understand rectangular coordinates, you need both algebra and geometry skills.

The rectangular coordinate system is a plane (a flat surface) divided by two perpendicular lines. The horizontal line is called the x-axis. The vertical line is called the y-axis. Each axis looks like the number line you studied earlier. On the x-axis, positive numbers are at the right of zero, and negative numbers are at the left. On the y-axis, positive numbers are above zero, and negative numbers are below. The lines intersect (cross) at zero on each axis. This point is called the **origin.**

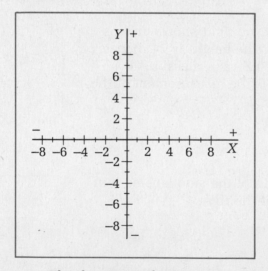

The location of every point on the plane can be described by two numbers called the **coordinates** of a point. The numbers, written inside a pair of parentheses, tell how far a point is from the origin. The first number, called the x-value, tells the distance to the right or left of the vertical axis. A positive x-value is to the right of the vertical axis, and a negative x-value is to the left. The second number, called the y-value, tells the distance above or below the horizontal axis. A positive y-value is above the horizontal axis, and a negative y-value is below. A comma (,) separates the coordinates.

Example: What are the coordinates for each lettered point on the rectangular coordinate system shown below.

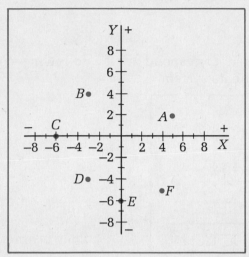

A = (+5, +2)   Point A is 5 spaces right of the y-axis and 2 spaces above the x-axis.

B = (−3, +4)   Point B is 3 spaces left of the y-axis and 4 spaces above the x-axis.

C = (−6, 0)   Point c is 6 spaces left of the y-axis and directly on the x-axis. Notice that the y-value is zero.

D = (−3, −4)   Point D is 3 spaces left of the y-axis and 4 spaces below the x-axis.

E = (0, −6)   Point E is directly on the y-axis and 6 spaces below the x-axis. Notice that the x-value is zero.

F = (+4, −5)   Point F is 4 spaces right of the y-axis and 5 spaces below the x-axis.

Lesson 4   Exercise

Directions: Solve each problem.

1. Write the coordinates of each lettered point that appears on the rectangular coordinate system.

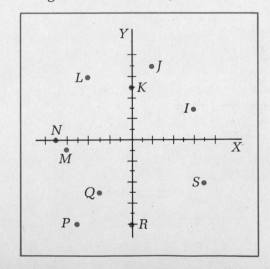

$I = ?$ $N = ?$
$J = ?$ $P = ?$
$K = ?$ $Q = ?$
$L = ?$ $R = ?$
$M = ?$ $S = ?$

2. Put the points that correspond to the following coordinates on the rectangular coordinate system.

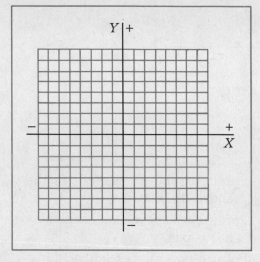

$A = (+6, +5)$ $E = (-4, -3)$
$B = (+1, +8)$ $F = (-2, -7)$
$C = (-3, +6)$ $G = (+4, -6)$
$D = (-5, 0)$ $H = (+5, 0)$

3. Write the coordinates of each lettered point that appears on the rectangular coordinate system.

$A = ?$ $F = ?$
$B = ?$ $G = ?$
$C = ?$ $H = ?$
$D = ?$ $I = ?$
$E = ?$ $J = ?$

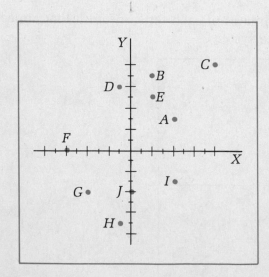

4. Put the points that correspond to the following coordinates on the rectangular coordinate system.

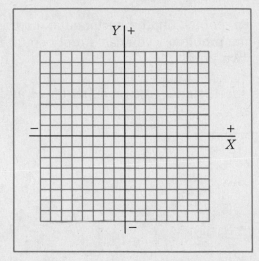

K = (−4, 0)   P = (6, 0)
L = (−5, 6)   Q = (3, −4)
M = (0, 8)   R = (0, −5)
N = (1, 7)   S = (−3, −4)
O = (3, 4)   T = (−8, −2)

5. Write the coordinates of each lettered point that appears on the rectangular coordinate system.

A = ?   F = ?
B = ?   G = ?
C = ?   H = ?
D = ?   I = ?
E = ?   J = ?

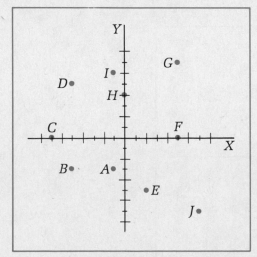

Answers are on page 355.

Lesson 5   Finding the Distance Between Points

The distances between points on the rectangular coordinate system are whole numbers. In some problems, you can simply count the spaces to find the distance between two points.

Example 1: What is the distance between points M and N on the graph below?

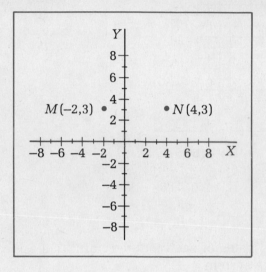

M is 2 spaces left of the y-axis. N is 4 spaces right of the y-axis. Distance MN is $2 + 4 = 6$.

Example 2: What is the distance between points P and Q on the graph at the right?

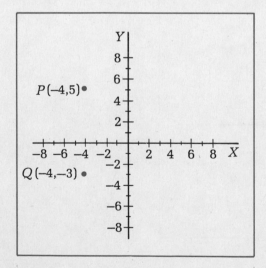

P is 5 spaces above the x-axis. Q is 3 spaces below the x-axis. Distance PQ is $5 + 3 = 8$.

When two points in a plane do not lie on a horizontal or vertical line, the distance (d) between the points is:

$$d = \sqrt{(x_2 - x_1)^2 + (y_2 - y_1)^2},$$

where (x_1, y_1) and (x_2, y_2) are two points.

Notice the small numbers in the expressions (x_1, y_1) and (x_2, y_2). These numbers, called **subscripts,** do not suggest any mathematical operation. The numbers simply distinguish between different x-values and different y-values.

Example 3: What is the distance between points C and D on the graph below?

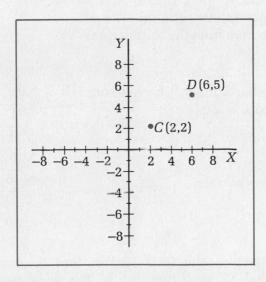

Step 1. Replace x_2, with 6, x_1 with 2, y_2 with 5, and y_1 with 2 in the formula for the distance between two points in a plane.

$C = (x_1, y_1) = (2, 2)$
$D = (x_2, y_2) = (6, 5)$

Step 2. Solve the formula for d. The distance between points C and D is 5.

$$\begin{aligned}
d &= \sqrt{(x_2 - x_1)^2 + (y_2 - y_1)^2} \\
&= \sqrt{(6 - 2)^2 + (5 - 2)^2} \\
&= \sqrt{(4)^2 + (3)^2} \\
&= \sqrt{16 + 9} \\
&= \sqrt{25} \\
&= 5
\end{aligned}$$

Example 4: What is the distance between points K and L on the graph below?

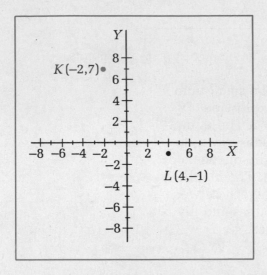

Step 1. Replace x_2 with 4, x_1 with -2, y_2 with -1, and y_1 with 7 in the formula for the distance between two points in a plane. Notice how the extra set of parentheses around -2 and 7 separates these numbers from the minus signs in the formula.

$K = (x_1, y_1) = (-2, 7)$
$L = (x_2, y_2) = (4, -1)$

Step 2. Solve the formula for d. The distance between points K and $L = 10$.

$$\begin{aligned}
d &= \sqrt{(x_2 - x_1)^2 + (y_2 - y_1)^2} \\
&= \sqrt{(4 - (-2))^2 + (-1 - (7))^2} \\
&= \sqrt{(4 + 2)^2 + (-1 - 7)^2} \\
&= \sqrt{(6)^2 + (-8)^2} \\
&= \sqrt{36 + 64} \\
&= \sqrt{100} \\
&= 10
\end{aligned}$$

You may be asked on the GED Test to find the coordinates of the midpoint of a line that connects two points. The formula for the midpoint M is:

$$M = \left(\frac{x_1 + x_2}{2}, \frac{y_1 + y_2}{2} \right),$$

where (x_1, y_1) and (x_2, y_2) are two points in a plane.

This formula tells you to find the average of the x-values and the average of the y-values for the two points.

Example 5: What are the coordinates of the midpoint M which lies half way between points R and S on the graph on the next page?

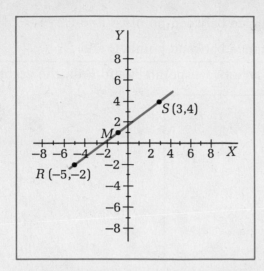

Step 1. Replace x_1, with -5, x_2 with 3, y_1 with -2, and y_2 with 4 in the formula for the coordinates of the midpoint between two points in a plane.

$R = (x_1, y_1) = (-5, -2)$
$S = (x_2, y_2) = (3, 4)$

Step 2. Simplify the expression. The coordinates for the midpoint between R and S are $(-1, +1)$.

$$M = \left(\frac{x_1 + x_2}{2}, \frac{y_1 + y_2}{2}\right)$$
$$= \left(\frac{-5 + 3}{2}, \frac{-2 + 4}{2}\right)$$
$$= \left(\frac{-2}{2}, \frac{+2}{2}\right)$$
$$= (-1, +1)$$

Lesson 5   Exercise

Directions: Solve each problem.

Items 1 to 3 are based on the following illustration.

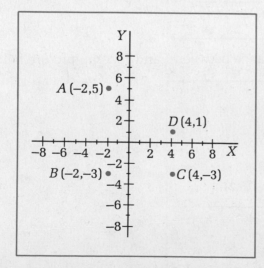

1.   What is the distance between points *A* and *B*?

2. What is the distance between points *B* and *C*?

3. What is the distance between points *C* and *D*?

4. Find the distance between points *E* and *F* on the graph below.

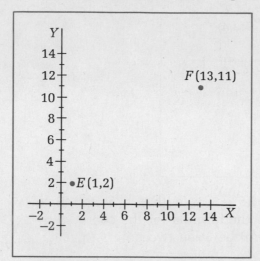

5. Find the distance between points *G* and *H* in the picture below.

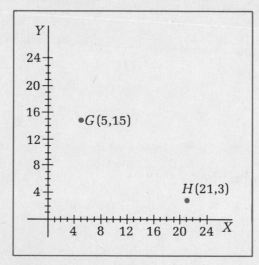

6. What is the distance between *I* and *J* in the picture below?

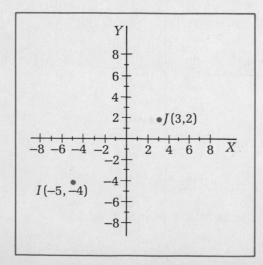

7. Find the distance between points K and L on the graph below.

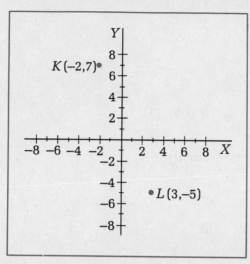

8. What are the coordinates of the midpoint between points P and N on the graph below?

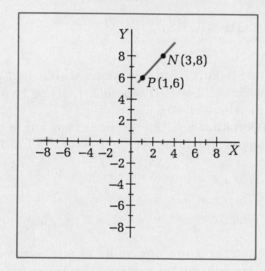

9. Find the coordinates of the midpoint between points Q and R on the graph below.

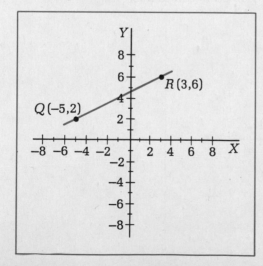

10. What are the coordinates of the midpoint between points S and T in the illustration below?

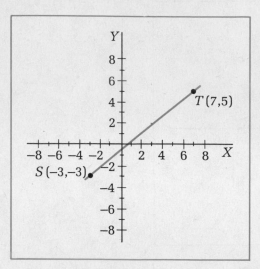

Answers are on pages 355–356.

Lesson 6 Graphs of Linear Equations

So far in this book, you have solved equations which had either one exact solution (for example, $3x + 2 = 14$, where $x = 4$) or two exact solutions (for example, $x^2 = 25$, where $x = +5$ or -5).

Think about the equation $y = x - 3$. For every value you can think of for x, there is a corresponding value for y. Look at the following example.

$$\text{When } x = 7, y = 7 - 3 = 4.$$
$$\text{When } x = 4, y = 4 - 3 = 1.$$
$$\text{When } x = -2, y = -2 - 3 = -5.$$

Each value of x and the corresponding value of y give the coordinates of a point on the rectangular coordinate system. The three points from the examples are $(7, 4)$, $(4, 1)$, and $(-2, -5)$. The picture below shows these three points on the rectangular coordinate system.

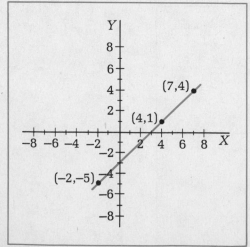

The three points lie in a straight line. The line is called the graph of the equation $y = x - 3$. An equation whose graph is a straight line is called a **linear equation.**

The coordinates of every point on the line are solutions to the equation.

Example 1: Draw a graph of the equation $y = 3x + 2$. Find values of y when $x = 2, -1$, and -3.

Step 1. Replace x with $2, -1$, and -3 to find the corresponding values of y.

$y = 3(2) + 2 = 6 + 2 = 8$

$y = 3(-1) + 2 = -3 + 2 = -1$

$y = 3(-3) + 2 = -9 + 2 = -7$

Step 2. Make a chart of each x-value and the corresponding y-value.

x	y
2	8
−1	−1
−3	−7

Step 3. Put each point on the graph, and connect the points with a line.

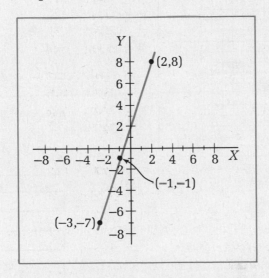

You can find out whether a point lies on the graph of an equation by using substitution. Replace the x-value of the coordinates into the equation. The result should equal the y-value.

Example 2: Does the point $(2, 4)$ lie on the graph of the equation $y = -x + 6$?

Step 1. The x-value of the coordinates is 2. Replace x with 2 in the equation $y = -x + 6$.

$y = -x + 6$
$y = -(2) + 6$

Step 2. Solve for y. Since $y = 4$, the point $(2, 4)$ does lie on the graph.

$y = -2 + 6$
$y = 4$
Point $(2, 4)$ is on the graph.

Example 3: Is point $(-3, 5)$ on the graph of the equation $y = -x + 6$?

Step 1. The x-value of the coordinate is -3. $y = -x + 6$
Replace x with -3 in $y = -x + 6$. $y = -(-3) + 6$

Step 2. Solve for y. Since $y = +9$, the point $y = +3 + 6$
$(-3, 5)$ is not on the graph. $y = +9$

Point $(-3, 5)$ is not on the graph.

Lesson 6   Exercise

Directions: *For items 1 to 4,* find a value of y for each value of x. Fill in the chart with the corresponding values of x and y. Then put each point on the graph, and connect the points with a line.

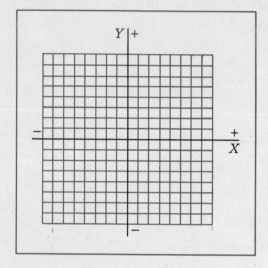

1. $y = x + 4$
    Let $x = 3, -2,$ and -5.

x	y

2. $y = \frac{x}{2} + 1$
    Let $x = 8, 4,$ and -6.

x	y

3. $y = -3x + 4$
    Let $x = 3, 1,$ and -2.

x	y

4. $y = -2x - 3$
    Let $x = 2, -3,$ and -4.

x	y

5. Is point (1, 8) a solution to the equation $y = 5x + 3$?

6. Is point (2, −4) a solution to the equation $y = -3x + 1$?

7. Does point (3, 5) lie on the graph of the equation $y = -x + 6$?

8. Does point (−8, −8) lie on the graph of the equation $y = \frac{3}{4}x - 2$?

9. Which of the following points is on the graph of the equation $y = x - 5$?

   (1) (4, 0)   **(4)** (−1, 5)
   (2) (3, −4)   **(5)** (−2, −8)
   (3) (2, −3)

10. Which of the following points lies on the graph of the equation $y = -2x - 3$?

    (1) (3, −9)   **(4)** (−3, + 4)
    (2) (2, −4)   **(5)** (−5, 9)
    (3) (−1, + 1)

Answers are on page 356.

Lesson 7 Graphs and Quadratic Equations

In the last section, you learned to recognize linear equations. The graphs of these equations are straight lines. Graphs of equations with x raised to the second power are not the same. Think about the equation $y = x^2 + 3$.

When $x = 2$, $y = (2)^2 + 3 = 4 + 3 = 7$.
When $x = 1$, $y = (1)^2 + 3 = 1 + 3 = 4$.
When $x = 0$, $y = (0)^2 + 3 = 0 + 3 = 3$.
When $x = -1$, $y = (-1)^2 + 3 = 1 + 3 = 4$.
When $x = -2$, $y = (-2)^2 + 3 = 4 + 3 = 7$.

Each value of x and the corresponding value of y give the coordinates of a point on the rectangular coordinate system. The five points from the examples are (2, 7), (1, 4), (0, 3), (−1, 4), and (−2, −7). The following picture shows these five points on the rectangular coordinate system.

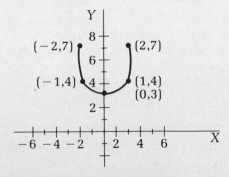

The five points lie in a curve called a **parabola.** An equation whose graph is a parabola is called a **quadratic equation.**

Example 1: Draw a graph of the equation $y = x^2 + x$. Find the values of y, when $x = 2, 1, 0, -1, -2,$ and -3.

Step 1. Replace x with $2, 1, 0, -1, -2,$ and -3 to find the corresponding values of y.

$y = (2)^2 + 2 = 4 + 2 = 6$

$y = (1)^2 + 1 = 1 + 1 = 2$

$y = (0)^2 + 0 = 0 + 0 = 0$

$y = (-1)^2 + (-1) = 1 - 1 = 0$

$y = (-2)^2 + (-2) = 4 - 2 = 2$

$y = (-3)^2 + (-3) = 9 - 3 = 6$

Step 2. Make a chart of each x-value and the corresponding y-value.

x	y
2	6
1	2
0	0
−1	0
−2	2
−3	6

Step 3. Put each point on the graph, and connect the points with a curved line.

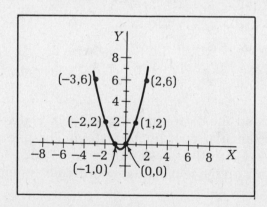

To find out whether a point lies on the graph of an equation, replace x in the equation with the x-value of the coordinates of the point. The result should equal the y-value.

Example 2: Does the point $(3, 12)$ lie on the graph of $y = x^2 + 3$?

Step 1. The x-value of the coordinates is 3. Replace x with 3 in the equation $y = x^2 + 3$.

$y = x^2 + 3$

$y = (3)^2 + 3$

Step 2. Solve for y. Since y = 12, the point (3, 12) does lie on the graph.

$y = 9 + 3 = 12$
Point (3, 12) is on the graph.

Example 3: Is point $(-2, 15)$ on the graph of the equation $y = x^2 - 4x + 5$?

Step 1. The x-value of the coordinates is -2. Replace x with -2 in the equation $y = x^2 - 4x + 5$.

$y = x^2 - 4x + 5$
$y = (-2)^2 - 4(-2) + 5$

Step 2. Solve for y. Since y = 17, the point $(-2, 15)$ is not on the graph.

$y = 4 + 8 + 5 = 17$
Point $(-2, 15)$ is not on the graph.

Lesson 7   Exercise

Directions: *For items 1 to 4,* find a value of y for each value of x. Fill in the chart with the corresponding values of x and y. Then put each point on the graph, and connect the points with a curved line.

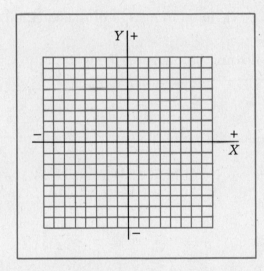

x	y

1. $y = x^2 + 2$
    Let x = 2, 1, 0, -1, and -2.

x	y

2. $y = x^2 - 2x$
    Let x = 4, 3, 2, 1, 0, -1, and -2.

3. $y = x^2 + x - 2$
    Let $x = 2, 1, 0, -1, -2,$ and -3.

x	y

4. $y = x^2 + x - 4$
    Let $x = 2, 1, 0, -1, -2,$ and -3.

x	y

5. Is point $(5, 4)$ on the graph of the equation $y = x^2 - 4x$?

6. Does point $(4, 23)$ lie on the graph of the equation $y = x^2 + x + 3$?

7. Is point $(-3, 11)$ on the graph of the equation $y = x^2 - x - 1$?

8. Does point $(-2, -4)$ lie on the graph of the equation $y = x^2 + 3x - 2$?

9. Which of the following points is on the graph of $y = x^2 + 2x$?
    (1) $(4, 20)$ **(4)** $(-2, 1)$
    (2) $(3, 15)$ **(5)** $(-4, 6)$
    (3) $(1, 2)$

10. Which of the following points lies on the graph of the equation $y = x^2 - 5x + 3$?
    (1) $(3, 3)$ **(4)** $(-2, 17)$
    (2) $(2, -4)$ **(5)** $(-3, 25)$
    (3) $(-1, 8)$

Answers are on page 357.

Lesson 8   Slope and Intercept

Slope and **intercept** are two words that describe the graph of a linear equation.

Slope tells how much a line slants or leans.

When a line rises from left to right, the slope is positive. The graph for the equation $y = x + 4$ shown at the right has positive slope.

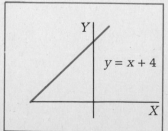

When a line falls from left to right, the slope is negative. The graph for the equation y = −x + 2 shown at the right has negative slope.

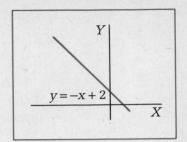

A horizontal line has zero slope. The graph of the equation y = 3 shown at the right has zero slope.

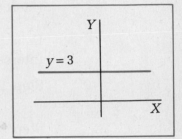

A vertical line has what is called undefined slope. The slope of the equation x = 5 shown below has undefined slope.

The formula for the slope (m) of a line is:

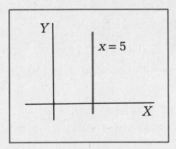

$$m = \frac{y_2 - y_1}{x_2 - x_1},$$

where (x_1, y_1) and (x_2, y_2) are two points in a plane.

Example 1: What is the slope of the line that passes through points *A* and *B* below?

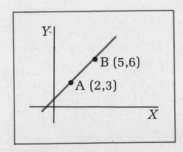

Step 1. Replace y_2 with 6, y_1 with 3, x_2 with 5, x_1 with 2 in the formula for the slope.

$A = (x_1, y_1) = (2, 3)$
$B = (x_2, y_2) = (5, 6)$

Step 2. Solve for m. The slope is +1.

$$m = \frac{y_2 - y_1}{x_2 - x_1} = \frac{6 - 3}{5 - 2}$$
$$= \frac{3}{3} = +1$$

An intercept is the point where two lines cross. For equations on the rectangular coordinate system, an intercept tells where the line of an equation crosses the x-axis or the y-axis.

The y-intercept tells where the line of an equation crosses the y-axis. The x-value of the coordinates of a y-intercept is always zero. To find a y-intercept, substitute 0 for x in an equation.

Example 2: What are the coordinates of the y-intercept for the equation y = −2x + 3?

Step 1. Replace x with 0 in y = −2x + 3.

$$y = -2x + 3$$
$$y = -2(0) + 3$$

Step 2. Solve for y. The value of y when x is zero is +3. The coordinates of the y-intercept are (0, 3).

$$y = 0 + 3 = 3$$
y-intercept = (0, 3)

The x-intercept tells where the line of an equation crosses the x-axis. The y-value of the coordinates of an x-intercept is always zero. To find an x-intercept, substitute 0 for y in an equation.

Example 3: What are the coordinates of the x-intercept for the equation y = 2x − 4.

Step 1. Replace y with 0 in y = 2x − 4.

$$y = 2x - 4$$
$$0 = 2x - 4$$

Step 2. Solve for x. The value of x when y is zero is +2. The coordinates of the x-intercept are (2, 0).

$$\begin{array}{cc} +4 & +4 \\ \hline \dfrac{4}{2} & = \dfrac{2x}{2} \end{array}$$

$$2 = x$$
x-intercept = (2, 0)

Lesson 8   Exercise

Directions: Solve each problem.

1.   What is the slope of a line that passes through points C and D at the right?

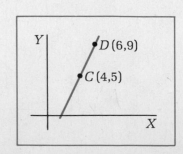

2. Find the slope of the line that passes through points *E* and *F* in the picture at the right.

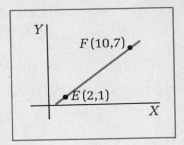

3. What is the slope of a line that passes through points *G* and *H* at the right?

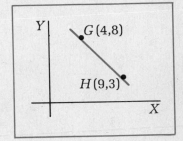

4. Find the slope of the line that passes through points *I* and *J* in the illustration at the right.

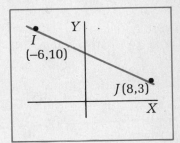

5. Which line at the right has a positive slope?

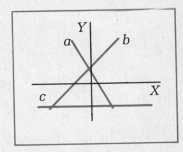

In items 6 to 8, find the coordinates of the y-intercept.

6. $y = 2x - 3$

7. $y = -2x + 1$

8. $y = 5x - 7$

In items 9 and 10, find the coordinates of the x-intercept.

9. $y = 3x - 9$

10. $y = 8x + 4$

Answers are on pages 357–358.

Level 2   Review

Directions: Solve each problem. Use the formulas on page 372.

1. The illustration at the right shows a diagram of two sails which are similar triangles. Find the height of the bigger sail.

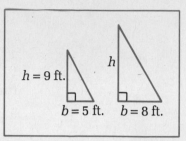

2. In the picture at the right, $MN = 12$, $NO = 20$, and $PQ = 16$. Find distance OP.

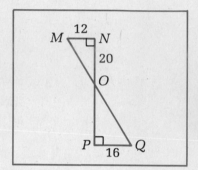

3. For the two triangles pictured at the right, $DE = GH$ and $\angle E = \angle H$. Along with this information, which of the following conditions is sufficient to make the two triangles congruent?

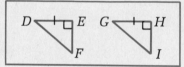

   (1) $\angle D = \angle I$ **(3)** $\angle F = \angle G$ **(5)** $\angle H = \angle F$
   (2) $\angle D = \angle G$ **(4)** $\angle E = \angle I$

4. Find the length of the side labeled b in the picture at the right.

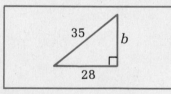

Items 5 and 6 are based on the following illustration.

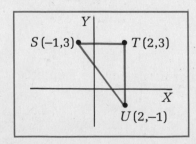

5. What is the distance between point S and point T?

6. Find the distance between point S and point U.

7. Which of the following points lies on the graph of the equation $y = 3x - 4$?
   (1) $(-3, -10)$ **(4)** $(4, 8)$
   (2) $(-1, -5)$ **(5)** $(6, 16)$
   (3) $(2, 3)$

8. Of the following points, which lies on the graph of the equation $y = x^2 + 2x - 1$?
   (1) $(5, 36)$ **(4)** $(-4, 6)$
   (2) $(3, 14)$ **(5)** $(-5, 12)$
   (3) $(-2, -2)$

9. What is the slope of a straight line that passes through points C and D in the figure at the right?
   (1) $+5$ **(4)** $+1$
   (2) -5 **(5)** 0
   (3) -1

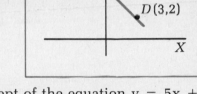

10. What are the coordinates of the y-intercept of the equation $y = 5x + 3$?
    (1) $(0, 5)$ **(4)** $(5, 0)$
    (2) $(0, 3)$ **(5)** $(0, -3)$
    (3) $(-3, 0)$

Check your answers. Correct answers are on page 358. If you have at least eight answers correct, go to the Level 3 Preview. If you have fewer than eight answers correct, go back to Lesson 1 and study Level 2.

Level 3   Problem Solving

Preview

Directions: Solve each problem.

1. The perimeter of the triangle pictured at the right is 48 inches. Find the length of the missing side.

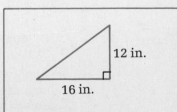

2. The distance around Silvia's rectangular vegetable garden is 40 feet. The garden is 9 feet wide. Find the length.

3. A rectangle has a width of 5 feet and an area of 95 square feet. What is the length of the rectangle?

4. Find the measurement of one side of a square which has the same area as the rectangle pictured at the right.

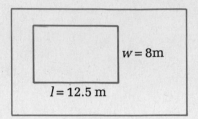

$w = 8$m

$l = 12.5$ m

5. The length of a rectangle is 8 inches more than the width. The perimeter is 100 inches. Find both the width and the length.

6. The ratio of the length to the width of the rectangle pictured at the right is 3:2. Find both the length and the width.

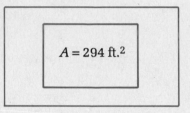

$A = 294$ ft.^{2}

7. The height of a triangle is twice the base. The area of the triangle is 64 square meters. Find both the base and the height.

8. The volume of the rectangular container pictured at the right is 162 cubic feet. The height of the container is one-third the length. Find both the length and the height of the container.

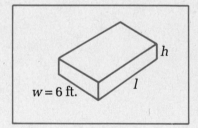

h

$w = 6$ ft.

l

9. Express in simplest form the hypotenuse of the triangle pictured at the right.

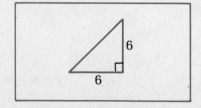

6

6

10. Jeff walked 4 miles east and then 8 miles north. Find the shortest distance from his starting point to his end point. Express your answer in simplest form.

Check your answers. Correct answers are on pages 358–359. If you have at least eight answers correct, do the Level 3 Review on page 262. If you have fewer than eight answers correct, study Level 3 beginning with Lesson 1.

Lesson 1 Using Algebra to Solve Formulas

In the algebra chapter, you learned to solve distance, cost, and interest problems when information you had was "backwards." You can use the same skills to solve many perimeter, area, and volume problems.

Example: Find the measurement of the side of a square which would have the same area as a rectangle which is 25 feet long and 16 feet wide.

Step 1. Find the area of the rectangle.

$$A = lw$$
$$A = 25 \cdot 16$$
$$A = 400 \text{ ft.}$$

Step 2. To find the side of the square, replace A with 400 in the formula $A = s^2$.

$$A = s^2$$
$$400 = s^2$$

Step 3. Solve for the side by finding the square root of 400. The side is 20 feet.

$$\sqrt{400} = s$$
$$20 \text{ ft.} = s$$

Lesson 1 Exercise

Directions: Solve each problem. Use the formulas on page 372.

1. The area of the rectangle at the right is 132 square feet. Find the width.

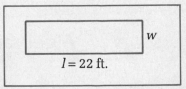

2. The distance around the garden in Mike's yard is 124 feet. The garden is a rectangle with a width of 16 feet. Find the length.

3. The perimeter of the triangle shown at the right is 55 inches. Find the measurement of the missing side.

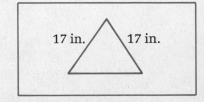

4. A circle has a circumference of 15.7 meters. Find the diameter of the circle.

5. What is the measurement of one side of a square which has an area of 81 square yards?

6. The perimeter of the square at the right is 150 feet. Find the measurement of one side.

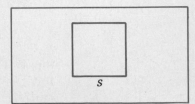

7. The area of the triangle shown at the right is 108 square feet. What is the height?

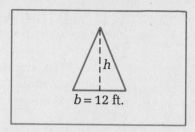

8. The volume of the concrete in a patio is 180 cubic feet. The patio is 10 feet wide and $\frac{1}{2}$ foot deep. What is the length?

9. The rectangle shown at the right has a perimeter of 46 centimeters. Find the width of the rectangle.

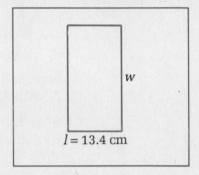

10. A triangle has an area of 84 square inches. The height of the triangle is 28 inches. What is the measurement of the base?

Answers are on page 359.

Lesson 2 Algebraic Expressions in Formulas

In the previous exercise, you substituted numbers into formulas and solved for the missing amounts. Sometimes the information in a problem is complicated. These problems require you to write algebraic expressions for pieces of information.

Example 1: A rectangle has a perimeter of 60 feet. The length of the rectangle is twice the width. Find both the length and the width.

Step 1. Let x stand for the width, and let 2x stand for the length.

$$w = x$$
$$l = 2x$$

Step 2. Replace P with 60, l with 2x, and w with x in the formula $P = 2l + 2w$.

$$P = 2l + 2w$$
$$60 = 2(2x) + 2x$$

Step 3. Solve for x.

$$60 = 4x + 2x$$
$$\frac{60}{6} = \frac{6x}{6}$$
$$10 = x$$

Step 4. The solution is 10, which is the width. Find the value of 2x, which is the length. The length is 20 feet and the width is 10 feet.

$$w = 10 \text{ ft.}$$
$$l = 2 \cdot 10 = 20 \text{ ft.}$$

Example 2: The volume of the rectangular container at the right is 160 cubic feet. The height of the container is twice the width. Find both the width and the height.

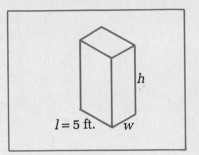

Step 1. Let x stand for the width, and let 2x stand for the height.

$w = x$
$h = 2x$

Step 2. Replace V with 160, l with 5, w with x, and h with 2x in the formula $V = lwh$.

$V = lwh$
$160 = 5 \cdot x \cdot 2x$

Step 3. Solve the equation for x. (Notice that we are interested in the positive value of x only.)

$\frac{160}{10} = \frac{10x^2}{10}$

$16 = x^2$

$\sqrt{16} = x$

$4 = x$

Step 4. The solution is 4, which is the width. Find the value of 2x, which is the height. The width is 4 feet and the height is 8 feet.

$w = 4$ ft.
$h = 2 \cdot 4 = 8$ ft.

Lesson 2   Exercise

Directions: Solve each problem. Use the formulas on page 372.

1. The distance around the kitchen in Miriam's house is 48 feet. The length of the kitchen is six feet more than the width. Find both the length and the width.

2. The length of a rectangle is three times the width. The perimeter is 64 feet. Find the length and the width.

3. In the triangle shown at the right, side b is 2 inches more than side a, and side c is 4 inches more than side a. The perimeter is 36 inches. Find the measurement of each side.

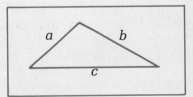

4. In the isosceles triangle shown at the right, each equal side is twice as long as the base. The perimeter is 45 meters. Find the measurement of each long side.

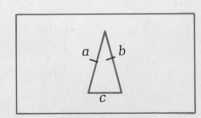

260   Chapter 7, Level 3   INSTRUCTION

5. The length of a rectangle is twice the width. The area of the rectangle is 98 square yards. Find the length and the width.

6. In another rectangle, the width is half the length. The area is 50 square inches. Find both the length and the width.

7. The ratio of the length to the width of the living room in Ann's house is 4:3. The area of the floor of the room is 300 square feet. Find both the length and the width.

8. The ratio of the base to the height of the triangle shown at the right is 2:3. The area of the triangle is 108 square inches. Find the base and the height.

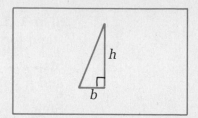

9. In another triangle, the height is one-third of the base, and the area is 24 square meters. Find both the height and the base.

10. The volume of the rectangular container pictured at the right is 1440 square inches. The width of the container is 5 inches. The height is eight times the length. Find both the length and the height.

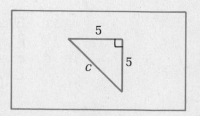

Answers are on pages 359–360.

Lesson 3   Factoring and the Pythagorean Relationship

In the algebra chapter of this book, you learned to simplify square roots by factoring. You can use this skill to express the answers to some right triangle problems on the GED Test.

Example 1: Find the hypotenuse of the right triangle pictured at the right.

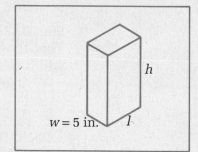

Step 1. Replace a with 5 and b with 5 in the formula for the Pythagorean relationship.

$c^2 = a^2 + b^2$
$c^2 = 5^2 + 5^2$

Step 2. Solve the equation for c.

$c^2 = 25 + 25$
$c^2 = 50$
$c = \sqrt{50}$

Step 3. The factors of 50 that include an exact square are 25 and 2. Put each factor inside a $\sqrt{}$ sign. Then write the square root of 25 next to $\sqrt{2}$.

$$c = \sqrt{25} \cdot \sqrt{2}$$
$$= 5\sqrt{2}$$

Sometimes there is no factor that has a perfect square.

Example 2: Find the hypotenuse of the triangle at the right.

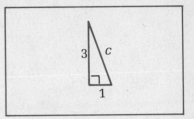

Step 1. Replace a with 1 and b with 3 in the formula for the Pythagorean relationship.

$$c^2 = a^2 + b^2$$
$$c^2 = 1^2 + 3^2$$

Step 2. Solve the equation for c. Since no factor of 10 has an exact square root, $\sqrt{10}$ is the simplest form for the answer.

$$c^2 = 1 + 9$$
$$c^2 = 10$$
$$c = \sqrt{10}$$

Lesson 3   Exercise

Directions: Solve each problem.

1. Express in simplest form the length of the hypotenuse of the triangle at the right.

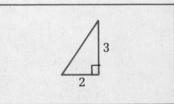

2. What is the hypotenuse of the triangle shown at the right? Express the answer in simplest form.

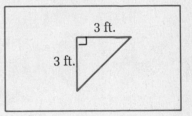

3. Find the diagonal distance AC of the square shown at the right. Express the answer in simplest form.

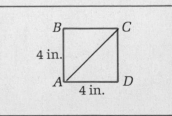

For items 4 and 5, first make a sketch of the distance each person travels in these problems.

4. Pete bicycled three miles west and then six miles south. Find the shortest distance from his starting point to his end point. Express the answer in simplest form.

5. Claire walked three miles north, then eight miles to the west, and finally five miles to the north. Find the shortest distance from her starting point to her end point. Express the answer in simplest form.

Answers are on page 360.

Level 3   Review

Directions: Solve each problem.

1. The distance around the dining room in the Johnson's house is 39 feet. The width of the room is 7.5 feet. Find the length.

2. A rectangle has a width of 3.4 meters and an area of 17 square meters. Find the length of the rectangle.

3. The volume of a rectangular container is 420 cubic feet. The length of the container is 7 feet and the height is 10 feet. What is the width?

4. Find the width of the rectangle at the right which would give the rectangle the same area as the square.

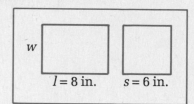

5. In the isosceles triangle pictured at the right, sides a and b are equal, and side c is 4 yards longer than side a or side b. The perimeter of the triangle is 28 yards. Find side c.

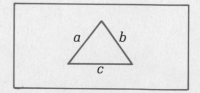

6. A rectangle has an area of 75 square feet. The width is one-third of the length. Find both the width and the length.

7. The ratio of the base to the height of the triangle at the right is 3:4. Find both the base and the height.

8. A rectangular container has a volume of 216 cubic inches. The height of the container is 12 inches. The length is twice the width. Find both the length and the width.

9. What is the diagonal distance AC on the square pictured at the right?

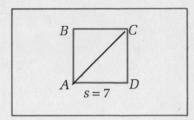

10. Ellen bicycled two miles south and then six miles east. Express in simplest form the shortest distance from her starting point to her end point.

Check your answers. Correct answers are on pages 360–361. If you have at least eight answers correct, go to the Quiz. If you have fewer than eight answers correct, go back to Lesson 1 and study Level 3.

Chapter 7 Quiz

Directions: Solve each problem. Use the formulas on page 372.

1. What type of angle is illustrated in the picture at the right?

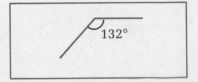

2. $\angle MON = 17°$ and $\angle MOP = 90°$. Find $\angle NOP$.

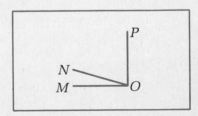

Items 3 and 4 are based on the following illustration below.

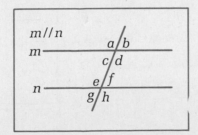

3. Which angle is an alternate interior angle with $\angle e$?

4. If ∠a = 111°, what is the measurement of ∠g?

---

5. In triangle RST, ∠R = 21° and ∠S = 69°. What kind of triangle is △RST?

6. In triangle JKL, ∠K = 55° and ∠L = 63°. Which side of △JKL is longest?

7. A 10-foot vertical pole casts a 7-foot shadow at the same time that a tree casts a 35-foot shadow. How tall is the tree?

8. ∠A and ∠E are both 90°. Also, AB = 24, AC = 32, and DE = 15. Find the length of CE.

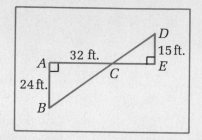

9. For the triangles pictured at the right, ∠K = ∠P and ∠M = ∠R. Along with this information, which of the following conditions is sufficient to guarantee that the two triangles are congruent?

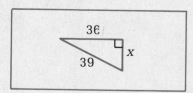

   A. ∠L = ∠P   C. KL = PQ
   B. KM = PR

   (1) A only   **(4)** A or B
   (2) B only   **(5)** B or C
   (3) C only

10. Find the length of side x in the triangle pictured at the right.

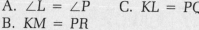

Items 11 and 12, are based on the illustration at the right.

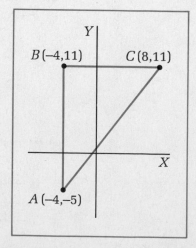

11. What is the distance from point A to point B?

12. What is the distance from point A to point C?

13. Of the following coordinates, which describes a point which lies on the graph of the equation y = 3x − 4?

    (1) (3, 6)   **(4)** (−3, −12)
    (2) (4, 8)   **(5)** (−4, −10)
    (3) (−1, −6)

14. Write the coordinates of the y-intercept of the equation y = 3x − 4.

15. What is the slope of a straight line that connects points P and Q in the picture at the right?

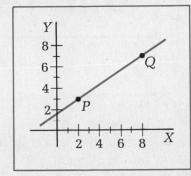

16. The volume of earth removed for the foundation of a new building was 120 cubic meters. Imagine that the hole was the shape of a rectangular container. If the hole was 8 meters long and 6 meters wide, how deep (high) was the hole?

17. The distance around a greenhouse is 56 feet. The greenhouse is a rectangle whose length is three times its width. Find the length of the greenhouse.

18. The ratio of the base to the height of the triangle pictured at the right is 6:5. The triangle has an area of 135 square feet. Find both the base and the height.

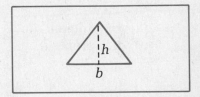

19. The volume of the figure shown at the right is 1500 cubic inches. The length is 10 inches, and the height is six times the width. Find both the width and the height.

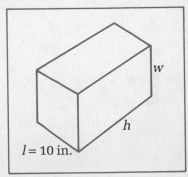

20. Arthur hiked six miles east, then ten miles south, and finally another four miles east. What was the shortest number of miles from his starting point to his finishing point?

    (1) 10   **(3)** $2\sqrt{10}$ **(5)** $4\sqrt{10}$
    (2) 20   **(4)** $10\sqrt{2}$

Check your answers. Correct answers are on pages 361–362. If you have at least sixteen answers correct, go to the simulated GED Test. If you have fewer than sixteen answers correct, study the Geometry chapter before you go on.

Unit Test

1. Janis works 35 hours per week and leaves her children at a day-care center during her working hours. If x = her hourly pay and y = the hourly day-care charge, which of the following expressions would show the difference between Janis' weekly salary and the weekly cost of day-care?

   (1) $35x - y$
   (2) $35(x - y)$
   (3) $35(x + y)$
   (4) $\dfrac{35x}{y}$
   (5) Insufficient data is given to solve the problem.

Item 2 is based on the following figure.

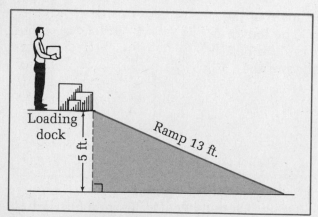

2. Stan uses a ramp to transfer goods from the loading dock. How many feet does the ramp extend from the base of the dock?

3. Since earning his GED, Mike earns $150 less than 3 times the amount of money he earned when he first dropped out of high school. If his current annual salary is $18,000, how much money did Mike earn when he dropped out of high school?

Items 4 to 6 are based on the following information.

The Sedleys are livestock and crop farmers. Each spring they borrow money that they pay back in the fall. This year they borrowed $21,000 at 8% interest for seed and fertilizer. Among their long-term loans they have a 30-year mortgage on their house at $7\frac{1}{2}$% interest. The principal of all the Sedleys' long-term loans combined is $300,000.

4. If the Sedleys repay the money they borrowed for seed and fertilizer in 9 months, how much interest will they have to pay?

   (1) $ 486
   (2) $ 547
   (3) $ 729
   (4) $1094
   (5) $1260

5. How much interest will the Sedleys have to pay during the life of their mortgage?

(1) $17,500
(2) $52,000
(3) $175,000
(4) $525,000
(5) Insufficient data is given to solve the problem.

6. The Sedleys can combine all their long-term loans in a new loan. The life of the new loan would be 25 years, and they would have to pay $450,000 interest. What would be the rate of the loan?

(1) 3%
(2) 4%
(3) 6%
(4) 12%
(5) Insufficient data is given to solve the problem.

---

7. One year, 16 pine trees were planted to reforest a tract of land. The next year, 16 saplings were taken from each of those trees and planted. The next year, this process was repeated with 16 saplings taken from each of the trees planted the previous year. If this process continued annually, which of the following expressions would show how many saplings would be planted in the twentieth year?

(1) 20(16)
(2) $\frac{16}{20}$
(3) $\frac{20}{16}$
(4) 20^{16}
(5) 16^{20}

Item 8 refers to the following figure.

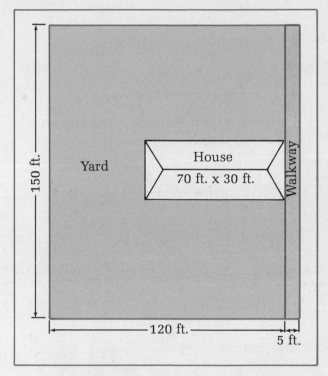

8. Except for the area taken up by the 5-foot walkway and the house, the Pucharskys' yard is being seeded with grass. What is the area of the lawn in square feet?

---

9. Kathy is writing a program for a computer game that requires the computer to select a random whole number between 50,001 and 75,000. If the computer won't pick the same number twice, what is the probability of its picking any unpicked number after 14 tries?

10. An airplane flight had 50 passengers that paid regular fares, plus 32 super-saver passengers that paid $\frac{3}{8}$ the regular fare. If $23,250 was paid for tickets, what is the cost of a regular ticket?

11. Quarterback Dunston Flowers had a season total completion-to-attempt pass ratio of 6:9. If he completed 120 passes, how many passes did he attempt?

12. If 6.023×10^{23} was written as a whole number, how many zeros would follow the digits 6023?

Items 13 and 14 are based on the following graph.

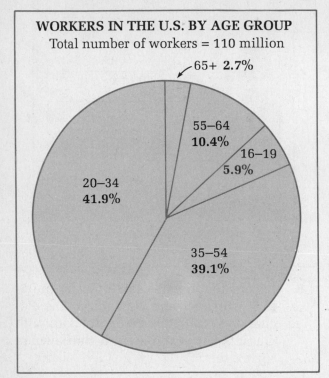

WORKERS IN THE U.S. BY AGE GROUP
Total number of workers = 110 million

65+ **2.7%**

55–64 **10.4%**

16–19 **5.9%**

20–34 **41.9%**

35–54 **39.1%**

13. If a company employs 600 workers, approximately how many of its employees would be between the ages of 35 and 54?

(1) 196
(2) 219
(3) 235
(4) 308
(5) Insufficient data is given to solve the problem.

14. How many million workers are age 35 or older?

(1) 14.41
(2) 41.14
(3) 57.42
(4) 63.91
(5) Insufficient data is given to solve the problem.

Items 15 and 16 are based on the following information.

Jona and Ed Bower have found a house that they would like to buy and must decide if they can afford it. Four times one's annual income is the amount recommended to spend. Ed earns 95% of Jona's salary. When the deal is closed, a 22% capital-gains tax will be applied to the profit the sellers earn on the sale.

15. If Jona's salary is b, which of the following expressions shows the recommended amount to spend for the house?

(1) $.95b$
(2) $4.95b$
(3) $4b + .95b$
(4) $4(b - .95b)$
(5) $4(b + .95b)$

16. The house the Bowers are buying has increased in value since the sellers purchased it. The sellers must therefore pay a capital-gains tax on $\frac{1}{2}$ of the profit they make on it. If d = the amount that the sellers paid for the house, and e = the price that the Bowers are paying for it, which of the following expressions would give the amount of capital-gains tax the sellers must pay?

(1) $2(e - d)$
(2) $.22(e - d)$
(3) $\dfrac{.22(e - d)}{2}$
(4) $\dfrac{2(e - d)}{.22}$
(5) $\dfrac{.22(2}{(e - d))}$

Item 17 refers to the following table.

1985 Tax Table

If line 37 (taxable income) is—		And you are—			
At least	But less than	Single	Married filing jointly *	Married filing sepa- rately	Head of a house- hold
			Your tax is—		
16,000					
16,000	16,050	2,178	1,711	2,623	2,055
16,050	16,100	2,190	1,719	2,637	2,065
16,100	16,150	2,201	1,727	2,651	2,075
16,150	16,200	2,213	1,735	2,665	2,085
16,200	16,250	2,224	1,743	2,679	2,095
16,250	16,300	2,236	1,751	2,693	2,105
16,300	16,350	2,247	1,759	2,707	2,115
16,350	16,400	2,259	1,767	2,721	2,125
16,400	16,450	2,270	1,775	2,735	2,135
16,450	16,500	2,282	1,783	2,749	2,145
16,500	16,550	2,293	1,791	2,763	2,155
16,550	16,600	2,305	1,799	2,777	2,165
16,600	16,650	2,316	1,807	2,791	2,175
16,650	16,700	2,328	1,816	2,805	2,185
16,700	16,750	2,339	1,825	2,819	2,195
16,750	16,800	2,351	1,834	2,833	2,205
16,800	16,850	2,362	1,843	2,847	2,215
16,850	16,900	2,374	1,852	2,861	2,225
16,900	16,950	2,385	1,861	2,875	2,235
16,950	17,000	2,397	1,870	2,889	2,245

* This column must also be used by a qualifying widow(er).

17. Dwight is single; his taxable income is $16,736, and he has paid $3002 in with-holding taxes. How much more was withheld than he actually had to pay?

Item 18 refers to the following figure.

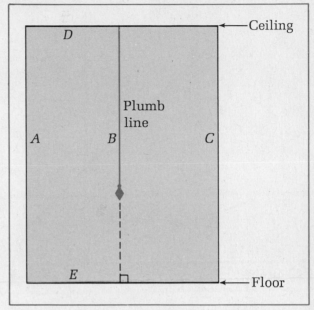

18. The figure above shows the plumb line, B, which is used to determine whether a wall is vertical. It is suspended from the ceiling in a room where a new wall will be built and is therefore perpendicular to the floor. If wall studs A and C are perpendicular to the floor, which of the following statements is true?

(1) A and D are parallel.
(2) A, B, and C are parallel.
(3) D and E are perpendicular.
(4) A and B are perpendicular.
(5) D, E, and C are parallel.

19. If $n^2 - 5n + 4 = 0$, what is the value of n?

20. Simplify $\dfrac{48y^2z^5}{12yz}$.

Items 21 and 22 refer to the following figure.

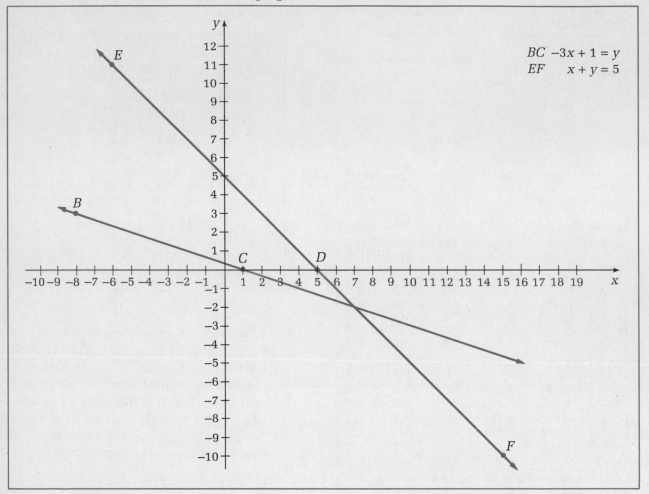

$$BC \quad -3x + 1 = y$$
$$EF \quad x + y = 5$$

21. What is the slope of line *EF*?

    (1) $\frac{-1}{3}$
    (2) 1
    (3) 3
    (4) −3
    (5) −1

22. What is the distance between points *B* and *E*?

    (1) $\sqrt{4}$
    (2) $\sqrt{64}$
    (3) $\sqrt{68}$
    (4) $\sqrt{256}$
    (5) Insufficient data is given to solve the problem.

Item 23 refers to the following figure.

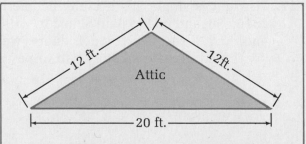

23. Barbara wants to install a triangular window in her attic. If the window will be similar to the dimensions of the attic and will have a base of 12 feet, what will each side measure in feet?

Item 24 is based on the following figure.

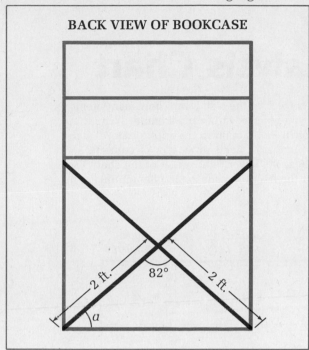

BACK VIEW OF BOOKCASE

2 ft.   82°   2 ft.

a

24. Jesse installed diagonal bracing to stabilize a bookcase he is building. What is the measure in degrees of angle a?

Item 25 refers to the following figure.

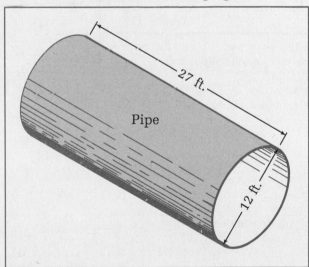

27 ft.

Pipe

12 ft.

25. Except for the 2 feet of pipe that extend out on each side of the road, how many cubic feet of earth will the culvert pipe replace?

26. A computer is programmed to reject cost items in an inventory that are greater than \$4300 and less than \$1900. Which of the following values of w will be retained by the computer?

    a. $1900 \leq w \leq 4300$
    b. $1900 \geq w \leq 4300$
    c. $4300 \geq w \geq 1900$
    d. $1900 \leq w \geq 4300$

(1) a
(2) c
(3) a and c
(4) d and a
(5) b

27. Kim has socket wrenches in the following sizes.

    a. $\frac{1}{4}$
    b. $\frac{5}{8}$
    c. $\frac{9}{16}$
    d. $\frac{1}{2}$
    e. $\frac{3}{4}$
    f. $\frac{11}{32}$

Which of the following sequences shows these socket wrenches from the smallest to the largest?

(1) d, a, b, c, e, f
(2) f, e, c, b, a, d
(3) e, a, f, b, c, d
(4) a, f, d, c, b, e
(5) a, d, b, e, c, f

28. Wanda has entered a 10,000-meter race. If she can run 6 kilometers per hour, how long will it take her to complete the course? (1 kilometer = 1000 meters)

Answers are on pages 363–364.

Performance Analysis Chart

Directions: Circle the number of each item you got correct on the test. Count how many items you got correct in each row; count how many items you got correct in each column. Write the amount correct per row and column as the numerator in the fraction in the appropriate "Total Correct" box. (The denominators represent the total number of items in the row or column.) Write the grand total correct over the denominator **28** at the lower right corner of the chart. (For example, if you got 24 items correct, write 24 so that the fraction reads 24/**28**.) Item numbers in color represent items based on graphic material.

Item Type	Arithmetic (page 21)	Algebra (page 174)	Geometry (page 213)	TOTAL CORRECT
Skills			18, 24	/2
Applications	5, 6, 7, 12, 17, 27, 28	19, 20	21, 22, 23	/12
Problem Solving	4, 8, 9, 13, 14, 25	1, 3, 10, 11, 15, 16, 26	2	/14
TOTAL CORRECT	/13	/9	/6	/28

The page numbers in parentheses indicate where in this book you can find the beginning of specific instruction about the various areas of mathematics and about the types of questions you encountered on the Unit Test.

On the chart, items are classified as Skills, Applications, or Problem Solving. In the chapters in this book, the three problem types are covered in different levels:

Skills items are covered in Level 1.

Applications items are covered in Level 2.

Problem Solving items are covered in Level 3.

For example, the skills needed to solve Item 22 are covered in Level 2 of Chapter 6. To locate in which chapter arithmetic items are addressed, reread the problem to see what kind of numbers (whole numbers, fractions, etc.) are used and then go to the designated level of the appropriate chapter.

Practice

Introduction

As you know, an orchestra is made up of many individual musicians. Before a performance, the conductor spends long hours rehearsing the musicians so that they all play well and play together. Sometimes during the practice, the conductor discovers problems with individual musicians or with the orchestra as a whole. The conductor sees to it that the musicians get extra practice in those weaker areas so, that when the time of the concert arrives, the orchestra is as well rehearsed and as comfortable with the material as possible. The activities in this GED Practice section will be your "rehearsals." By completing them, you will get valuable practice on answering GED-type questions and taking GED-type tests. And when it comes time to take your test, you will be as "well rehearsed" and "as comfortable with the material" as you can be.

This section is filled with GED-type test questions, or *items*. It provides valuable practice on the kinds of items found on the Mathematics Test. Arranged in two groups, the practices make it easy to test your ability at arithmetic, algebra, or geometry items separately as well as to test your ability with a collection of items structured like the actual test.

On the pages that follow, you will find:

- **PRACTICE ITEMS**—This practice contains 56 simulated GED test items, grouped according to the types of mathematics. You will find arithmetic items grouped together, geometry items grouped together, and so on.

- **PRACTICE TEST**—This is a 56-item test structured like the actual Mathematics Test. The items are *not* grouped together according to type. Rather, the types vary throughout the test.

As on the actual Mathematics Test, all the items are multiple choice. By completing the Practice Items and the Practice Test, you will discover your strong points and weak points in mathematics. And if you discover any weak points, *don't worry*—you will be shown how to strengthen them. The answer section not only provides the correct answer to each practice item, it also shows how to solve each problem. The Performance Analysis Chart following each practice will direct you to parts of the book where you can review the skills or subjects that give your trouble.

You can use the Practice Items and the Practice Test in a number of different ways. The introductions that precede the practices will provide you with choices for using them to your best advantage. You also may wish to talk with your teacher in order to get suggestions about how best to make use of the Practice Items and Practice Test.

Practice Items

These Practice Items are similar to a real Mathematics Test in many ways, but there is one major difference. The items that follow are grouped according to types of mathematics: arithmetic, algebra, and geometry. As you work on the Practice Items, you will focus on one skill area at a time. (The actual Mathematics Test presents problems that require various types of mathematics in mixed order.)

The whole group of 56 Practice Items is the same length as an actual test; the items are similarly as challenging as the actual test items. Your results will help you determine which skills you have mastered and which you should study further.

Using the Practice Items to Your Best Advantage

You can use the Practice Items in the following ways:

- After you finish Chapters 1 through 5 of the instruction part of this book, you can test your skill by completing the arithmetic section of the Practice Items. You can do the algebra section after you finish Chapter 6 and the geometry section after you finish Chapter 7. Or you may prefer to save the Practice Items until you've completed all the instruction. You can do all the Practice Items at once, like a test, or you can do them one group at a time. Afterward, review the chapters for the areas in which you have difficulty.

- You can use the Practice Items as a practice test. To do this, complete all 56 items in one sitting. Since the actual test allows you 90 minutes, you may want to time yourself. If 90 minutes elapse and you have not finished, circle the last item you finished and then continue. This way, you can learn what score you'd earn within the time limit as well as your total score counting the untimed portion of the practice. This will give you a rough idea of how you would perform on the actual Mathematics Test.

Keep an accurate record of your performance. Write your answers neatly on a sheet of paper or use an answer sheet provided by your teacher.

Using the Answers and Solutions

The answer section can be a very helpful study tool. Compare your answers to the correct answers beginning on page 364, and check each item you answered correctly. Whether you answer an item correctly or not, you should read through the solutions in the answer section. Doing this will reinforce your ability with mathematics and develop your test-taking skills.

How to Use Your Score

Regardless of how you use these Practice Items, you will gain valuable experience with GED-type items. After scoring your work with the answer section, fill in the Performance Analysis Chart on page 285. The chart will help you determine which skills and problem types you are strongest in and will direct you to parts of the book where you can review areas in which you will need additional work.

PRACTICE ITEMS

Directions: *Choose the one best answer to each question.*

Arithmetic Problems

1. Of the 150 members in the First Street Tenants' Association, 85 of them voted to go on a rent strike. What is the ratio of the number of association members who voted to strike to the number of members who did NOT vote to strike?

    (1) 13:30   **(3)** 1:85   **(5)** 85:1
    (2) 17:13   **(4)** 17:30

2. Which of the following is equal to $(\frac{2}{3})^3$?

    (1) 2   **(3)** $\frac{2}{3}$ **(5)** $\frac{8}{27}$
    (2) $\frac{8}{9}$ **(4)** $\frac{4}{9}$

3. The three highest mountains in Alaska are McKinley (20,320 ft.), St. Elias (18,008 ft.), and Foraker (17,400 ft.). Find the mean height of these three mountains.

    (1) 17,704 ft.   **(3)** 18,576 ft.   **(5)** 19,164 ft.
    (3) 18,008 ft.   **(4)** 18,860 ft.

4. Three families who are friends with each other bought raffle tickets. The Johnsons bought 8 tickets, the Millers bought 12, and the Smiths bought 10. Altogether, 500 raffle tickets were sold. What is the probability that the Smiths will win?

    (1) $\frac{3}{500}$ **(3)** $\frac{1}{10}$ **(5)** $\frac{1}{2}$
    (2) $\frac{1}{50}$ **(4)** $\frac{1}{3}$

5. What is the chance that any of the three families in problem 4 will win the raffle?

    (1) $\frac{3}{500}$ **(3)** $\frac{3}{50}$ **(5)** $\frac{1}{3}$
    (2) $\frac{1}{30}$ **(4)** $\frac{3}{25}$

6. Which of the following expressions is the same as 9×10^4?

    (1) $9 \times 10 \times 4$
    (2) $10 \times 9 \times 9 \times 9 \times 9$
    (3) $9 \times 4 \times 4 \times 4 \times 4$
    (4) $9 \times 10 \times 10 \times 10 \times 10$
    (5) $4 \times 9 \times 9 \times 9 \times 9$

Items 7 and 8 are based on the following figure.

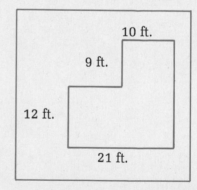

7. What is the total distance in feet around the room?

    (1) 96   **(3)** 73   **(5)** 52
    (2) 84   **(4)** 63

8. Find the total number of square feet of the floor of the room.

    (1) 84   **(3)** 318   **(5)** 441
    (2) 309   **(4)** 342

9. What is the area in square inches of a square with a side that measures $4\frac{1}{2}$ inches?

(1) $24\frac{1}{4}$ (4) $16\frac{1}{4}$

(2) $20\frac{1}{4}$ (5) 9

(3) 18

10. The picture below shows a 5-centimeter scale. What is the distance from point X to point Y?

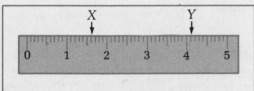

(1) 2.5 cm   (4) 3.5 cm
(2) 2.7 cm   (5) 3.7 cm
(3) 2.9 cm

11. What is the volume in cubic centimeters of the rectangular container shown below?

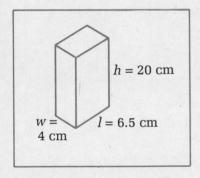

(1) 46   (4) 502
(2) 134   (5) 520
(3) 460

12. Arrange the following lengths of metal tubing in order from shortest to longest.

Tube A is 0.85 m long.
Tube B is 0.095 m long.
Tube C is 1.2 m long
Tube D is 0.9 m long.
Tube E is 1.07 m long.

(1) E, B, C, A, D
(2) B, A, D, E, C
(3) D, A, C, B, E
(4) C, E, B, A, D
(5) C, A, E, B, D

Items 13 to 16 are based on the following circle graph.

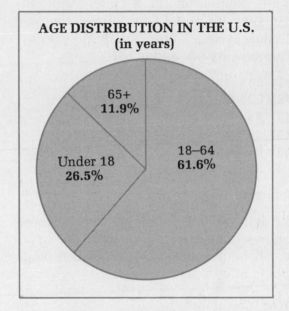

13. The category "under 18" is about what fraction of the total?

(1) $\frac{1}{4}$ (4) $\frac{3}{5}$

(2) $\frac{1}{3}$ (5) $\frac{2}{3}$

(3) $\frac{1}{2}$

14. The combined percent of the "over 65" and "under 18" group is how many percentage points less than the "18–64" group?

(1) 49.7%   (4) 23.2%
(2) 47.0%   (5) 14.6%
(3) 35.1%

15. People from ages 18 to 40 are about what fraction of the total?

(1) $\frac{4}{5}$ (4) $\frac{1}{4}$

(2) $\frac{3}{5}$ (5) Insufficient data is given to solve the problem.

(3) $\frac{1}{3}$

16. The population of Capital City is 300,000. If Capital City follows the national average, how many people in Capital City are under 18?

(1) 184,800
(2) 92,400
(3) 79,500
(4) 35,700
(5) Insufficient data is given to solve the problem.

---

17. Which of the following expressions shows the total amount Joe paid for a car if he put $600 down and then paid $80 a month for five years?

(1) 600 + 60 × 80
(2) 12 × 600 × 80
(3) 5 × 500 × 80
(4) 60(600 + 80)
(5) 60 × 80 × 600

18. Blanca works two days each week and earns a total of $150. How many weeks does she have to work in order to earn $3000.

(1) 10   (4) 20
(2) 12.5   (5) Insufficient data is
(3) 17.5   given to solve the problem.

Items 19 to 22 are based on the following situation.

In Capital City, which has a population of 300,000, the estimated number of jobs is 120,000. Elk Electronics plans to build a new factory in Capital City, which should increase the number of jobs by 5%. The new jobs will be phased in gradually over a three-year period. The first third of the jobs will be open at the end of a year, the next third at the end of two years, and the last third at the end of three years. In addition, Paulson's Plastics plans to open a new shop in one year. Paulson's will increase the current number of jobs by 2%.

19. How many new jobs will be available at Elk Electronics at the end of one year?

(1) 1200   (4) 4000
(2) 2000   (5) 6000
(3) 2400

20. By about what percent will the population of Capital City have grown two years from now?

(1) 2%   (4) 10%
(2) 5%   (5) Insufficient data is
(3) 7%   given to solve the problem.

21. Assuming no other changes, what will be the total number of jobs in Capital City at the end of two years if both Elk Electronics and Paulson's Plastics complete their plans?

(1) 124,000
(2) 124,400
(3) 126,400
(4) 128,400
(5) Insufficient data is given to solve the problem.

22. Suppose after one year that Elk Electronics sticks to their plan, but Paulson's decides not to build a shop, and a bottling plant closes and lays off 700 people. Assuming no other changes, what will be the net change to the current number of jobs in the city?

(1) 2000 more
(2) 1300 more
(3) 700 more
(4) 300 fewer
(5) Insufficient data is given to solve the problem.

23. Nick can paint an average of three rooms a day. Which of the following tells the number of work days Nick needs to paint the interior of 20 houses if each house has an average of six rooms?

(1) $3(20 + 6)$

(2) $\dfrac{20 + 6}{3}$

(3) $20 \times 6 \times 3$

(4) $\dfrac{3}{20 \times 6}$

(5) $\dfrac{20 \times 6}{3}$

24. Janet weighs 160 pounds, which is 80% of what she weighed one year ago. How many pounds did she lose?

(1) 20   **(4)** 40
(2) 25   **(5)** 60
(3) 30

25. The ratio of wins to losses for Mike's softball team last season was 5:2. The team played a total of 49 games. How many did they win, and how many did they lose?

(1) won 40, lost 9
(2) won 35, lost 14
(3) won 30, lost 19
(4) won 25, lost 24
(5) won 33, lost 16

26. Ellen drove for four hours at an average speed of 40 mph. Then Terry took over and drove five more hours at an average speed of 55 mph. Which of the following expressions tells the total distance they traveled?

(1) $4 \times 40 \times 5 \times 55$

(2) $(4 + 5) + \dfrac{40 + 55}{2}$

(3) $4(40) + 5(55)$

(4) $4(5) + 40(55)$

(5) $9(40 + 55)$

Items 27 and 28 are based on the following chart.

Postal Rates	
First-Class Letters	
1 oz.	$ .22
Each additional oz.	.17
Certified Mail	
(In addition to postage)	.75
Registry	
$0.01 to $100.00	3.55

27. Find the cost of sending a first-class letter that weighs 4 ounces.

(1) $.56   **(4)** $.88
(2) $.68   **(5)** $.90
(3) $.73

28. Find the cost of sending a two-ounce first-class letter by certified mail.

(1) $1.09   **(4)** $1.31
(2) $1.14   **(5)** $1.41
(3) $1.19

Algebra Problems

29. Which point on the number line corresponds to $\frac{-6}{5}$?

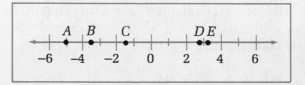

(1) A   **(4)** D
(2) B   **(5)** E
(3) C

30. Solve for a in $8a - 2 = 6$.

(1) 8   **(4)** 1
(2) 4   **(5)** $\frac{1}{2}$
(3) 2

31. Which of the following expressions is equal to $18m + 24$?

(1) $6(3m + 4)$ (4) $18(m + 3)$
(2) $18m(24)$ (5) $6(3m + 8)$
(3) $6(m + 4)$

32. Write an equation which expresses the following statement: Two more than four times a number is the same as eleven more than three times the same number.

(1) $4x + 3 = 2x + 11$
(2) $4x + 2 = 3x + 11$
(3) $4x + 11 = 3x + 2$
(4) $2x + 4 = 11x + 3$
(5) $2x + 3 = 4x + 11$

33. Solve for y in $5y - 8 = 4y + 1$.

(1) $\frac{4}{9}$ (4) 7
(2) $\frac{5}{9}$ (5) 9
(3) $\frac{4}{5}$

34. Simplify $\frac{-8a^2}{-12a}$.

(1) $20a^3$ (4) $\frac{-20a}{-2}$
(2) $-\frac{2}{3}a^3$ (3) $\frac{2a}{3}$ (5) $\frac{-2}{-3a}$

35. George makes x dollars an hour. His supervisor, Sam, makes \$1 more than twice as much as George in an hour. Which of the following expressions shows Sam's hourly wage?

(1) $2x - 1$ (4) $\frac{2x + 1}{2}$
(2) $\frac{x - 1}{2}$ (5) $2x + 1$
(3) $\frac{x + 1}{2}$

36. Which of the following expressions equals $x^2 + 12x + 35$?

(1) $(x + 3)(x + 5)$
(2) $(x + 35)(x + 1)$
(3) $(x - 5)(x - 7)$
(4) $(x + 7)(x + 5)$
(5) $(x - 7)(x + 5)$

37. Choose the expression that represents twice the quantity of a number decreased by six.

(1) $6(x - 2)$ (4) $2x - 6$
(2) $2(x - 6)$ (5) $\frac{x - 6}{2}$
(3) $2x + 6$

38. Solve for d in $-2 = 4(3d - 2)$.

(1) 3 (4) $\frac{1}{3}$
(2) 2 (5) $\frac{1}{4}$
(3) $\frac{1}{2}$

39. Write an equation that expresses the following statement: Half a number decreased by one equals one-third of the same number increased by five.

(1) $\frac{x}{2} - 1 = \frac{x}{3} + 5$
(2) $x - \frac{1}{2} = x - \frac{1}{3}$
(3) $\frac{x}{2} + 1 = \frac{x}{3} + 5$
(4) $x - 1 = \frac{x}{5} + 3$
(5) $1 - \frac{x}{2} = \frac{x}{3} + 5$

40. Which of the following expressions is the same as $12x - (-3x)$?

(1) $36x$ (4) $36x$
(2) $9x$ (5) $-9x$
(3) $15x$

41. One less than three times a number is the same as two times the quantity of the same number increased by five. Find the number.

(1) 2 (4) 9
(2) 4 (5) 11
(3) 5

42. Solve for p in $3p - 2 > p + 8$.

(1) $p > 2$ (4) $p > 5$
(2) $p > 3$ (5) $p > 8$
(3) $p > 4$

43. Deborah, Jeff, and Caroline worked together painting Caroline's apartment. Jeff worked four more hours than Deborah, and Caroline worked twice as many hours as Deborah. Altogether, the three of them worked 28 hours. How many hours did Caroline work?

(1) 6   (4) 12
(2) 8   (5) 14
(3) 10

44. Solve for x in $x^2 - 36 = 0$.

(1) $x = +3$ and $+12$
(2) $x = +4$ and -9
(3) $x = -6$ and $+6$
(4) $x = +18$ and -2
(5) $x = -4$ and -9

Geometry Problems

45. In the picture below, $m//n$ and $\angle s = 84°$. Find the measurement of $\angle x$.

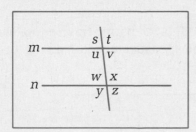

(1)   6°   (4) 84°
(2) 16°   (5) 96°
(3) 76°

46. In the diagram below, $\angle AOB = 62°$ and $\angle COD = 79°$. Find $\angle BOC$.

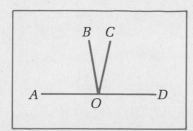

(1) 78°   (4) 22°
(2) 39°   (5) 11°
(3) 28°

47. In triangle PQR, $\angle P = 59°$ and $\angle Q = 64°$. Which side of the triangle is longest?

(1) PR (4) All three sides are
(2) PQ equal.
(3) QR (5) Insufficient data is given to solve the problem.

48. Find the measurement of the missing side of the triangle pictured below.

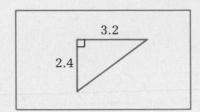

(1) 5.6   (4) 3
(2) 4.2   (5) 2.8
(3) 4

49. The two triangles shown below are similar. Find the height of the triangle at the right.

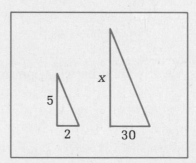

(1) 12   (4) 75
(2) 18   (5) 150
(3) 60

50. The vertex angle of an isosceles triangle equals 48°. Find the measurement of each base angle.

(1) 42°   (4) 84°
(2) 64°   (5) 96°
(3) 66°

51. What is the slope of the line that passes through points F and G on the graph at the right?

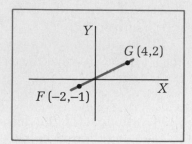

(1) -1 **(4)** $\frac{-1}{2}$

(2) -2

(3) $+2$ **(5)** $\frac{+1}{2}$

Items 52 and 53 are based on the following diagram.

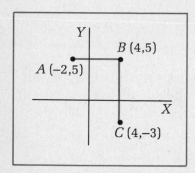

52. What is the distance between points A and B?

(1) 3   **(4)** 8

(2) 6   **(5)** 10

(3) 7

53. The midpoint M between two points in a plane is given by the expression

$$M = \left(\frac{x_1 + x_2}{2}, \frac{y_1 + y_2}{2}\right),$$

where (x_1, y_1) and (x_2, y_2) are the coordinates of the two points. Find the midpoint between points A and C.

(1) $(2, 1)$ **(4)** $(1, 1)$

(2) $(-1, -1)$ **(5)** $(2, 3)$

(3) $(-2, -2)$

54. What are the coordinates of the y-intercept of the graph of the equation $y = -x + 3$?

(1) $(0, 3)$ **(4)** $(0, 1)$

(2) $(3, 0)$ **(5)** $(0, -3)$

(3) $(-3, 0)$

55. For the rectangle shown below, the length is 7 feet more than the width, and the perimeter is 66 feet. Find the length of the rectangle in feet.

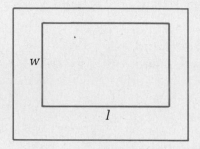

(1) 7   **(4)** 22

(2) 13   **(5)** 33

(3) 20

56. The two figures below have equal areas. Find the measurement of a side of the square.

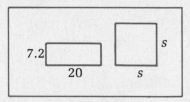

(1) 10   **(4)** 27.2

(2) 12   **(5)** 36

(3) 14.4

Answers are on pages 364–366.

PRACTICE ITEMS
Performance Analysis Chart

Directions: Circle the number of each item you got correct on the Practice Items. Count how many items you got correct in each row; count how many items you got correct in each column. Write the amount correct per row and column as the numerator in the fraction in the appropriate "Total Correct" box. (The denominators represent the total number of items in the row or column.) Write the grand total correct over the denominator **56** at the lower right corner of the chart. (For example, if you got 50 items correct, write 50 so that the fraction reads *50/56*.) Item numbers in color represent items based on graphic material.

Item Type	Arithmetic (page 21)	Algebra (page 174)	Geometry (page 213)	TOTAL CORRECT
Skills		29, 34, 40	45, 46, 47, 50	/7
Applications	1, 2, 3, 4, 5, 6, 9, 10, 11, 12, 13, 15, 16, 27, 28	30, 31, 33, 35, 36, 37, 38, 42, 44	48, 49, 51, 52, 53, 54	/30
Problem Solving	7, 8, 14, 17, 18, 19, 20, 21, 22, 23, 24, 25, 26	32, 39, 41, 43	55, 56	/19
TOTAL CORRECT	/28	/16	/12	/56

The page numbers in parentheses indicate where in this book you can find the beginning of specific instruction about the various areas of mathematics you encountered on the Practice Items.

On the chart, items are classified as Skills, Applications, or Problem Solving. In the chapters in this book, the three problem types are covered in different levels:

Skills items are covered in Level 1.

Applications items are covered in Level 2.

Problem Solving items are covered in Level 3.

For example, the skills needed to solve Item 48 are covered in Level 2 of Chapter 6. To locate in which chapter arithmetic items are addressed, reread the problem to see what kind of numbers (whole numbers, fractions, etc.) are used and then go to the designated level of the appropriate chapter.

Practice Test

Like the actual Mathematics Test, the items on the following Practice Test appear in a mixed order. The test provides you with two kinds of practice necessary for the GED: practice on the items themselves and practice on switching from one type of mathematics to another.

This Practice Test is the same length (56 items) as the actual test, and it is similarly challenging. By taking the Practice Test, you can gain valuable test-taking experience and will know what to expect when you sit down to take the actual Mathematics Test.

You can use the Practice Test in the following ways: (1) Try to answer all the questions within a 90-minute time limit. If time runs out before you finish, circle the last question you have answered. Then continue with the test. This way, you can learn your score within the time limit as well as your total score on the test. Comparing the two scores will give you an idea about how much speed can affect your score. (2) If you want, you can take the Practice Test in sections. While this does not simulate the actual testing situation, your results will still give you a pretty good idea of how well you would do on the real test.

Write your answers on a sheet of paper or use an answer sheet. If a question gives you trouble, take an educated guess and then move on.

Compare your answers to those in the answer key on page 366. Whether you answer an item correctly or not, look over the solution in the answer key. This will help you reinforce your ability with mathematics and develop your test-taking skills.

However you decide to take the Practice Test, your final score will point out your strengths and weaknesses in mathematics. The Performance Analysis Chart at the end of the test will help you identify those strengths and weaknesses.

PRACTICE TEST

Directions: *Choose the one best answer to each question.*

1. What is the ratio of 21 inches to one yard?

   (1) 1:7   (4) 1:21
   (2) 7:1   (5) 7:12
   (3) 5:12

2. Which point on the line below corresponds to $\frac{5}{2}$?

   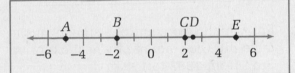

   (1) A   (4) D
   (2) B   (5) E
   (3) C

3. Margaret tutors five students—Mary, age 56; Kevin, 23; Ramona, 41; James, 29; and Laura, 32. Find their median age.

   (1) 23   (4) 36
   (2) 29   (5) 41
   (3) 32

4. In an isosceles triangle, each base angle contains 72°. Find the measurement of the vertex angle.

   (1) 36°   (4) 72°
   (2) 54°   (5) 108°
   (3) 64°

5. Suppose n represents the number of men in Paul's night-school class. The number of women in the class is seven less than twice the number of men. Which expression represents the number of women?

   (1) n + 7   (4) 7n − 2
   (2) 2n − 7   (5) 7n + 2
   (3) 2n + 7

6. Which of the following expressions equals 7×10^3?

   (1) 7 × 300
   (2) 7 × 10 × 3
   (3) 7 × 10 × 10
   (4) 3 × 7 × 7 × 7
   (5) 7 × 10 × 10 × 10

7. In the diagram below, ∠x = 121°. Find the measurement of ∠w.

   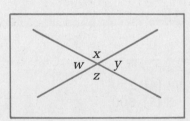

   (1) 121°   (4) 59°
   (2) 118°   (5) 31°
   (3) 62°

8. George takes home $380 a week, and he saves $60 each week. How many weeks will it take him to save $750?

(1) 6.25
(2) 10
(3) 12.5
(4) 15
(5) Insufficient data is given to solve the problem.

9. Which of the following expressions tells Fernando's net income for the year if his gross monthly pay is $1800, and his employer deducts $400 each month for taxes and social security?

(1) $12(1800 - 400)$
(2) $\dfrac{1800 + 400}{12}$
(3) $12 \times 1800 \times 400$
(4) $52 \times 1800 \times 400$
(5) $52(1800 - 400)$

10. The picture below shows the plan of a swimming pool and a walkway around the pool. The shaded portion of the drawing represents the walkway. Find the distance in feet around the pool.

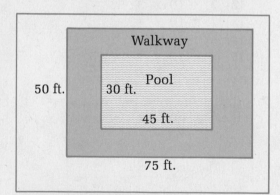

(1) 135
(2) 150
(3) 180
(4) 250
(5) 300

11. What is the area in square feet of the walkway around the pool in problem 10?

(1) 1200
(2) 2400
(3) 2500
(4) 2550
(5) 5100

Items 12 and 13 are based on the following table.

Major Wheat Exporters		
Country	*millions of metric tons*	
	1982	1983
United States	40.5	38.5
Canada	19.2	21.8
France	10.7	13.4
Argentina	3.8	10.2
Australia	10.9	8.2

Source: U.S. Department of Agriculture.

12. By how many million metric tons did the amount of wheat France exported in 1983 exceed the amount France exported in 1982?

(1) 2.0
(2) 2.6
(3) 2.7
(4) 3.7
(5) 6.4

13. For which two countries shown in the table did the amount of wheat exported drop from 1982 to 1983?

(1) Canada and France
(2) United States and Canada
(3) Argentina and Australia
(4) United States and Australia
(5) France and Argentina

14. Of the registered voters in Greenport, the ratio of those who voted to those who did not vote in the last election was 3:2. 5400 people voted. How many registered voters are there in Greenport?

(1) 3240
(2) 3600
(3) 5400
(4) 9000
(5) 12,000

15. Which of the following expressions shows the total distance a hiker traveled if he walked for two hours at 4.5 mph and then for four hours at 3.5 mph?

(1) $\frac{2 \times 4.5}{4 \times 3.5}$

(2) $2 \times 4.5 \times 4 \times 3.5$

(3) $4(4.5) + 2(3.5)$

(4) $(2 + 4) + (4.5 + 3.5)$

(5) $2(4.5) + 4(3.5)$

Items 16 to 19 are based on the following situation.

Tom works in the shipping department of Elk Electronics, where he makes $1350 a month. His boss offered him a choice of two promotions. With choice A, Tom would become the head of the shipping department in the plant where he works, and he would get a raise of 8%. With choice B, Tom would head the shipping department in another plant. For this job, Tom would get a raise of 10%. Tom estimates that additional monthly travel expenses to the other plant would be $40 a month.

16. How much would Tom make in a year if he accepts choice A?

(1) $ 9350   (4) $17,496

(2) $12,000   (5) $17,820

(3) $14,580

17. With choice B, Tom's monthly salary would be how much more than his current monthly salary?

(1) $ 95   (4) $125

(2) $100   (5) $135

(3) $110

18. If Tom chooses job A, how much would he have at the end of a year in the company's pension plan?

(1) $ 850   (4) $1485

(2) $1100   (5) Insufficient data is

(3) $1350   given to solve the
                   problem.

19. Find Tom's yearly salary with job B, minus the estimated travel expenses.

(1) $17,016   (4) $17,780

(2) $17,340   (5) Insufficient data is

(3) $17,420   given to solve the
                     problem.

20. Which of the following expressions equals $4ab - 6ab + 7ab$?

(1) $-6ab$ (4) $-17ab$

(2) $17ab$ (5) $9ab$

(3) $5ab$

21. In the picture below, $\angle COE = 90°$, and $\angle DOE = 23°$. Find $\angle COD$.

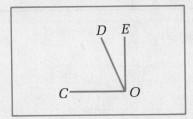

(1) 23°   (4) 77°

(2) 57°   (5) 157°

(3) 67°

22. Solve for c in $\frac{3}{5}c + 1 = 13$.

(1) 1   (3) 15   (5) 25

(2) 5   (4) 20

23. What is the area of a triangle with a base of $2\frac{3}{4}$ inches and a height of $1\frac{1}{2}$ inches?

(1) $4\frac{3}{4}$ in.^{2} (4) $2\frac{1}{16}$ in.^{2}

(2) $3\frac{1}{8}$ in.^{2} (5) $1\frac{15}{16}$ in.^{2}

(3) $2\frac{7}{8}$ in.^{2}

24. Which expression below is equal to the quantity of one more than five times a number, all divided by four?

(1) $\frac{4x + 1}{5}$ (4) $\frac{x + 4}{4}$

(2) $\frac{5x + 1}{4}$ (5) $\frac{x + 5}{4}$

(3) $5x + \frac{1}{4}$

25. Simplify the expression $\frac{-4xy}{12x}$.

    (1) $-3x^2y$ **(4)** $\frac{-x^2y}{3}$

    (2) $8xy$ **(5)** $\frac{-y}{3}$

    (3) $-3y$

Items 26 to 29 are based on the following line graph.

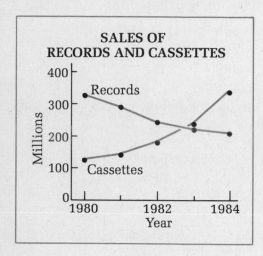

26. By what year was the sale of cassettes greater than the sale of long-playing records?

    (1) 1980   **(4)** 1983
    (2) 1981   **(5)** 1984
    (3) 1982

27. What was the total value of the cassettes sold in 1983?

    (1) \$335 million
    (2) \$240 million
    (3) \$140 million
    (4) \$100 million
    (5) Insufficient data is given to solve the problem.

28. Which of the following statements describes the change in the sales of cassettes from 1980 to 1984?

    (1) The number of cassettes sold in 1984 was about three times the number sold in 1980.
    (2) The number of cassettes sold in 1984 was about twice the number sold in 1980.
    (3) The number of cassettes sold in 1984 was about one-third of the number sold in 1980.
    (4) The number of cassettes sold in 1984 was about one-half the number sold in 1980.
    (5) The number of cassettes sold in 1984 was about the same as the number sold in 1980.

29. In 1980, about how many more long-playing records were sold than cassettes?

    (1) 50 million
    (2) 100 million
    (3) 150 million
    (4) 200 million
    (5) Insufficient data is given to solve the problem.

30. The figure below is a 5-inch scale. Find the distance between points C and D.

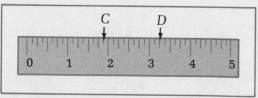

    (1) $1\frac{3}{8}$ in.   **(4)** $4\frac{1}{8}$ in.

    (2) $1\frac{5}{8}$ in.   **(5)** $4\frac{7}{8}$ in.

    (3) $1\frac{7}{8}$ in.

31. Which of the following expressions is equal to $24y - 32$?

    (1) $8(2y - 4y)$ **(4)** $24(y - 32)$
    (2) $8(3y - 4)$ **(5)** $24y(-32)$
    (3) $12(2y - 3)$

32. Find the volume in cubic meters of the cylinder shown below.

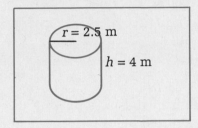

(1) 31.4   (4) 78.5
(2) 39.3   (5) 157.0
(3) 62.8

33. Write an equation which expresses the following statement: Twice the quantity of a number decreased by five is 12.

(1) 2x − 5 = 12
(2) 2x + 5 = 12
(3) 2(x − 5) = 12
(4) 2(x + 5) = 12
(5) 5(x − 2) = 12

34. On a test, Yoshiko got 48 problems right and 12 problems wrong. What percent of the problems did she get right?

(1) 20%   (4) 80%
(2) 25%   (5) 85%
(3) 75%

35. Solve for e in 3(e + 4) = 6.

(1) $\frac{2}{3}$ (4) 6
(2) −2   (5) −6
(3) −3

36. An 8-foot-tall vertical pole casts a shadow 3 feet long at the same time that a tree casts a shadow 48 feet long. Find the height of the tree.

(1) 16 ft.   (4) 128 ft.
(2) 32 ft.   (5) 256 ft.
(3) 64 ft.

37. Which of the following expressions is equal to x² − 100?

(1) (x + 25) (x − 4)
(2) (x + 50) (x − 50)
(3) (x + 10)²
(4) (x − 20) (x + 5)
(5) (x + 10) (x − 10)

38. Solve for t in 5(t + 1) < 20.

(1) $t < 2\frac{1}{2}$ (4) $t < 5$
(2) $t < 3$ (5) $t < 7\frac{1}{2}$
(3) $t < 4$

39. Arrange the following packages in order from heaviest to lightest.

Package A weighs $1\frac{15}{16}$ pounds.
Package B weighs $2\frac{7}{16}$ pounds.
Package C weighs $1\frac{3}{4}$ pounds.
Package D weighs $2\frac{1}{2}$ pounds.
Package E weighs $2\frac{3}{8}$ pounds.

(1) D, B, E, A, C
(2) A, C, D, B, E
(3) D, E, C, B, A
(4) C, A, E, D, B
(5) A, D, B, E, C

40. Simplify the expression (.4)² + (.01)².

(1) .81   (4) .15
(2) .161   (5) .009
(3) .1601

41. A grocery store donated 60 cans of tomato soup, 48 cans of vegetable soup, and 36 cans of chicken soup to the Community Day Care Center. The cans were unlabeled. What is the probability that the first can the cook opens will be tomato soup?

(1) $\frac{1}{60}$ (4) $\frac{5}{12}$
(2) $\frac{1}{4}$ (5) $\frac{7}{12}$
(3) $\frac{1}{3}$

42. Of the first 12 cans that the cook in problem 41 opened, three were tomato soup, three were vegetable soup, and six were chicken soup. What is the probability that the next can the cook opens will be chicken soup?

(1) $\frac{1}{30}$ (3) $\frac{5}{22}$ (5) $\frac{5}{6}$

(2) $\frac{1}{6}$ (4) $\frac{17}{22}$

43. Write an equation which expresses the following statement: 6 more than one-third of a number is four.

(1) $\frac{x}{3} - 6 = 4$ (4) $\frac{x}{3} + 6 = 4$

(2) $\frac{x}{6} + 3 = 4$ (5) $\frac{x}{6} + 4 = 6$

(3) $\frac{x}{3} + 4 = 6$

44. In the rectangle pictured below, the length is twice the width, and the area is 800. Find the width.

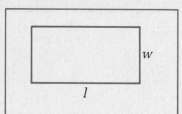

(1) 20   (3) 80   (5) 160
(2) 40   (4) 120

45. Solve for s in $7s + 1 = 3s + 3$.

(1) $\frac{1}{3}$ (4) 3

(2) $\frac{1}{2}$ (5) 4

(3) 2

46. In triangle XYZ, side $XY = 9$ inches, side $YZ = 12$ inches, and side $XZ = 15$ inches. Which angle of the triangle is largest?

(1) $\angle X$ (4) All three angles are the
(2) $\angle Y$ same.
(3) $\angle Z$ (5) Insufficient data is
                       given to solve the prob-
                       lem.

47. Two more than five times a number equals the same number increased by 30. Find the number.

(1) 5   (4) 8
(2) 6   (5) 10
(3) 7

48. Find the length of the missing side of the triangle pictured below.

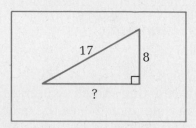

(1) 25   (4) 14
(2) 20   (5) 12
(3) 15

49. For the two triangles below, $DE = GH$ and $DF = GI$. Along with the information given, which of the following conditions is enough to guarantee that the triangles are congruent?

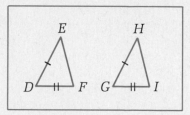

A. $EF = HI$
B. $\angle D = \angle E$
C. $\angle G = \angle I$

(1) A only   (4) both A and C
(2) B only   (5) both B and C
(3) C only

50. Solve for y in $y^2 + y - 12 = 0$.

(1) $+ 12$ and -1 (4) -12 and -1
(2) -4 and $+3$ (5) -6 and $+2$
(3) $+6$ and -2

51. What is the slope of the line that passes through points *A* and *B* on the graph at the right?

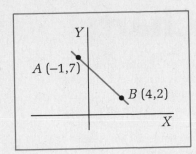

(1) $+1$ (4) $+\frac{1}{2}$
(2) -1 (5) -2
(3) $+2$

52. A machine can label 15 envelopes per minute. Which expression tells the number of hours the machine needs to label 3000 envelopes?

(1) $3000 \times 15 \times 60$ (4) $\dfrac{3000}{15 \times 60}$

(2) $\dfrac{15 \times 60}{3000}$ (5) $\dfrac{60 \times 3000}{15}$

(3) $15 + 60 + 3000$

53. For every dollar the Chung family spends in a month on car payments, they spend $2 on food, and $3 on mortgage payments. The Chungs spend a total of $696 a month for these three items. How much do they spend each month on mortgage payments?

(1) $174
(2) $232
(3) $261
(4) $348
(5) $464

Items 54 to 56 are based on the following graph.

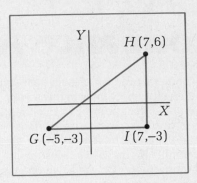

54. What is the distance between points *G* and *H*?

(1) 8 (3) 12 (5) 15
(2) 9 (4) 13

55. The coordinates of the midpoint M between two points in a plane is given by the expression

$$M = \left(\frac{x_1 + x_2}{2}, \frac{y_1 + y_2}{2} \right),$$

where (x_1, y_1) and (x_2, y_2) are two points in the plane. What are the coordinates of the midpoint between *G* and *H*?

(1) $(1, 2)$ (4) $(2\frac{1}{2}, 1)$

(2) $(2, 3)$ (5) $(1\frac{1}{2}, 1)$

(3) $(1, 1\frac{1}{2})$

56. What is the distance between points *H* and *I*?

(1) 12 (4) 3
(2) 9 (5) 2
(3) 7

Answers are on pages 366–368.

PRACTICE TEST
Performance Analysis Chart

Directions: Circle the number of each item you got correct on the Practice Test. Count how many items you got correct in each row; count how many items you got correct in each column. Write the amount correct per row and column as the numerator in the fraction in the appropriate "Total Correct" box. (The denominators represent the total number of items in the row or column.) Write the grand total correct over the denominator **56** at the lower right corner of the chart. (For example, if you got 50 items correct, write *50* so that the fraction reads *50*/**56**.) Item numbers in color represent items based on graphic material.

Item Type	Arithmetic (page 21)	Algebra (page 174)	Geometry (page 213)	TOTAL CORRECT
Skills		2, 20, 25	4, 7, 21, 46	/7
Applications	1, 3, 6, 23, 26, 27, 28, 29, 30, 32, 39, 40, 41, 42	5, 22, 24, 31, 35, 37, 38, 45, 50	36, 48, 49, 51, 54, 55, 56	/30
Problem Solving	8, 9, 10, 11, 12, 13, 14, 15, 16, 17, 18, 19, 34, 52	33, 43, 47, 53	44	/19
TOTAL CORRECT	/28	/16	/12	/56

The page numbers in parentheses indicate where in this book you can find the beginning of specific instruction about the various areas of mathematics you encountered on the Practice Test.

On the chart, items are classified as Skills, Applications, or Problem Solving. In the chapters in this book, the three problem types are covered in different levels:

Skills items are covered in Level 1.

Applications items are covered in Level 2.

Problem Solving items are covered in Level 3.

For example, the skills needed to solve Item 36 are covered in Level 2 of Chapter 6. To locate in which chapter arithmetic items are addressed, reread the problem to see what kind of numbers (whole numbers, fractions, etc.) are used and then go to the designated level of the appropriate chapter.

Simulation

Simulated Test

This test is as much like the real Mathematics Test as possible. The number of items and their degree of difficulty are the same as on the real test. The time limit and the mixed order of the test items are also the same. By taking the Simulated Test, you will gain valuable test-taking experience and get a better idea about how prepared you are to take the actual test.

Unlike the Practice activities, there is only one way you should take the Simulated Test. You should take the test under the same conditions you will have when you take the real test. You will have 90 minutes to complete the test. Set aside at least 90 minutes so you can work without interruption. Do not talk to anyone or consult any books as you take the test. If you have questions, ask the instructor. If you are not sure of an answer, make an educated guess. On the real GED you are not penalized for wrong answers.

As you take the Simulated Test, write your answers on a sheet of paper or use an answer sheet. When time is up, you may wish to circle the item that you answered last and then continue with the test. This way, when you score your test, you can see how much of a factor time was in your performance.

Whether you answered an item correctly or not, you should look over each problem solution in the answer section. This will reinforce your test-taking skills and your understanding of the material. If you get 45 items or more correct, you will have done 80% work or better. This shows that you should do well on the actual Mathematics Test. If you get less than 45 items correct, you probably need to do some light reviewing. If your score is far below the 80% mark, you should spend additional time reviewing lessons that will strengthen your weak areas. The Performance Analysis Chart at the end of the test will help you identify your stronger and weaker areas.

Simulated Test

TIME: *90 minutes*
Directions: *Choose the one best answer to each question.*

1. Simplify the expression $10^4 - 5^2$.

   (1)   15
   (2)   30
   (3)   975
   (4)   9,975
   (5)   10,025

2. Shirley works three days a week and earns $180 a week. How many weeks does she have to work in order to earn $4500?

   (1)   40
   (2)   25
   (3)   20
   (4)   15
   (5)   Insufficient data is given to solve the problem.

3. Which point on the line below corresponds to $\frac{-7}{2}$?

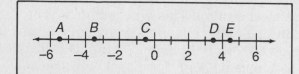

   (1)   *A*
   (2)   *B*
   (3)   *C*
   (4)   *D*
   (5)   *E*

4. The ratio of green paint to white paint in a mixture used on the walls of the neighborhood community center is 3:4. A dealer told the volunteers who are painting the center that they will need 56 gallons of paint altogether to finish the job. How many gallons of green paint will they need?

   (1) 12   (4) 20
   (2) 15   (5) 24
   (3) 18

5. Write an expression for the quantity of eight more than three times a number, all divided by six.

   (1) $\frac{3x + 8}{6}$ (4) $\frac{8x + 3}{6}$

   (2) $\frac{3}{6}x + 8$ (5) $\frac{8x - 3}{6}$

   (3) $\frac{6x + 8}{3}$

6. Find the sales tax on three $15 shirts if the tax rate is 6%.

   (1) $0.90
   (2) $1.35
   (3) $1.80
   (4) $2.70
   (5) Insufficient data is given to solve the problem.

7. Let *c* stand for the number of cases of cola Bill sells in a week in his store. The number of cases of orange drink he sells in a week is 20 less than half the number of cases of cola. Which expression tells the number of cases of orange drink he sells in a week?

(1) $\dfrac{c + 2}{20}$

(2) $\dfrac{c - 20}{2}$

(3) $\frac{1}{2}c - 20$

(4) $\dfrac{c + 20}{2}$

(5) $\frac{1}{2}c + 20$

8. Solve for *b* in $\frac{2}{3}b - 2 = 8$.

(1) 2
(2) 4
(3) 6
(4) 9
(5) 15

9. In triangle *MNO*, $\angle M = 61°$ and $\angle N = 52°$. Which side of the triangle is longest?

(1) *NO*
(2) *MO*
(3) *MN*
(4) All three sides are the same.
(5) Insufficient data is given to solve the problem.

10. Simplify the expression $\dfrac{15c^2d}{-25cd}$.

(1) $\dfrac{-5d}{3}$

(2) $\dfrac{3c^3d^2}{5}$

(3) $\dfrac{-3cd}{5}$

(4) $\dfrac{-3c}{5}$

(5) $\dfrac{3c^2d}{5}$

11. Which of the following expressions is equal to $(-5m^3)(-2m)$?

(1) $-10m^2$
(2) $+10m^4$
(3) $+7m^4$
(4) $-7m^2$
(5) $+3m^2$

12. Seven less than twice a number equals the same number increased by three. Find the number.

(1) 2
(2) 6
(3) 10
(4) 14
(5) 20

13. The rainfall in Capital City during June was 3.6 inches the first week, 2.45 inches the second week, 4.63 inches the third week, and 3.84 inches the fourth week. Find the mean weekly rainfall in inches for that period.

(1) 3.48
(2) 3.54
(3) 3.63
(4) 3.68
(5) 3.72

14. John received a shipment of jackets to sell in his store. There were 15 small jackets, 25 medium ones, and 20 large ones. The sizes were mixed together. What is the probability that the first jacket John took from the box was large?

(1) $\frac{1}{2}$

(2) $\frac{1}{3}$

(3) $\frac{1}{6}$

(4) $\frac{1}{20}$

(5) $\frac{1}{30}$

15. In fact, the first two jackets John took from the box in problem 14 were medium, and the next three were large. What is the probability that the sixth jacket he takes from the box will be small?

(1) $\frac{15}{16}$

(2) $\frac{1}{3}$

(3) $\frac{3}{11}$

(4) $\frac{1}{4}$

(5) $\frac{1}{15}$

16. Jeff makes $6 an hour for the first seven hours of the work day and then $9 an hour for each additional hour. Which of the following expressions shows the amount he makes in a ten-hour work day?

(1) $7(9) + 3(6)$
(2) $7(6) + 3(9)$
(3) $7 \times 9 \times 3 \times 6$
(4) 10×7
(5) 10×9

17. In an isosceles triangle, the vertex angle measures 55°. Find the measurement of each base angle.

(1) 55°
(2) 62.5°
(3) 70°
(4) 110°
(5) 140°

Items 18 and 19 are based on the following table.

Number of Newspapers and Periodicals (in thousands)					
	1965	1970	1975	1980	1985
Newspapers	11.4	11.4	11.4	9.6	9.1
Periodicals	9.0	9.6	9.7	10.2	11.1

Source: Statistical Abstract of the United States, 1986.

18. For which year shown in the table was the number of periodicals first more than the number of newspapers?

(1) 1965
(2) 1970
(3) 1975
(4) 1980
(5) 1985

19. Which of the following statements describes the trend shown in the table?

(1) While the number of newspapers gradually decreased, the number of periodicals increased.
(2) While the number of periodicals decreased, the number of newspapers increased.
(3) The number of newspapers and the number of periodicals both increased.
(4) The number of newspapers and the number of periodicals both decreased.
(5) The number of newspapers and the number of periodicals both remained the same.

20. Which of the following expressions equals 4×10^6?

(1) $4 \times 6 \times 10$
(2) 4×6000
(3) $4 \times 10 \times 10 \times 10 \times 10 \times 10 \times 10$
(4) $10 \times 4 \times 4 \times 4 \times 4 \times 4 \times 4$
(5) $4 \times 10 \times 10$

21. A machine produces 100 motor parts per hour and runs for eight hours a day. Which of the following expressions shows the number of days needed to produce 12,000 motor parts?

(1) $8 \times 100 \times 12{,}000$

(2) $\dfrac{12{,}000}{(8 \times 100)}$

(3) $\dfrac{(100 \times 12{,}000)}{8}$

(4) $\dfrac{(8 + 100)}{12{,}000}$

(5) $\dfrac{100}{(8 \times 12{,}000)}$

Items 22 to 25 are based on the following situation.

José wants to buy a video recorder. The model he likes is for sale at Sav-a-Lot for $389. Sales tax is 5%, and the delivery charge is $10. Sav-a-lot offers a time payment plan of $100 down and $16 a month for 24 months.

    The same model of video recorder is for sale at Al's Appliances in a nearby state for $439. Sales tax is 6%, and the delivery charge is $8. The time payment plan at Al's is $60 down and $36 a month for 12 months.

22. What is the total cost of the recorder at Sav-a-lot with the time payment plan?

(1) $384
(2) $444
(3) $484
(4) $524
(5) Insufficient data is given to solve the problem.

23. The total cost of the recorder at Al's with the time payment plan is how much more than Al's list price for the recorder?

(1) $ 53
(2) $ 60
(3) $ 70

(4) $100
(5) Insufficient data is given to solve the problem.

24. Find the price of the recorder, including an additional three-year guarantee at Al's.

(1) $449
(2) $459
(3) $476
(4) $496
(5) Insufficient data is given to solve the problem.

25. What is the total price of the recorder at Sav-a-Lot, including tax and delivery?

(1) $418.45
(2) $460.95
(3) $468.95
(4) $578.45
(5) Insufficient data is given to solve the problem.

26. Solve for n in $2n - 7 \geq 5$.

(1) $n \geq 1$
(2) $n \geq 2$
(3) $n \geq 4$
(4) $n \geq 6$
(5) $n \geq 12$

27. Arrange the following boxes in order from heaviest to lightest.

    Box A weighs 0.65 kilogram.
    Box B weighs 0.5 kilogram.
    Box C weighs 1.09 kilogram.
    Box D weighs 1.45 kilogram.
    Box E weighs 0.505 kilogram.

(1) D, C, A, E, B
(2) C, E, A, B, D
(3) A, E, B, D, C
(4) A, D, C, B, E
(5) D, E, B, C, A

28. Choose the equation which expresses the following statement: One-half the quantity of a number decreased by five is seven.

(1) $2x - 5 = 7$

(2) $\frac{x}{2} + 5 = 7$

(3) $\frac{x - 5}{2} = 7$

(4) $\frac{x + 5}{2} = 7$

(5) $\frac{x - 7}{2} = 5$

29. Geraldine drove for two hours at 35 mph and then for three hours at 55 mph. Which expression shows the total distance she drove?

(1) 5×90

(2) $2(55) + 3(35)$

(3) $2 \times 35 \times 3 \times 55$

(4) $2(35) + 3(55)$

(5) $(35 + 55) \times (3 + 2)$

30. Solve for z in $8z - 1 = 6z + 4$

(1) $\frac{1}{2}$ (4) $1\frac{1}{2}$

(2) 1 (5) $2\frac{1}{2}$

(3) 2

Items 31 to 34 are based on the following graph.

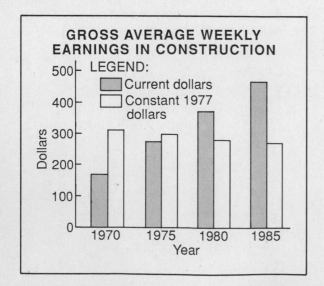

GROSS AVERAGE WEEKLY EARNINGS IN CONSTRUCTION

LEGEND:
- Current dollars
- Constant 1977 dollars

31. What were the approximate gross average weekly earnings in construction in current dollars in 1975?

(1) $195 (4) $370

(2) $270 (5) $460

(3) $300

32. What percent of current dollar gross average weekly earnings in construction in 1980 went to taxes and social security?

(1) 10%

(2) 12%

(3) 15%

(4) 20%

(5) Insufficient data is given to solve the problem.

33. Which of the following statements describes the change in *current* dollar earnings in construction from 1970 to 1985?

(1) Gross earnings were about half as much in 1985 as in 1970.

(2) Gross earnings steadily dropped from 1970 to 1985.

(3) Gross earnings more than doubled from 1970 to 1985.

(4) Gross earnings were about the same in 1985 as in 1970.

(5) Gross earnings were about one-quarter as much in 1985 as in 1970.

34. Which of the following statements describes the change in *constant* dollar earnings in construction from 1970 to 1985?

(1) 1985 earnings were less than 1970 earnings.

(2) 1985 earnings were three times 1970 earnings.

(3) 1985 earnings were the same as in 1970.

(4) 1985 earnings were twice as much as 1970 earnings.

(5) Insufficient data is given to solve the problem.

35. The figure shown below is a 6-centi-meter scale. What is the distance from *M* to *N*?

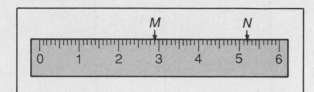

(1) 2.3 cm
(2) 2.9 cm
(3) 3.3 cm
(4) 3.7 cm
(5) 8.1 cm

36. The figure below shows the plan of a rectangular lot. Find the number of yards of fencing needed to enclose the lot.

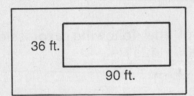

36 ft.
90 ft.

(1) 84
(2) 126
(3) 168
(4) 252
(5) 324

37. A packet of grass seed makes 300 square feet of grass. How many packets are needed to cover the lot in probblem 36 with grass?

(1) 3.2
(2) 10.8
(3) 12.4
(4) 21.6
(5) 32.4

38. Solve for *t* in $5(t - 3) = 2(t + 6)$.

(1) 2
(2) $6\frac{1}{2}$
(3) 9
(4) 13
(5) $13\frac{1}{2}$

39. In the diagram below, $k/\!/l$ and $\angle b = 106°$. Find $\angle g$.

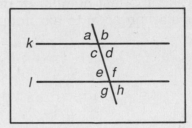

(1) 16°
(2) 26°
(3) 74°
(4) 84°
(5) 106°

40. Find, to the nearest tenth of a cubic meter, the volume of the cube shown at the right.

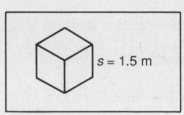

$s = 1.5$ m

(1) 2.4
(2) 3.4
(3) 4.5
(4) 6.0
(5) 33.6

41. The Nieves family has paid off $16,000 of their $40,000 mortgage. What is the ratio of the amount they have paid to the amount they still owe?

(1) 1:2
(2) 2:5
(3) 3:5
(4) 2:3
(5) 3:4

42. Frank drove 36 miles east, then 36 miles north, and finally 12 miles east. What is the shortest number of miles from his starting point to his stopping point?

(1) 36
(2) 40
(3) 50
(4) 56
(5) 60

43. Which of the following expressions equals $x^2 + 2x - 48$?

(1) $(x + 24)(x - 24)$
(2) $(x + 24)(x - 2)$
(3) $(x - 16)(x + 3)$
(4) $(x + 8)(x - 6)$
(5) $(x + 4)(x - 12)$

44. Find, to the nearest tenth of a square meter, the area of a circle with a radius of 6.5 meters.

(1) 40.8
(2) 81.6
(3) 132.7
(4) 265.4
(5) 408.2

45. In the figure below, $\angle POS = 90°$, $\angle POQ = 27°$, and $\angle QOR = 29°$. Find $\angle ROS$.

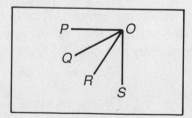

(1) 24°
(2) 34°
(3) 61°
(4) 72°
(5) 124°

46. Find the length of TS in the picture below.

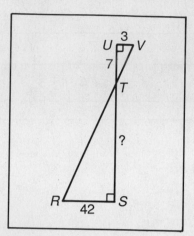

(1) 18
(2) 21
(3) 35
(4) 84
(5) 98

47. Which of the following expressions equals $a^2 - 12a$?

(1) $a(a - 12)$
(2) $a^2(-12a)$
(3) $a(a - 4)$
(4) $(a - 4)(a + 4)$
(5) $(a - 6)(a + 2)$

48. What is the slope of the line that passes through points S and T on the graph below?

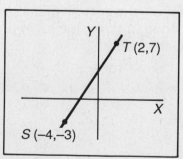

(1) $\frac{-3}{5}$
(2) $\frac{+5}{3}$
(3) -5
(4) $+3$
(5) $\frac{+3}{5}$

49. The triangle shown below has an area of 54 square inches. The height is one-third of the base. Find the measurement of the base in inches.

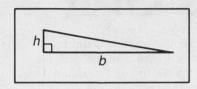

- **(1)** 4
- **(2)** 6
- **(3)** 9
- **(4)** 18
- **(5)** 27

50. For the triangles pictured below, *MN* = *PQ* and ∠*M* = ∠*P*. Along with the information given, which of the following conditions is enough to guarantee that the triangles are congruent?

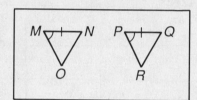

- A. ∠*N* = ∠*Q*
- B. *MO* = *PR*
- C. *MN* = *MO*

- **(1)** only A
- **(2)** only B
- **(3)** only C
- **(4)** both A and B
- **(5)** both B and C

51. Choose the equation which expresses the following statement: Four times the quantity of a number decreased by one is equal to twice the quantity of the same number increased by nine.

- **(1)** $4(x - 2) = x + 9$
- **(2)** $4(x + 1) = 2(x - 9)$
- **(3)** $4(x - 1) = 2(x + 9)$
- **(4)** $4x + 1 = 2x - 9$
- **(5)** $4x - 1 = 2x + 9$

52. What are the coordinates of the *y*-intercept of the graph of the equation $y = +3x - 4$?

- **(1)** (0, 3)
- **(2)** (0, −3)
- **(3)** (−3, 0)
- **(4)** (−4, 0)
- **(5)** (0, −4)

53. Solve for *c* in $c^2 + 3c - 18 = 0$.

- **(1)** +6 and −3
- **(2)** −6 and +3
- **(3)** +9 and −2
- **(4)** +18 and −1
- **(5)** −18 and +1

Items 54 and 55 are based on the following figure.

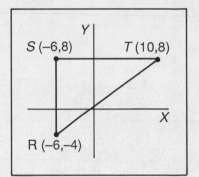

54. What is the distance from *S* to *T*?

- **(1)** 6
- **(2)** 8
- **(3)** 10
- **(4)** 16
- **(5)** 18

55. What is the distance from *R* to *T*?

- **(1)** 10
- **(2)** 12
- **(3)** 14
- **(4)** 18
- **(5)** 20

56. Fred is an electrician. His assistant Gordon makes $5 an hour less than Fred. Together they made $760 on a job they each worked on for 40 hours. Find both their hourly rates.

- **(1)** Fred gets $12, and Gordon gets $7.
- **(2)** Fred gets $13, and Gordon gets $8.
- **(3)** Fred gets $14, and Gordon gets $9.
- **(4)** Fred gets $15, and Gordon gets $10.
- **(5)** Fred gets $16, and Gordon gets $11.

Answers are on page 369.

SIMULATED TEST
Performance Analysis Chart

Directions: Circle the number of each item you got correct on the Simulated Test. Count how many items you got correct in each row; count how many items you got correct in each column. Write the amount correct per row and column as the numerator in the fraction in the appropriate "Total Correct" box. (The denominators represent the total number of items in the row or column.) Write the grand total correct over the denominator **56** at the lower right corner of the chart. (For example, if you got 50 items correct, write 50 so that the fraction reads 50/**56**.) Item numbers in color represent items based on graphic material.

Item Type	Arithmetic (page 21)	Algebra (page 174)	Geometry (page 213)	TOTAL CORRECT
Skills		3, 10, 11	9, 17, 39, 45	/7
Applications	1, 6, 13, 14, 15, 20, 27, 31, 32, 33, 34, 35, 40, 41, 44	5, 7, 8, 26, 30, 38, 43, 47, 53	42, 46, 48, 50, 52, 54, 55	/31
Problem Solving	2, 4, 16, 18, 19, 21, 22, 23, 24, 25, 29, 36, 37	12, 28, 51, 56	49	/18
TOTAL CORRECT	/28	/16	/12	/56

The page numbers in parentheses indicate where in this book you can find the beginning of specific instruction about the various areas of mathematics you encountered on the Simulated Test.

On the chart, items are classified as Skills, Applications, or Problem Solving. In the chapters in this book, the three problem types are covered in different levels:

Skills items are covered in Level 1.

Applications items are covered in Level 2.

Problem Solving items are covered in Level 3.

For example, the skills needed to solve Item 42 are covered in Level 2 of Chapter 6. To locate in which chapter arithmetic items are addressed, reread the problem to see what kind of numbers (whole numbers, fractions, etc.) are used and then go to the designated level of the appropriate chapter.

Answers
and Solutions

Answers and Solutions

In the Answers and Solutions section, you will find answers to all the questions in these sections of the book:

- Previews
- Lesson Exercises
- Level Reviews
- Chapter Quizzes
- The Unit Test
- The Practice Items
- The Practice Test
- The Simulated Test

You will discover that the Answers and Solutions section is a valuable study tool. It not only tells you the correct answer, it shows how each problem is solved. It also points out the type of math that is required to solve each problem successfully.

Even if you get a question right, it will help to review the solution. The solution will reinforce your understanding of the problem and the method for its solution. Because you might have guessed a correct answer or answered correctly for the wrong reason, it can't hurt to review solutions. It may help a lot.

INSTRUCTION

Chapter 1   Whole Numbers

Level 1   Whole Number Skills

Preview

1.
```
  9000
-  496
  8504
```

2.
```
   473
 ×  90
 42,570
```

3.
```
      509
15)7635
   75
   13
    0
   135
   135
```

4.
```
    49
   207
  5653
+   28
  5937
```

5.
```
     470 r 9
21)9879
   84
   147
   147
    09
     0
     9
```

Lesson 1

1. c.
```
  9015
   493
+   76
  9584
```

2. b.
```
  22,500
-  6,087
  16,413
```

3.
```
  5030
-  763
  4267
```

4.
```
   9,704
      86
 +10,471
  20,261
```

5.
```
    78
  4062
 + 529
  4669
```

6.
```
  42,003
-  8,346
  33,657
```

7.
```
  18,206
 -11,954
   6,252
```

8.
```
   428
    61
   593
 +   7
  1089
```

9.
```
  30,005
 -19,472
  10,533
```

10.
```
   365
 +7048
  7413
```

Lesson 2

1. c.
```
     803
12)9636
   96
   03
    0
   36
   36
```

2. a.
```
   704
 ×  18
 5 632
 7 04
 12,672
```

3.
```
    536
 ×800
 428,800
```

4.
```
     907
8)7256
  72
  056
   56
```

5.
```
      78
46)3588
   322
   368
   368
```

6.
```
      39
78)3042
   234
   702
   702
```

7.
```
   230
 × 34
   920
   690
  7820
```

8.
```
   40,570
 ×     19
  365 130
  405 70
  770,830
```

9.
```
      14
316)4424
    316
   1264
   1264
```

10.
```
       4 r 300
523)2392
    2092
     300
```

Level 1   Review

1.
```
   890
    23
  4017
 + 605
  5535
```

2.
```
       920 r 3
12)11,043
   10 8
   24
   24
   03
   00
    3
```

3.
```
  12,050
 - 9,947
   2,103
```

4.
```
    308
 ×  76
  1 848
 21 56
 23,408
```

5.
```
      3,096
8)24,768
  24
   0 76
   72
   48
   48
```

6.
```
  108,270
   33,580
 + 6,095
  147,945
```

7.
```
    675
 ×208
  5 400
 135 00
 140,400
```

8.
```
  306,471
 - 28,295
  278,176
```

9.
```
  508,000
 - 29,460
  478,540
```

10.
```
      804
32)25,728
   25 6
    12
     0
    128
    128
```

Level 2   Whole Number Applications

Preview

1. **500,000**

2. $d = rt$
   $d = 435 \times 7$
   $= 3045$ mi.

3. $A = \frac{1}{2}bh$
   $A = \frac{(40 \times 24)}{2}$
   $A = \frac{960}{2} = 480$

4. In order: $1950 $2155
   $3080 $6075
   $8470
   The median is **$3080.**

5. (3) $(7 \times 20) - (7 \times 1)$

Lesson 1

1. **80   160   3200   2430**
2. **800   1300   6600   400**
3. **3000   42,000   29,000   150,000**
4. **800,000   300,000   600,000   3,500,000**
5. **6,000,000   12,000,000   32,000,000   19,000,000**

Lesson 2

1. $d = rt$
   $d = 4 \times 3 = $ **12 mi.**

2. $d = rt$
   $d = 65 \times 4 = $ **260 mi.**

3. $d = rt$
   $d = 475 \times 5$
   $= $ **2375 mi.**

4. $d = rt$
   $d = 15 \times 3 = $ **45 mi.**

5. $c = nr$
   $c = 3 \times 18 = $ **$54**

6. $c = nr$
   $c = 5 \times 3.60 = $ **$18.00**

7. $c = nr$
   $c = 12 \times 6 = $ **$72**

8. $c = nr$
   $c = 30 \times 65 = $ **$1950**

9. $d = rt$
   $d = 55 \times 6 = $ **330 mi.**

10. $c = nr$
    $c = 6 \times 1.51 = $ **$9.06**

Lesson 3

1. $2^4 = 2 \times 2 \times 2 \times 2 = $ **16**
   $3^3 = 3 \times 3 \times 3 = $ **27**
   $9^2 = 9 \times 9 = $ **81**
   $8^1 = $ **8**

2. $13^2 = 13 \times 13 = $ **169**
   $6^3 = 6 \times 6 \times 6 = $ **216**
   $50^2 = 50 \times 50 = $ **2500**
   $12^0 = $ **1**

3. $2^5 = 2 \times 2 \times 2 \times 2 \times 2 = $ **32**
   $16^2 = 16 \times 16 = $ **256**
   $1^5 = $ **1**
   $40^2 = 40 \times 40 = $ **1600**

4. $5^2 - 2^3 = $
   $5 \times 5 - 2 \times 2 \times 2 = $
   $25 - 8 = $ **17**

   $8^2 + 3^3 = $
   $8 \times 8 + 3 \times 3 \times 3 = $
   $64 + 27 = $ **91**

   $10^2 - 4^2 + 5^2 = $
   $10 \times 10 - 4 \times 4 + 5 \times 5 = $
   $100 - 16 + 25 = $ **109**

5. $4^3 + 6^1 - 2^4 = $
   $4 \times 4 \times 4 + 6 - 2 \times 2 \times 2 \times 2 = $
   $64 + 6 - 16 = $ **54**

   $12^2 - 5^0 - 3^2 = $
   $12 \times 12 - 1 - 3 \times 3 = $
   $144 - 1 - 9 = $ **134**

   $10^3 - 10^2 = $
   $10 \times 10 \times 10 - 10 \times 10 = $
   $1000 - 100 = $ **900**

Lesson 4

1. $\sqrt{289} = 17$
   Guess 20.

   ```
       14         14        17
   20)289       +20       2)34
       20         34
       89
       80
   ```

   $\sqrt{784} = 28$
   Guess 30.

   ```
       26         26        28
   30)784       +30       2)56
       60         56
      184
      180
   ```

   $\sqrt{1444} = 38$
   Guess 40.

   ```
       36         36        38
   40)1444      +40       2)76
      120         76
      244
      240
   ```

   $\sqrt{484} = 22$
   Guess 20.

   ```
       24         24        22
   20)484       +20       2)44
       40         44
       84
       80
   ```

2. $\sqrt{1521} = 39$
   Guess 40.

   ```
       38         38        39
   40)1521      +40       2)78
      120         78
      321
      320
   ```

   $\sqrt{1849} = 43$
   Guess 40.

   ```
       46         46        43
   40)1849      +40       2)86
      160         86
      249
      240
   ```

   $\sqrt{529} = 23$
   Guess 20.

   ```
       26         26        23
   20)529       +20       2)46
       40         46
      129
      120
   ```

   $\sqrt{2704} = 52$
   Guess 50.

   ```
       54         54        52
   50)2704      +50       2)104
      250        104
      204
      200
   ```

3. $\sqrt{2025} = 45$
   Guess 50.

   ```
       40         40        45
   50)2025      +50       2)90
      200         90
       25
        0
   ```

   $\sqrt{961} = 31$
   Guess 30.

   ```
       32         32        31
   30)961       +30       2)62
       90         62
       61
       60
   ```

   $\sqrt{4761} = 69$
   Guess 70.

   ```
       68         68        69
   70)4761      +70       2)138
      420        138
      561
      560
   ```

   $\sqrt{3481} = 59$
   Guess 60.

   ```
       58         58        59
   60)3481      +60       2)118
      300        118
      481
      480
   ```

4. $\sqrt{3844} = 62$

Guess 60.

$$
\begin{array}{r}
64 \\
60\overline{)3844} \\
360 \\
\hline
244 \\
240 \\
\end{array}
\quad
\begin{array}{r}
64 \\
+60 \\
\hline
124 \\
\end{array}
\quad
\begin{array}{r}
62 \\
2\overline{)124} \\
\end{array}
$$

$\sqrt{8649} = 93$

Guess 90.

$$
\begin{array}{r}
96 \\
90\overline{)8649} \\
810 \\
\hline
549 \\
540 \\
\end{array}
\quad
\begin{array}{r}
96 \\
+90 \\
\hline
186 \\
\end{array}
\quad
\begin{array}{r}
93 \\
2\overline{)186} \\
\end{array}
$$

$\sqrt{5929} = 77$

Guess 80.

$$
\begin{array}{r}
74 \\
80\overline{)5929} \\
560 \\
\hline
329 \\
320 \\
\end{array}
\quad
\begin{array}{r}
74 \\
+80 \\
\hline
154 \\
\end{array}
\quad
\begin{array}{r}
77 \\
2\overline{)154} \\
\end{array}
$$

$\sqrt{2401} = 49$

Guess 50.

$$
\begin{array}{r}
48 \\
50\overline{)2401} \\
200 \\
\hline
401 \\
400 \\
\end{array}
\quad
\begin{array}{r}
48 \\
+50 \\
\hline
98 \\
\end{array}
\quad
\begin{array}{r}
49 \\
2\overline{)98} \\
\end{array}
$$

5. $\sqrt{4096} = 64$

Guess 60.

$$
\begin{array}{r}
68 \\
60\overline{)4096} \\
360 \\
\hline
496 \\
480 \\
\end{array}
\quad
\begin{array}{r}
68 \\
+60 \\
\hline
128 \\
\end{array}
\quad
\begin{array}{r}
64 \\
2\overline{)128} \\
\end{array}
$$

$\sqrt{8836} = 94$

Guess 90.

$$
\begin{array}{r}
98 \\
90\overline{)8836} \\
810 \\
\hline
736 \\
720 \\
\end{array}
\quad
\begin{array}{r}
98 \\
+90 \\
\hline
188 \\
\end{array}
\quad
\begin{array}{r}
94 \\
2\overline{)188} \\
\end{array}
$$

$\sqrt{6084} = 78$

Guess 80.

$$
\begin{array}{r}
76 \\
80\overline{)6084} \\
560 \\
\hline
484 \\
480 \\
\end{array}
\quad
\begin{array}{r}
76 \\
+80 \\
\hline
156 \\
\end{array}
\quad
\begin{array}{r}
78 \\
2\overline{)156} \\
\end{array}
$$

$\sqrt{7056} = 84$

Guess 80.

$$
\begin{array}{r}
88 \\
80\overline{)7056} \\
640 \\
\hline
656 \\
640 \\
\end{array}
\quad
\begin{array}{r}
88 \\
+80 \\
\hline
168 \\
\end{array}
\quad
\begin{array}{r}
84 \\
2\overline{)168} \\
\end{array}
$$

Lesson 5

1. $P = 4s$
$P = 4 \times 11 = $ **44 ft.**

2. $P = 2l + 2w$
$P = 2 \times 9 + 2 \times 7$
$P = 18 + 14 = $ **32 m**

3. $P = 2l + 2w$
$P = 2 \times 26 + 2 \times 13$
$P = 52 + 26 = $ **78 ft.**

4. $P = 4s$
$P = 4 \times 20 = $ **80 in.**

5. $P = a + b + c$
$P = 14 + 14 + 14$
$= $ **42 in.**

6. $P = 4s$
$P = 4 \times 18 = $ **72 yd.**

7. $P = 2l + 2w$
$P = 2 \times 10 + 2 \times 2$
$P = 20 + 4 = $ **24**

8. $P = a + b + c$
$P = 12 + 16 + 20$
$= $ **48**

9. $P = a + b + c$
$P = 7 + 8 + 9 = $ **24 cm**

10. $P = 2l + 2w$
$P = 2 \times 21 + 2 \times 15$
$P = 42 + 30 = $ **72 in.**

Lesson 6

1. $A = lw$
$A = 16 \times 9 = $ **144 in.²**

2. $A = s^2$
$A = 8^2 = $ **64 ft.²**

3. $A = \frac{1}{2}bh$

$A = \frac{(10 \times 8)}{2}$

$A = \frac{80}{2} = $ **40 m²**

4. $A = bh$
$A = 20 \times 14 = $ **280**

5. $A = s^2$
$A = 15^2 = $ **225 in.²**

6. $A = \frac{1}{2}bh$

$A = \frac{(30 \times 18)}{2}$

$A = \frac{540}{2} = $ **270 ft.²**

7. $A = lw$
$A = 13 \times 3 = $ **39**

8. $A = bh$
$A = 14 \times 8$
$A = $ **112 yd.²**

9. $A = \frac{1}{2}bh$

$A = \frac{(12 \times 22)}{2}$

$A = \frac{264}{2} = $ **132 ft.²**

10. $A = s^2$
$A = 16^2 = $ **256 m²**

Lesson 7

1. $V = s^3$
$V = 6^3$
$V = 6 \times 6 \times 6$
$= $ **216 ft.³**

2. $V = lwh$
$V = 12 \times 8 \times 5$
$V = $ **480 in.³**

3. $V = lwh$
$V = 20 \times 5 \times 6$
$V = $ **600 ft.³**

4. $V = s^3$
$V = 8^3$
$V = 8 \times 8 \times 8$
$= $ **512 cm³**

5. $V = s^3$
$V = 1^3$
$V = 1 \times 1 \times 1 = $ **1 yd.³**

6. $V = lwh$
$V = 100 \times 24 \times 8$
$V = $ **19,200 ft.³**

7. $V = lwh$
$V = 50 \times 4 \times 1$
$V = $ **200 ft.³**

8. $V = s^3$
$V = 3^3$
$V = 3 \times 3 \times 3$
$= $ **27 ft.³**

9. $V = lwh$
$V = 18 \times 11 \times 30$
$V = $ **5940 yd.³**

10. $V = s^3$
$V = 27^3$
$V = $ **19,683 cm³**

Lesson 8

1. **(3) 6 × 9**
2. **(2) (3 × 12) × 8**
3. **(5) 15 + (4 + 10)**
4. **(1) (7 × 8) − (7 × 1)**
5. **(4) 30 + 20**
6. **(3) 2(19 + 7)**
7. **(5) 6 + (2 + 12)**
8. **(3) (20 × 12) + (20 × 3)**
9. **(4) 50(3 + 4)**
10. **(1) 5(24 × .70)**

Lesson 9

1.
$$
\begin{array}{r}
\$14 \\
22 \\
+\ 18 \\
\hline
\$54 \\
\end{array}
\qquad
\begin{array}{r}
\mathbf{\$18} \\
3\overline{)\$54} \\
\end{array}
$$

2. In order: $14 $18
           $22
$18 is the median.

3.
$$
\begin{array}{r}
185 \\
138 \\
97 \\
+\ 88 \\
\hline
508 \\
\end{array}
\qquad
\begin{array}{r}
\mathbf{127\ lb.} \\
4\overline{)508} \\
\end{array}
$$

4.
$$
\begin{array}{r}
65 \\
86 \\
79 \\
92 \\
+88 \\
\hline
410 \\
\end{array}
\qquad
\begin{array}{r}
\mathbf{82} \\
5\overline{)410} \\
\end{array}
$$

5.
$$
\begin{array}{r}
284 \\
191 \\
297 \\
162 \\
+256 \\
\hline
1190 \\
\end{array}
\qquad
\begin{array}{r}
\mathbf{238\ mi.} \\
5\overline{)1190} \\
\end{array}
$$

6.
$$
\begin{array}{r}
11 \\
9 \\
+13 \\
\hline
33 \\
\end{array}
\qquad
\begin{array}{r}
\mathbf{11\ gal.} \\
3\overline{)33} \\
\end{array}
$$

7. 213   **249** people   8. In order: 191 213 289
    191   4)996   303
    289
  + 303   Find the mean of 213
  ─────   and 289.
    996
                                          213   **251** people
                                        + 289   2)502
                                        ─────
                                          502

9. In order: $1.16 $1.25   10. $24,800   **$14,375**
              $1.29 $1.39   22,500   4)$57,500
              $1.49   6,000
    The median is **$1.29**   + 4,200
                                       ───────
                                        $57,500

Lesson 10

1. **New York**

2. $1504
   − 856
   ──────
   $ 648

3. The two middle values   4. $1221
   are $1068 and $1440.   −   898
   Find the mean of these   ──────
   two values.   **$ 323**

    $1068   **$1254**
  + 1440   2)$2508
  ──────   2
    $2508   ──
               05
                4
               ──
               10
               10
               ──
               08
                8

5. $1440   6. $1164   **$933**
   − 1221   1137   6)5598
   ──────   694   54
   $ 219   800   ──
                                        705   19
                                     + 1098   18
                                      ──────   ──
                                       $5598   18
                                                    18
                                                    ──

7. $1,221   8. $856
   ×   300,000   − 694
   ───────────   ──────
   $366,300,000 to the   **$162**
   nearest $10 million =
   $370,000,000

9. $902   10. $1164
   ×   15,000   − 694
   ──────────   ──────
   4 510 000   **$ 470**
   9 02
   ──────
   $13,530,000 to the
   nearest hundred
   thousand = **$13,500,000**

Level 2   Review

1. **2,400,000**

2. $c = nr$
   $c = 12 \times 1.20$
   $= $ **$14.40**

3. $15^2 - 10^2 + 25^1 =$
   $(15 \times 15) - (10 \times 10)$
   $+ 25 =$
   $225 - 100 + 25 =$ **150**

4. $\sqrt{3481} =$ **59**

   Guess 60.

         58   58   59
      60)3481   + 60   2)118
         300   ────   118
         ───   118
         481
         480

5. $P = a + b + c$
   $P = 8 + 11 + 14$
   $= $ **33 m**

6. $A = lw$
   $A = 36 \times 20$
   $= $ **720 yd.²**

7. $V = lwh$
   $V = 25 \times 12 \times 8$
   $V = $ **2400 ft.³**

8. 12   **23 lb.**
    36   4)92
    19
  + 25
  ────
    92

9. In order: 12 19 25 36

   Find the mean of 19 and
   25.

    19   **22 lb.**
  + 25   2)44
  ────
    44

10. **(2) 9(15 + 20)**

Level 3   Whole Number Problem Solving

Preview

1.   **$13,500**   2.   **19 mi.**
    5)$67,500   23)437
      5   23
     ──   ───
     17   207
     15   207
     ──   ───
      2 5
      2 5

3. 4 yr. = 4 × 12 = 48 mo.
         48 × $350 = $16,800
   2 yr. = 2 × 12 = 24 mo.
         24 × $250 = $6,000

   Total: $16,800
             5,000
          + 6,000
          ────────
            $27,800

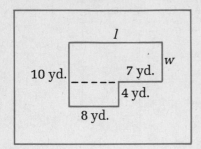

4. $l = 8 + 7 = 15$ yd.
   $w = 10 - 4 = 6$ yd.
   Area of top: $A = lw =$
   $15 \times 6 = 90$ yd.^{2}
   Area of bottom: $A = lw$
   $= 8 \times 4 = 32$ yd.^{2}
   Sum: $90 + 32 = 122$ yd.^{2}

```
   122
 ×$25
   610
   244
 $3050
```

5. (1) $\dfrac{(24 + 8 + 31 + 13)}{4}$

Lesson 1

1.
```
 $16,593,650
 -  1,108,212
  $15,485,438
```

2.
```
      $ 6,984
  4)$27,936
      24
       3 9
       3 6
        33
        32
        16
        16
```

3.
```
  $16,456
   11,294
 +  3,367
  $31,117
```

4.
```
    $1850
 ×   230
   55 500
   370 0
  $425,500
```

5.
```
   14,273
 +  8,498
  22,771 people
```

6.
```
      17 in.
  6)102
      6
      42
      42
```

7. $V = lwh$
   $V = 80 \times 20 \times 6$
   $V = $ **9600 ft.^{3}**

8.
```
   1987
 -1859
    128 yr.
```

9. $P = 4s$
   $P = 4 \times 6 = $ **24 mi.**

10.
```
    $462
     436
     194
     323
     245
   $1660
```

Lesson 2

1.
```
    $52        $780
  × 15        - 650
   260        $130
    52
   780
```

2. $3 \times 12 = \quad 36$
   $4 \times 12 = \quad 48$
   $2 \times 12 = +24$
   $\qquad\qquad$ **108 cans**

3.
```
   $235        $475
 +  240      ×    4
   $475       $1900
```

4.
```
    $290       $1300
 +  400        - 690
    $690       $ 610
```

5.
```
    $25     $12      $ 8     $500
  × 20    ×   8    ×  6       96
   $500     $96     $48    +  48
                            $644
```

6.
```
    $6.50      $9.75      $227.50
  ×   35     ×    8     +  78.00
    32 50     $78.00      $305.50
   195 0
   $227.50
```

7.
```
    $260       $290      $3120        $ 270
  ×  12      ×   6     + 1740    18)$4860
    520       $1740      $4860         36
    260                               126
   $3120                              126
```

8. $d = rt = 55 \times 4 = \quad 220$ mi.
   $d = rt = 30 \times 2 = +60$
   $\qquad\qquad\qquad$ **280 mi.**

9.
```
   $60,000      $ 35,000
 +  45,000   3)$105,000
   $105,000
```

10.
```
    $1500        $1150
  -   350      ×   12
    $1150       2 300
               11 50
              $13,800
```

Lesson 3

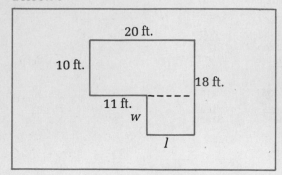

1. $l = 20 - 11 = 9$ Area of top rectangle:
   $w = 18 - 10 = 8$ $A = lw = 20 \times 10$
                 $= 200 \text{ ft.}^2$
        Area of bottom rectangle:
          $A = lw = 9 \times 8 = 72 \text{ ft.}^2$
        Sum: $200 + 72 = \textbf{272 ft.}^2$

2. $P = 20 + 18 + 9 + 8 + 11 + 10 = \textbf{76 ft.}$

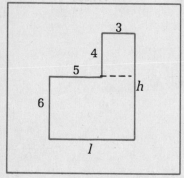

3. $l = 5 + 3 = 8$ Area of top rectangle:
   $h = 6 + 4 = 10$ $A = lw = 4 \times 3 = 12 \text{ in.}^2$
        Area of bottom rectangle:
          $A = lw = 8 \times 6 = 48 \text{ in.}^2$
        Sum: $12 + 48 = \textbf{60 in.}^2$

4. $P = 6 + 5 + 4 + 3 + 10 + 8 = \textbf{36 in.}$

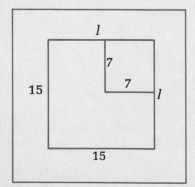

5. $l = 15 - 7 = 8$ Area of large square:
   $l = 15 - 7 = 8$ $A = s^2 = 15^2 = 225$
        Area of small square:
          $A = s^2 = 7^2 = 49$
        Difference: $225 - 49 = \textbf{176}$

6. $P = 8 + 7 + 7 + 8 + 15 + 15 = \textbf{60}$

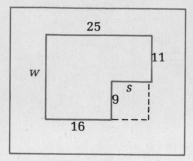

7. $s = 25 - 16 = 9$ Area of large rectangle:
   $w = 11 + 9 = 20$ $A = lw = 25 \times 20 = 500$
        Area of small square:
          $A = s^2 = 9^2 = 81$
        Difference: $500 - 81 = \textbf{419}$

8. $P = 25 + 11 + 9 + 9 + 16 + 20 = \textbf{90}$

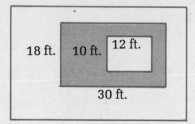

9. Area of large rectangle:
        $A = lw = 30 \times 18 = 540 \text{ ft.}^2$
      Area of small rectangle:
        $A = lw = 12 \times 10 = 120 \text{ ft.}^2$
      Difference: $540 - 120 = \textbf{420 ft.}^2$

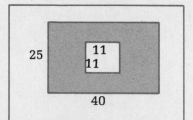

10. Area of large rectangle:
        $A = lw = 40 \times 25 = 1000$
      Area of small square:
        $A = s^2 = 11^2 = 121$
      Difference: $1000 - 121 = \textbf{879}$

Lesson 4

1. **(1) about $300 million**

$$\begin{array}{r} \$306{,}783{,}000 \\ -\ \ 15{,}487{,}000 \\ \hline \$291{,}296{,}000 \end{array}$$

2. **(3) a little less than $20,000**

$$\begin{array}{r} \$19{,}658 \\ 3)\overline{\$58{,}974} \\ \underline{3} \\ 28 \\ \underline{27} \\ 1\ 9 \\ \underline{1\ 8} \\ 17 \\ \underline{15} \\ 24 \\ \underline{24} \end{array}$$

3. **(3) nearly $20,000**

$$\begin{array}{r} \$23{,}496 \\ -\ \ 3{,}984 \\ \hline \$19{,}512 \end{array}$$

4. **(5) a little less than $60,000**

$$\begin{array}{r} \$29{,}260 \\ 18{,}420 \\ +\ \ 9{,}455 \\ \hline \$57{,}135 \end{array}$$

5. **(3) a little more than $25**

$$\begin{array}{r} \$\ 28 \\ 4)\overline{\$112} \\ \underline{8} \\ 32 \\ \underline{32} \end{array}$$

6. **(1) about 30 inches**

$$\begin{array}{r} 29\ \text{in.} \\ 3)\overline{87} \\ \underline{6} \\ 27 \\ \underline{27} \end{array}$$

7. **(3) around $60,000**

$$\begin{array}{r} \$\ 2{,}160 \\ \times\ \ \ \ \ 30 \\ \hline \$64{,}800 \end{array}$$

8. **(3) about 1400**

$$\begin{array}{r} 328 \\ 217 \\ 421 \\ 186 \\ +313 \\ \hline 1465 \end{array}$$

9. **(3) between $60 and $70**

$$\begin{array}{r} \$\ 65.50 \\ 5)\overline{\$327.50} \\ \underline{30} \\ 27 \\ \underline{25} \\ 25 \\ \underline{25} \\ 0 \end{array}$$

10. **(4) between 1200 and 1400**

$A = lw = 62 \times 21 = 1302\ \text{ft.}^2$

Lesson 5

1.
$$\begin{array}{r} \$950 \\ +\ 550 \\ \hline \$1500 \end{array} \qquad \begin{array}{r} \$1500 \\ \times\ \ \ \ 12 \\ \hline 3\ 000 \\ 15\ 00 \\ \hline \$18{,}000 \end{array}$$

2.
$$\begin{array}{r} \$950 \\ \times\ \ \ \ 6 \\ \hline \$5700 \end{array} \qquad \begin{array}{r} \$1500 \\ \times\ \ \ \ 6 \\ \hline \$9000 \end{array} \qquad \begin{array}{r} \$5{,}700 \\ +\ 9{,}000 \\ \hline \$14{,}700 \end{array}$$

3.
$$\begin{array}{r} \$\ 1{,}225 \\ 12)\overline{\$14{,}700} \\ \underline{12} \\ 27 \\ \underline{24} \\ 30 \\ \underline{24} \\ 60 \\ \underline{60} \end{array}$$

4.
$$\begin{array}{r} \$1500 \\ +\ 250 \\ \hline \$1750 \end{array} \qquad \begin{array}{r} \$1{,}750 \\ \times\ \ \ \ 12 \\ \hline 3\ 500 \\ 17\ 50 \\ \hline \$21{,}000 \end{array}$$

5.
$$\begin{array}{r} 8 \\ 35 \\ +10 \\ \hline \textbf{53 employees} \end{array}$$

6.
$$\begin{array}{r} \$\ 216{,}000 \\ 756{,}000 \\ +\ 144{,}000 \\ \hline \$1{,}116{,}000 \end{array}$$

7.
$$\begin{array}{r} \$\ \ \ 93{,}000 \\ 12)\overline{\$1{,}116{,}000} \\ \underline{1\ 08} \\ 36 \\ \underline{36} \end{array}$$

8.
$$\begin{array}{r} \$\ \ 21{,}600 \\ 35)\overline{\$756{,}000} \\ \underline{70} \\ 56 \\ \underline{35} \\ 21\ 0 \\ \underline{21\ 0} \end{array}$$

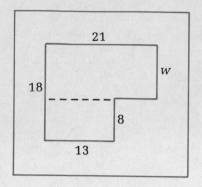

9.
$$\begin{array}{r} \$\ \ 14{,}400 \\ 10)\overline{\$144{,}000} \end{array}$$
$$\begin{array}{r} \$\ \ 1{,}200 \\ 12)\overline{\$14{,}400} \\ \underline{12} \\ 24 \\ \underline{24} \end{array}$$

10.
$$\begin{array}{r} \$21{,}600 \\ -\ \ 14{,}400 \\ \hline \$\ 7{,}200 \end{array}$$

Lesson 6

1. (4) $(2 \times 20) + (3 \times 60)$

2. (2) $\dfrac{(80 + 95 + 74)}{3}$

3. (1) $(4 \times 7.99) + 1.92$

4. (3) $(8 \times 350) + (6 \times 425)$

5. (3) $12(1800 - 360)$

6. (1) $\dfrac{(2500 + 4850 + 4200)}{5}$

7. (2) $12(20 - 2)$

8. (4) $(2 \times 100) + (2 \times 30)$

9. (2) $3(35 + 9)$

10. (3) $\dfrac{(3 \times 2 \times 4)}{2}$

Level 3   Review

1.
$$\begin{array}{r} 16\ \text{lb.} \\ 48)\overline{768} \\ \underline{48} \\ 288 \\ \underline{288} \end{array}$$

2.
$$\begin{array}{r} \$315 \\ +\ 135 \\ \hline \$450 \end{array}$$
$$\begin{array}{r} \$450 \\ \times\ \ \ 12 \\ \hline 900 \\ 450 \\ \hline \$5400 \end{array}$$
$$\begin{array}{r} \$12{,}360 \\ -\ \ 5{,}400 \\ \hline \$\ 6{,}960 \end{array}$$

3.
$$\begin{array}{r} 760 \\ \times\ \ \ 6 \\ \hline 4560 \end{array}$$
$$\begin{array}{r} 1250 \\ \times\ \ \ 3 \\ \hline 3750 \end{array}$$
$$\begin{array}{r} 410 \\ \times\ \ \ 3 \\ \hline 1230 \end{array}$$
$$\begin{array}{r} 4560 \\ 3750 \\ +1230 \\ \hline 9540 \end{array}$$

4. $w = 18 - 8 = 10$

Area of top rectangle:
$A = lw = 21 \times 10 = 210$
Area of bottom rectangle:
$A = lw = 13 \times 8 = 104$
Sum: $210 + 104 = 314$

5.
$$\begin{array}{r} \$18.00 \\ +\ 12.50 \\ \hline \$30.50 \end{array}$$
$$\begin{array}{r} \$30.50 \\ \times\ \ \ \ \ \ 4 \\ \hline \mathbf{\$122.00} \end{array}$$

6.
$$\begin{array}{r} \$12.00 \\ +\ 10.25 \\ \hline \$22.25 \end{array}$$
$$\begin{array}{r} \$22.25 \\ \times\ \ \ \ \ \ 4 \\ \hline \mathbf{\$89.00} \end{array}$$

7.
$$\begin{array}{r} 100 \\ \times .20 \\ \hline \$20.00 \end{array}$$
$$\begin{array}{r} \$89.00 \\ +\ 20.00 \\ \hline \mathbf{\$109.00} \end{array}$$

8.
$$\begin{array}{r} 200 \\ \times .20 \\ \hline \$40.00 \end{array}$$
$$\begin{array}{r} \$89.00 \\ +\ 40.00 \\ \hline \mathbf{\$129.00} \end{array}$$

9. (4) $\dfrac{(33 + 25 + 47 + 19)}{4}$

10. (2) $12(8 + 6 + 5)$

Chapter 1   Quiz

1. **987,000**

2. $d = rt$
   $d = 48 \times 6 =$ **288 mi.**

3. $c = nr$
   $c = 12 \times \$29.50$
   $= $ **\$354.00**

4. $14^2 - 8^0 + 16^1 =$
   $(14 \times 14) - 1 + 16 =$
   $196 - 1 + 16 =$ **211**

5. $\sqrt{2401} =$ **49**

   Guess 50.

$$
\begin{array}{r}
48 \\
50\overline{)2401} \\
200 \\
\hline
401 \\
400 \\
\end{array}
\quad
\begin{array}{r}
48 \\
+50 \\
\hline
98 \\
\end{array}
\quad
\begin{array}{r}
49 \\
2\overline{)98} \\
98 \\
\end{array}
$$

6. $P = 2l + 2w$
   $P = 2 \times 22 + 2 \times 14$
   $P = 44 + 28 =$ **72 m**

7. $A = lw$
   $A = 22 \times 14 =$ **308 m^2**

8. $V = s^3$
   $V = 9^3 = 9 \times 9 \times 9$
   $=$ **729 in.^{3}**

9.
$$
\begin{array}{r}
\$\ 32,400 \\
46,600 \\
+\ \ 29,300 \\
\hline
\$108,300 \\
\end{array}
\quad
\begin{array}{r}
\mathbf{\$\ 36,100} \\
3\overline{)\$108,300} \\
9 \\
\hline
18 \\
18 \\
\hline
0\ \ 3 \\
3 \\
\end{array}
$$

10. In order:
    $29,300 $32,400
    $46,600
    $32,400 is the median.

11. **(3) \$182**

$$
\begin{array}{r}
\$553 \\
-\ 371 \\
\hline
\$182 \\
\end{array}
$$

12. **(5) \$327**

$$
\begin{array}{r}
\$371 \\
278 \\
+\ \ 69 \\
\hline
\$718 \\
\end{array}
\quad
\begin{array}{r}
\$500 \\
392 \\
+\ 153 \\
\hline
\$1045 \\
\end{array}
\quad
\begin{array}{r}
\$1045 \\
-\ 718 \\
\hline
\$\ 327 \\
\end{array}
$$

13. **(4) 14**

$$
\begin{array}{r}
14\text{ yd.} \\
6\overline{)84} \\
6 \\
\hline
24 \\
24 \\
\end{array}
$$

14. **(4) \$5724**

$$
\begin{array}{r}
\$112 \\
+\ 365 \\
\hline
\$477 \\
\end{array}
\quad
\begin{array}{r}
\$477 \\
\times\ \ 12 \\
\hline
954 \\
477 \\
\hline
\$5724 \\
\end{array}
$$

15. **(3) 1184**

$$
\begin{array}{r}
78 \\
\times\ 8 \\
\hline
624 \\
\end{array}
\quad
\begin{array}{r}
112 \\
\times\ 5 \\
\hline
560 \\
\end{array}
\quad
\begin{array}{r}
624 \\
+560 \\
\hline
1184 \\
\end{array}
$$

16. **(3) 775**

    Area of rectangle:
    $A = lw = 40 \times 25$
    $= 1000$ ft.^{2}
    Area of square:
    $A = s^2 = 15^2$
    $= 225$ ft.^{2}
    Difference:
    $1000 - 225 =$
    775 ft.^{2}

17. **(3) 13**

$$
\begin{array}{r}
38,734 \\
-38,526 \\
\hline
208 \\
\end{array}
\quad
\begin{array}{r}
13\text{ mi.} \\
16\overline{)208} \\
16 \\
\hline
48 \\
48 \\
\end{array}
$$

18. **(2) 21**

$$
\begin{array}{r}
39,028 \\
38,734 \\
\hline
294 \\
\end{array}
\quad
\begin{array}{r}
21\text{ mi.} \\
14\overline{)294} \\
28 \\
\hline
14 \\
14 \\
\end{array}
$$

19. **(4) \$34.50**

$$
\begin{array}{r}
16 \\
+14 \\
\hline
30 \\
\end{array}
\quad
\begin{array}{r}
\$1.15 \\
\times\ \ 30 \\
\hline
\$34.50 \\
\end{array}
$$

20. **(4) (35 × 8.50) +
    (6 × 12.75)**

Chapter 2   Fractions

Level 1   Fraction Skills

Preview

1. $\dfrac{24 \div 8}{64 \div 8} = \dfrac{3}{8}$

2. $4\frac{7}{20} = 4\frac{7}{20}$
   $+9\frac{1}{2} = 9\frac{10}{20}$
   $\overline{\phantom{+9\frac{1}{2} = }13\frac{17}{20}}$

3. $9\frac{1}{2} = 9\frac{4}{8}$
   $-4\frac{3}{8} = 4\frac{3}{8}$
   $\overline{\phantom{-4\frac{3}{8} = }5\frac{1}{8}}$

4. $12 \times 4\frac{1}{2} \times 2\frac{2}{3} =$

   $\dfrac{\overset{6}{\cancel{12}}}{1} \times \dfrac{\overset{3}{\cancel{9}}}{\underset{1}{\cancel{2}}} \times \dfrac{8}{\underset{1}{\cancel{3}}} = \dfrac{144}{1} =$ **144**

5. $6\frac{1}{4} \div 1\frac{7}{8} =$
   $\dfrac{25}{4} \div \dfrac{15}{8} =$

   $\dfrac{\overset{5}{\cancel{25}}}{\underset{1}{\cancel{4}}} \times \dfrac{\overset{2}{\cancel{8}}}{\underset{3}{\cancel{15}}} = \dfrac{10}{3} = 3\frac{1}{3}$

Lesson 1

1. $\dfrac{5 \div 5}{20 \div 5} = \dfrac{1}{4}$ $\dfrac{32 \div 8}{56 \div 8} = \dfrac{4}{7}$

   $\dfrac{25 \div 5}{30 \div 5} = \dfrac{5}{6}$ $\dfrac{90 \div 10}{200 \div 10} = \dfrac{9}{20}$

2. $\dfrac{40 \div 5}{55 \div 5} = \dfrac{8}{11}$ $\dfrac{6 \div 6}{18 \div 6} = \dfrac{1}{3}$

   $\dfrac{45 \div 15}{75 \div 15} = \dfrac{3}{5}$ $\dfrac{22 \div 2}{24 \div 2} = \dfrac{11}{12}$

3. $\dfrac{24 \div 8}{40 \div 8} = \dfrac{3}{5}$ $\dfrac{19 \div 19}{38 \div 19} = \dfrac{1}{2}$

   $\dfrac{60 \div 12}{144 \div 12} = \dfrac{5}{12}$ $\dfrac{50 \div 50}{1000 \div 50} = \dfrac{1}{20}$

4. $\dfrac{25 \div 5}{45 \div 5} = \dfrac{5}{9}$ $\dfrac{48 \div 16}{64 \div 16} = \dfrac{3}{4}$

   $\dfrac{24 \div 24}{72 \div 24} = \dfrac{1}{3}$ $\dfrac{9 \div 9}{54 \div 9} = \dfrac{1}{6}$

5. $\dfrac{12 \div 12}{60 \div 12} = \dfrac{1}{5}$ $\dfrac{20 \div 4}{84 \div 4} = \dfrac{5}{21}$

   $\dfrac{36 \div 4}{76 \div 4} = \dfrac{9}{19}$ $\dfrac{68 \div 34}{102 \div 34} = \dfrac{2}{3}$

6. $\dfrac{9 \times 4}{10 \times 4} = \dfrac{36}{40}$ $\dfrac{5 \times 3}{12 \times 3} = \dfrac{15}{36}$

   $\dfrac{2 \times 9}{5 \times 9} = \dfrac{18}{45}$ $\dfrac{2 \times 4}{3 \times 4} = \dfrac{8}{12}$

7. $\dfrac{4 \times 5}{9 \times 5} = \dfrac{20}{45}$ $\dfrac{3 \times 10}{20 \times 10} = \dfrac{30}{200}$

   $\dfrac{3 \times 7}{5 \times 7} = \dfrac{21}{35}$ $\dfrac{9 \times 2}{50 \times 2} = \dfrac{18}{100}$

8. $\dfrac{7 \times 3}{25 \times 3} = \dfrac{21}{75}$ $\dfrac{1 \times 8}{4 \times 8} = \dfrac{8}{32}$

   $\dfrac{2 \times 4}{9 \times 4} = \dfrac{8}{36}$ $\dfrac{5 \times 2}{8 \times 2} = \dfrac{10}{16}$

9. $\dfrac{6 \times 6}{7 \times 6} = \dfrac{36}{42}$ $\dfrac{7 \times 4}{12 \times 4} = \dfrac{28}{48}$

   $\dfrac{5 \times 9}{8 \times 9} = \dfrac{45}{72}$ $\dfrac{1 \times 18}{2 \times 18} = \dfrac{18}{36}$

10. $\dfrac{7 \times 3}{8 \times 3} = \dfrac{21}{24}$

$\dfrac{4 \times 5}{5 \times 5} = \dfrac{20}{25}$

$\dfrac{3 \times 4}{16 \times 4} = \dfrac{12}{64}$

$\dfrac{4 \times 16}{5 \times 16} = \dfrac{64}{80}$

Lesson 2

1. $\dfrac{11}{2} = 2\overline{)11}^{\;5\frac{1}{2}}$

$\dfrac{17}{3} = 3\overline{)17}^{\;5\frac{2}{3}}$

$\dfrac{25}{8} = 8\overline{)25}^{\;3\frac{1}{8}}$

$\dfrac{36}{9} = 9\overline{)36}^{\;4}$

2. $\dfrac{52}{8} = 8\overline{)52}^{\;6\frac{4}{8}} = 6\frac{1}{2}$

$\dfrac{35}{10} = 10\overline{)35}^{\;3\frac{5}{10}} = 3\frac{1}{2}$

$\dfrac{28}{6} = 6\overline{)28}^{\;4\frac{4}{6}} = 4\frac{2}{3}$

$\dfrac{40}{12} = 12\overline{)40}^{\;3\frac{4}{12}} = 3\frac{1}{3}$

3. $\dfrac{36}{3} = 3\overline{)36}^{\;12}$

$\dfrac{48}{5} = 5\overline{)48}^{\;9\frac{3}{5}}$

$\dfrac{17}{8} = 8\overline{)17}^{\;2\frac{1}{8}}$

$\dfrac{28}{19} = 19\overline{)28}^{\;1\frac{9}{19}}$

4. $\dfrac{7}{2} = 2\overline{)7}^{\;3\frac{1}{2}}$

$\dfrac{14}{10} = 10\overline{)14}^{\;1\frac{4}{10}} = 1\frac{2}{5}$

$\dfrac{38}{8} = 8\overline{)38}^{\;4\frac{6}{8}} = 4\frac{3}{4}$

$\dfrac{27}{4} = 4\overline{)27}^{\;6\frac{3}{4}}$

5. $\dfrac{54}{9} = 9\overline{)54}^{\;6}$

$\dfrac{13}{2} = 2\overline{)13}^{\;6\frac{1}{2}}$

$\dfrac{16}{5} = 5\overline{)16}^{\;3\frac{1}{5}}$

$\dfrac{98}{15} = 15\overline{)98}^{\;6\frac{8}{15}}$

6. $3\frac{7}{10} = \dfrac{37}{10}$

$7\frac{1}{2} = \dfrac{15}{2}$

$2\frac{2}{3} = \dfrac{8}{3}$

$1\frac{5}{6} = \dfrac{11}{6}$

7. $10\frac{3}{8} = \dfrac{83}{8}$

$7\frac{2}{5} = \dfrac{37}{5}$

$2\frac{3}{4} = \dfrac{11}{4}$

$12\frac{7}{20} = \dfrac{247}{20}$

8. $4\frac{5}{8} = \dfrac{37}{8}$

$9\frac{4}{5} = \dfrac{49}{5}$

$6\frac{1}{2} = \dfrac{13}{2}$

$1\frac{1}{3} = \dfrac{4}{3}$

9. $5\frac{3}{5} = \dfrac{28}{5}$

$3\frac{1}{2} = \dfrac{7}{2}$

$8\frac{2}{3} = \dfrac{26}{3}$

$9\frac{7}{8} = \dfrac{79}{8}$

10. $1\frac{8}{9} = \dfrac{17}{9}$

$5\frac{2}{7} = \dfrac{37}{7}$

$4\frac{7}{10} = \dfrac{47}{10}$

$7\frac{1}{8} = \dfrac{57}{8}$

Lesson 3

1. $\begin{array}{r} \frac{5}{8} \\ + \frac{7}{8} \\ \hline \frac{12}{8} = 1\frac{4}{8} = 1\frac{1}{2} \end{array}$

$\begin{array}{r} 3\frac{1}{6} \\ + 8\frac{5}{6} \\ \hline 11\frac{6}{6} = 12 \end{array}$

$\begin{array}{r} 2\frac{3}{5} \\ + 1\frac{4}{5} \\ \hline 3\frac{7}{5} = 4\frac{2}{5} \end{array}$

$\begin{array}{r} 4\frac{11}{12} \\ + 4\frac{7}{12} \\ \hline 8\frac{18}{12} = 9\frac{6}{12} = 9\frac{1}{2} \end{array}$

2. $\begin{array}{r} \frac{1}{4} = \frac{5}{20} \\ + \frac{3}{5} = \frac{12}{20} \\ \hline \frac{17}{20} \end{array}$

$\begin{array}{r} \frac{5}{12} = \frac{5}{12} \\ + \frac{3}{4} = \frac{9}{12} \\ \hline \frac{14}{12} = 1\frac{2}{12} = 1\frac{1}{6} \end{array}$

$\begin{array}{r} 8\frac{2}{5} = 8\frac{6}{15} \\ + 1\frac{7}{15} = 1\frac{7}{15} \\ \hline 9\frac{13}{15} \end{array}$

$\begin{array}{r} 6\frac{1}{2} = 6\frac{4}{8} \\ + 5\frac{3}{8} = 5\frac{3}{8} \\ \hline 11\frac{7}{8} \end{array}$

3. $\begin{array}{r} \frac{7}{8} = \frac{21}{24} \\ + \frac{2}{3} = \frac{16}{24} \\ \hline \frac{37}{24} = 1\frac{13}{24} \end{array}$

$\begin{array}{r} \frac{1}{6} = \frac{3}{18} \\ + \frac{5}{9} = \frac{10}{18} \\ \hline \frac{13}{18} \end{array}$

$\begin{array}{r} 5\frac{3}{4} = 5\frac{27}{36} \\ + 2\frac{2}{9} = 2\frac{8}{36} \\ \hline 7\frac{35}{36} \end{array}$

$\begin{array}{r} 4\frac{3}{5} = 4\frac{18}{30} \\ + 4\frac{1}{6} = 4\frac{5}{30} \\ \hline 8\frac{23}{30} \end{array}$

4. $\begin{array}{r} \frac{3}{10} = \frac{6}{20} \\ \frac{1}{2} = \frac{10}{20} \\ + \frac{3}{4} = \frac{15}{20} \\ \hline \frac{31}{20} = 1\frac{11}{20} \end{array}$

$\begin{array}{r} \frac{3}{8} = \frac{9}{24} \\ \frac{2}{3} = \frac{16}{24} \\ + \frac{1}{12} = \frac{2}{24} \\ \hline \frac{27}{24} = 1\frac{3}{24} = 1\frac{1}{8} \end{array}$

$\begin{array}{r} 2\frac{5}{6} = 2\frac{20}{24} \\ 5\frac{3}{8} = 5\frac{9}{24} \\ + 3\frac{1}{4} = 3\frac{6}{24} \\ \hline 10\frac{35}{24} = 11\frac{11}{24} \end{array}$

$\begin{array}{r} 1\frac{5}{9} = 1\frac{10}{18} \\ 4\frac{1}{2} = 4\frac{9}{18} \\ + 3\frac{2}{3} = 3\frac{12}{18} \\ \hline 8\frac{31}{18} = 9\frac{13}{18} \end{array}$

5. $\begin{array}{r} \frac{11}{12} \\ - \frac{7}{12} \\ \hline \frac{4}{12} = \frac{1}{3} \end{array}$

$\begin{array}{r} \frac{7}{8} \\ - \frac{3}{8} \\ \hline \frac{4}{8} = \frac{1}{2} \end{array}$

$\begin{array}{r} 6\frac{9}{10} \\ - 5\frac{7}{10} \\ \hline 1\frac{2}{10} = 1\frac{1}{5} \end{array}$

$\begin{array}{r} 9\frac{15}{16} \\ - 3\frac{3}{16} \\ \hline 6\frac{12}{16} = 6\frac{3}{4} \end{array}$

6. $\begin{array}{r} \frac{3}{4} = \frac{6}{8} \\ - \frac{3}{8} = \frac{3}{8} \\ \hline \frac{3}{8} \end{array}$

$\begin{array}{r} \frac{4}{5} = \frac{16}{20} \\ - \frac{3}{4} = \frac{15}{20} \\ \hline \frac{1}{20} \end{array}$

$\begin{array}{r} 7\frac{1}{2} = 7\frac{5}{10} \\ - 2\frac{1}{5} = 2\frac{2}{10} \\ \hline 5\frac{3}{10} \end{array}$

$\begin{array}{r} 4\frac{7}{9} = 4\frac{14}{18} \\ - 1\frac{1}{6} = 1\frac{3}{18} \\ \hline 3\frac{11}{18} \end{array}$

7. $\begin{array}{r} 7 = 6\frac{8}{8} \\ - 3\frac{5}{8} = 3\frac{5}{8} \\ \hline 3\frac{3}{8} \end{array}$

$\begin{array}{r} 9 = 8\frac{12}{12} \\ - 4\frac{7}{12} = 4\frac{7}{12} \\ \hline 4\frac{5}{12} \end{array}$

$\begin{array}{r} 8 = 7\frac{10}{10} \\ - 2\frac{7}{10} = 2\frac{7}{10} \\ \hline 5\frac{3}{10} \end{array}$

$\begin{array}{r} 12 = 11\frac{16}{16} \\ - 5\frac{13}{16} = 5\frac{13}{16} \\ \hline 6\frac{3}{16} \end{array}$

8. $5\frac{1}{6} = 4\frac{1}{6} + \frac{6}{6} = 4\frac{7}{6}$
$\phantom{5\frac{1}{6}} -2\frac{5}{6} \phantom{+ \frac{6}{6}} = 2\frac{5}{6}$
$\phantom{5\frac{1}{6} = } 2\frac{2}{6} = 2\frac{1}{3}$

   $9\frac{1}{3} = 8\frac{1}{3} + \frac{3}{3} = 8\frac{4}{3}$
$\phantom{9\frac{1}{3}} -3\frac{2}{3} \phantom{+ \frac{3}{3}} = 3\frac{2}{3}$
$\phantom{9\frac{1}{3} = 8\frac{1}{3} +} 5\frac{2}{3}$

$8\frac{5}{12} = 7\frac{5}{12} + \frac{12}{12} = 7\frac{17}{12}$
$\phantom{8\frac{5}{12}} -7\frac{7}{12} \phantom{+ \frac{12}{12}} = 7\frac{7}{12}$
$\phantom{8\frac{5}{12} = 7\frac{5}{12}} \frac{10}{12} = \frac{5}{6}$

   $4\frac{1}{5} = 3\frac{1}{5} + \frac{5}{5} = 3\frac{6}{5}$
$\phantom{4\frac{1}{5}} -1\frac{4}{5} \phantom{+\frac{5}{5}} = 1\frac{4}{5}$
$\phantom{4\frac{1}{5} = 3\frac{1}{5} +} 2\frac{2}{5}$

9. $6\frac{1}{5} = 6\frac{4}{20} = 5\frac{4}{20} + \frac{20}{20} = 5\frac{24}{20}$
$\phantom{6\frac{1}{5}} -3\frac{3}{4} = 3\frac{15}{20} \phantom{+ \frac{20}{20}} = 3\frac{15}{20}$
$\phantom{6\frac{1}{5} = 6\frac{4}{20} = 5\frac{4}{20} +} 2\frac{9}{20}$

$7\frac{1}{2} = 7\frac{4}{8} = 6\frac{4}{8} + \frac{8}{8} = 6\frac{12}{8}$
$\phantom{7\frac{1}{2}} -1\frac{5}{8} = 1\frac{5}{8} \phantom{+ \frac{8}{8}} = 1\frac{5}{8}$
$\phantom{7\frac{1}{2} = 7\frac{4}{8} = 6\frac{4}{8} +} 5\frac{7}{8}$

$9\frac{1}{3} = 9\frac{4}{12} = 8\frac{4}{12} + \frac{12}{12} = 8\frac{16}{12}$
$\phantom{9\frac{1}{3}} -5\frac{3}{4} = 5\frac{9}{12} \phantom{+ \frac{12}{12}} = 5\frac{9}{12}$
$\phantom{9\frac{1}{3} = 9\frac{4}{12} = 8\frac{4}{12} +} 3\frac{7}{12}$

   $2\frac{1}{2} = 2\frac{3}{6} = 1\frac{3}{6} + \frac{6}{6} = 1\frac{9}{6}$
$\phantom{2\frac{1}{2}} -1\frac{2}{3} = 1\frac{4}{6} \phantom{+ \frac{6}{6}} = 1\frac{4}{6}$
$\phantom{2\frac{1}{2} = 2\frac{3}{6} = 1\frac{3}{6} +} \frac{5}{6}$

10. $8\frac{2}{3} = 8\frac{16}{24}$
$\phantom{8\frac{2}{3}} -1\frac{5}{8} = 1\frac{15}{24}$
$\phantom{8\frac{2}{3} = } 7\frac{1}{24}$

   $5\frac{1}{4} = 4\frac{1}{4} + \frac{4}{4} = 4\frac{5}{4}$
$\phantom{5\frac{1}{4}} -2\frac{3}{4} \phantom{+ \frac{4}{4}} = 2\frac{3}{4}$
$\phantom{5\frac{1}{4} = 4\frac{1}{4} +} 2\frac{2}{4} = 2\frac{1}{2}$

$3\frac{1}{9} = 3\frac{1}{9} = 2\frac{1}{9} + \frac{9}{9} = 2\frac{10}{9}$
$\phantom{3\frac{1}{9}} -1\frac{2}{3} = 1\frac{6}{9} \phantom{+ \frac{9}{9}} = 1\frac{6}{9}$
$\phantom{3\frac{1}{9} = 3\frac{1}{9} = 2\frac{1}{9} +} 1\frac{4}{9}$

   $5\frac{1}{8} = 4\frac{1}{8} + \frac{8}{8} = 4\frac{9}{8}$
$\phantom{5\frac{1}{8}} -4\frac{3}{8} \phantom{+ \frac{8}{8}} = 4\frac{3}{8}$
$\phantom{5\frac{1}{8} = 4\frac{1}{8} +} \frac{6}{8} = \frac{3}{4}$

Lesson 4

1. $\frac{2}{3} \times \frac{4}{5} = \frac{8}{15}$ $\frac{5}{8} \times \frac{3}{4} = \frac{15}{32}$ $\frac{3}{4} \times \frac{1}{2} \times \frac{1}{5} = \frac{3}{10}$

2. $\frac{9}{10} \times \frac{2}{3} = \frac{3}{5}$ $\frac{1}{20} \times \frac{5}{12} = \frac{1}{16}$ $\frac{2}{3} \times \frac{1}{6} \times \frac{7}{10} = \frac{7}{18}$

3. $6 \times \frac{3}{4} =$ $\frac{5}{6} \times 15 =$ $8 \times \frac{11}{12} =$
$\frac{6}{1} \times \frac{3}{4} = \frac{9}{2} = 4\frac{1}{2}$ $\frac{5}{6} \times \frac{15}{1} = \frac{25}{2} = 12\frac{1}{2}$ $\frac{8}{1} \times \frac{11}{12} = \frac{22}{3} = 7\frac{1}{3}$

4. $\frac{3}{4} \times 3\frac{1}{5} =$ $2\frac{1}{4} \times 1\frac{2}{3} =$ $5\frac{1}{3} \times 1\frac{5}{16} =$
$\frac{3}{4} \times \frac{16}{5} = \frac{12}{5}$ $\frac{9}{4} \times \frac{5}{3} = \frac{15}{4} = 3\frac{3}{4}$ $\frac{16}{3} \times \frac{21}{16} = \frac{7}{1} = 7$
$\phantom{\frac{3}{4} \times \frac{16}{5}} = 2\frac{2}{5}$

5. $\frac{2}{3} \div \frac{4}{9} =$ $\frac{9}{10} \div \frac{2}{5} =$ $\frac{3}{4} \div \frac{1}{8} =$
$\frac{2}{3} \times \frac{9}{4} = \frac{3}{2} = 1\frac{1}{2}$ $\frac{9}{10} \times \frac{5}{2} = \frac{9}{4} = 2\frac{1}{4}$ $\frac{3}{4} \times \frac{8}{1} = \frac{6}{1} = 6$

6. $4 \div \frac{2}{3} =$ $8 \div \frac{4}{5} =$ $5 \div \frac{3}{4} =$
$\frac{4}{1} \times \frac{3}{2} = \frac{6}{1} = 6$ $\frac{8}{1} \times \frac{5}{4} = \frac{10}{1} = 10$ $\frac{5}{1} \times \frac{4}{3} = \frac{20}{3} = 6\frac{2}{3}$

7. $1\frac{1}{9} \div \frac{5}{6} =$ $2\frac{5}{8} \div \frac{1}{4} =$ $6\frac{1}{2} \div \frac{3}{8} =$
$\frac{10}{9} \times \frac{6}{5} = \frac{4}{3} = 1\frac{1}{3}$ $\frac{21}{8} \times \frac{4}{1} = \frac{21}{2} = 10\frac{1}{2}$ $\frac{13}{2} \times \frac{8}{3} = \frac{52}{3} = 17\frac{1}{3}$

8. $4\frac{1}{3} \div 5 =$ $10\frac{1}{2} \div 8 =$ $\frac{3}{8} \div 4 =$
$\frac{13}{3} \div \frac{5}{1} =$ $\frac{21}{2} \div \frac{8}{1} =$ $\frac{3}{8} \div \frac{4}{1} =$
$\frac{13}{3} \times \frac{1}{5} = \frac{13}{15}$ $\frac{21}{2} \times \frac{1}{8} = \frac{21}{16} = 1\frac{5}{16}$ $\frac{3}{8} \times \frac{1}{4} = \frac{3}{32}$

9. $\frac{5}{9} \div 1\frac{1}{3} =$ $4 \div 1\frac{3}{4} =$ $\frac{3}{8} \div 2\frac{2}{3} =$
$\frac{5}{9} \div \frac{4}{3} =$ $\frac{4}{1} \div \frac{7}{4} =$ $\frac{3}{8} \div \frac{8}{3} =$
$\frac{5}{9} \times \frac{3}{4} = \frac{5}{12}$ $\frac{4}{1} \times \frac{4}{7} = \frac{16}{7} = 2\frac{2}{7}$ $\frac{3}{8} \times \frac{3}{8} = \frac{9}{64}$

10. $1\frac{1}{4} \div 2\frac{1}{2} =$ $2\frac{3}{4} \div 1\frac{5}{8} =$ $7\frac{1}{2} \div 3\frac{3}{4} =$
$\frac{5}{4} \div \frac{5}{2} =$ $\frac{11}{4} \div \frac{13}{8} =$ $\frac{15}{2} \div \frac{15}{4} =$
$\frac{5}{4} \times \frac{2}{5} = \frac{1}{2}$ $\frac{11}{4} \times \frac{8}{13} = \frac{22}{13} = 1\frac{9}{13}$ $\frac{15}{2} \times \frac{4}{15} = \frac{2}{1} = 2$

Level 1   Review

1. $\frac{35 \div 7}{84 \div 7} = \frac{5}{12}$ 2. $8\frac{4}{5} = \frac{44}{5}$

3. $8\frac{3}{4} = 8\frac{12}{16}$
$+7\frac{9}{16} = 7\frac{9}{16}$
$15\frac{21}{16} = 16\frac{5}{16}$

4. $2\frac{3}{8} = 2\frac{15}{40}$
$ 4\frac{1}{2} = 4\frac{20}{40}$
$+6\frac{3}{5} = 6\frac{24}{40}$
$12\frac{59}{40} = 13\frac{19}{40}$

5. $7\frac{2}{3} = 7\frac{8}{12}$
$-3\frac{5}{12} = 3\frac{5}{12}$
$4\frac{3}{12} = 4\frac{1}{4}$

6. $8\frac{1}{5} = 8\frac{4}{20} = 7\frac{4}{20} + \frac{20}{20} = 7\frac{24}{20}$
$-3\frac{3}{4} = 3\frac{15}{20} \phantom{= 7\frac{4}{20} +} = 3\frac{15}{20}$
$\phantom{8\frac{1}{5} = 8\frac{4}{20} = 7\frac{4}{20} +} 4\frac{9}{20}$

7. $3\frac{3}{5} \times 10 =$ 8. $2\frac{1}{3} \times 4\frac{1}{8} \times \frac{1}{2} =$
$\frac{18}{5} \times \frac{10}{1} = \frac{36}{1} = 36$ $\frac{7}{3} \times \frac{33}{8} \times \frac{1}{2} = \frac{77}{16} = 4\frac{13}{16}$

9. $6\frac{2}{3} \div 8 =$ 10. $9\frac{1}{3} \div 3\frac{1}{5} =$
$\frac{20}{3} \div \frac{8}{1} =$ $\frac{28}{3} \div \frac{16}{5} =$
$\frac{20}{3} \times \frac{1}{8} = \frac{5}{6}$ $\frac{28}{3} \times \frac{5}{16} = \frac{35}{12} = 2\frac{11}{12}$

Level 2   Fraction Applications

Preview

1. Total: $8 + 6 + 6 = 20$
$\frac{8}{20} = \frac{2}{5}$

2. $d = rt$
$d = 52 \times 3\frac{1}{2}$
$d = \frac{52}{1} \times \frac{7}{2} = 182$ mi.

3. $3\frac{4}{12} = 3\frac{1}{3}$ yr.
$12\overline{)40}$
$\frac{36}{4}$

4. $A = lw$
$A = 6\frac{5}{8} \times 4$
$A = \frac{53}{8} \times \frac{4}{1} = \frac{53}{2}$
$= 26\frac{1}{2}$ in.²

5. Total: $4 + 6 = 10$
women: total = 4:10 = **2:5**

Lesson 1

1. $15 - 9 = 6$ women
   a. men $\frac{9}{15} = \frac{3}{5}$
     total
   b. women $\frac{6}{15} = \frac{2}{5}$
     total

2. $45 + 36 = 81$ total
   a. union $\frac{45}{81} = \frac{5}{9}$
     total
   b. nonunion $\frac{36}{81} = \frac{4}{9}$
     total

3. $33 - 15 = 18$ used cars
  a. new $\frac{15}{33} = \frac{5}{11}$
    total
  b. used $\frac{18}{33} = \frac{6}{11}$
    total

4. $12 + 2 + 6 = 20$ total
  a. cola $\frac{12}{20} = \frac{3}{5}$
    total
  b. lime $\frac{2}{20} = \frac{1}{10}$
    total
  c. orange $\frac{6}{20} = \frac{3}{10}$
    total

9. $c = nr$
  $c = 1\frac{1}{4} \times \$3.00$
  $c = \frac{5}{4} \times \frac{\overset{.75}{\$3.00}}{1} = \$3.75$

10. $d = rt$
  $d = 600 \times 5\frac{3}{4}$
  $d = \frac{\overset{150}{600}}{1} \times \frac{23}{4} = 3450$ mi.

5. $45 + 35 + 20 = 100$ total
  a. for $\frac{45}{100} = \frac{9}{20}$
    total
  b. against $\frac{35}{100} = \frac{7}{20}$
    total
  c. undecided $\frac{20}{100} = \frac{1}{5}$
    total

Lesson 2

1. $\frac{3}{8} \times \frac{\overset{1200}{9600}}{1} = 3600$

2. $\begin{array}{r} 9600 \\ -3600 \\ \hline \textbf{6000 people voted} \end{array}$

3. $209 \div 9\frac{1}{2} =$
$\frac{209}{1} \div \frac{19}{2} =$
$\frac{\overset{11}{209}}{1} \times \frac{2}{19} = 22$ mi.

4. $2\frac{1}{2} = 2\frac{5}{10} = 1\frac{5}{10} + \frac{10}{10} = 1\frac{15}{10}$
$\underline{-1\frac{9}{10} = 1\frac{9}{10} \qquad\qquad = 1\frac{9}{10}}$
$\qquad\qquad\qquad\qquad\qquad \frac{6}{10} = \frac{3}{5}$

5. $52 \div 3\frac{1}{4} =$
$\frac{52}{1} \div \frac{13}{4} =$
$\frac{\overset{4}{52}}{1} \times \frac{4}{13} = 16$ pieces

6. Total drop: $\frac{3}{8} + \frac{1}{8} = \frac{4}{8} = \frac{1}{2}$
$24\frac{1}{4} = 24\frac{1}{4} = 23\frac{1}{4} + \frac{4}{4} = 23\frac{5}{4}$
$\underline{-\frac{1}{2} = \frac{2}{4} \qquad\qquad = \frac{2}{4}}$
$\qquad\qquad\qquad\qquad\qquad 23\frac{3}{4}$

7. $\frac{1}{4} \times \frac{\overset{18}{72}}{1} = \$18$
$\begin{array}{r} \$72 \\ -18 \\ \hline \$54 \end{array}$

8. $24 \div \frac{3}{4} =$
$\frac{24}{1} \div \frac{3}{4} =$
$\frac{\overset{8}{24}}{1} \times \frac{4}{3} = 32$ lots

9. $2\frac{1}{2} \times \$12.50 =$
$\frac{5}{2} \times \frac{\overset{6.25}{\$12.50}}{1} = \$31.25$

10. $2\frac{3}{4} = 2\frac{3}{4}$
$2\frac{1}{2} = 2\frac{2}{4}$
$\underline{+3\frac{1}{4} = 3\frac{1}{4}}$
$7\frac{6}{4} = 8\frac{2}{4} = 8\frac{1}{2}$ hr.

Lesson 3

1. $d = rt$
$d = 40 \times 3\frac{1}{2}$
$d = \frac{\overset{20}{40}}{1} \times \frac{7}{2} = 140$ mi.

2. $d = rt$
$d = 520 \times 4\frac{1}{4}$
$d = \frac{\overset{130}{520}}{1} \times \frac{17}{4} = 2210$ mi.

3. $c = nr$
$c = \frac{3}{4} \times \$2.80$
$c = \frac{3}{4} \times \frac{\overset{.70}{\$2.80}}{1} = \$2.10$

4. $c = nr$
$c = 2\frac{1}{2} \times \$.84$
$c = \frac{5}{2} \times \frac{\overset{.42}{\$.84}}{1} = \$2.10$

5. $d = rt$
$d = 5 \times 1\frac{1}{2}$
$d = \frac{5}{1} \times \frac{3}{2} = \frac{15}{2} = 7\frac{1}{2}$ mi.

6. $c = nr$
$c = 9\frac{1}{2} \times \$.36$
$c = \frac{19}{2} \times \frac{\overset{.18}{\$.36}}{1} = \$3.42$

7. $d = rt$
$d = 22 \times 2\frac{3}{4}$
$d = \frac{\overset{11}{22}}{1} \times \frac{11}{4} = \frac{121}{2} = 60\frac{1}{2}$ mi.

8. $c = nr$
$c = 2\frac{1}{4} \times \$.48$
$c = \frac{9}{4} \times \frac{\overset{.12}{\$.48}}{1} = \$1.08$

Lesson 4

1. $\left(\frac{1}{2}\right)^2 = \frac{1}{2} \times \frac{1}{2} = \frac{1}{4}$
  $\left(\frac{2}{3}\right)^3 = \frac{2}{3} \times \frac{2}{3} \times \frac{2}{3} = \frac{8}{27}$
  $\left(\frac{1}{9}\right)^2 = \frac{1}{9} \times \frac{1}{9} = \frac{1}{81}$
  $\left(\frac{1}{10}\right)^4 = \frac{1}{10} \times \frac{1}{10} \times \frac{1}{10} \times \frac{1}{10}$
$= \frac{1}{10,000}$

2. $\left(\frac{3}{10}\right)^3 = \frac{3}{10} \times \frac{3}{10} \times \frac{3}{10}$
$= \frac{27}{1000}$
  $\left(\frac{7}{12}\right)^2 = \frac{7}{12} \times \frac{7}{12} = \frac{49}{144}$
  $\left(\frac{5}{6}\right)^2 = \frac{5}{6} \times \frac{5}{6} = \frac{25}{36}$
  $\left(\frac{3}{4}\right)^3 = \frac{3}{4} \times \frac{3}{4} \times \frac{3}{4} = \frac{27}{64}$

3. $\left(\frac{1}{6}\right)^4 = \frac{1}{6} \times \frac{1}{6} \times \frac{1}{6} \times \frac{1}{6}$
$= \frac{1}{1296}$
$\left(\frac{1}{8}\right)^3 = \frac{1}{8} \times \frac{1}{8} \times \frac{1}{8} = \frac{1}{512}$
  $\left(\frac{1}{2}\right)^5 = \frac{1}{2} \times \frac{1}{2} \times \frac{1}{2} \times \frac{1}{2} \times \frac{1}{2}$
$= \frac{1}{32}$
$\left(\frac{2}{7}\right)^3 = \frac{2}{7} \times \frac{2}{7} \times \frac{2}{7} = \frac{8}{343}$

4. $\sqrt{\frac{25}{36}} = \frac{5}{6}$ $\sqrt{\frac{4}{49}} = \frac{2}{7}$
$\sqrt{\frac{1}{81}} = \frac{1}{9}$ $\sqrt{\frac{9}{100}} = \frac{3}{10}$

5. $\sqrt{\frac{64}{81}} = \frac{8}{9}$ $\sqrt{\frac{1}{144}} = \frac{1}{12}$
$\sqrt{\frac{9}{16}} = \frac{3}{4}$ $\sqrt{\frac{36}{49}} = \frac{6}{7}$

Lesson 5

1. a. $\begin{array}{r} 2\frac{2}{3} \text{ yd.} \\ 3\overline{)8} \\ 6 \\ \hline 2 \end{array}$
  b. $\begin{array}{r} \textbf{2 yd. 2 ft.} \\ 3\overline{)8} \\ 6 \\ \hline 2 \end{array}$

2. a. $\begin{array}{r} 2\frac{15}{60} = 2\frac{1}{4} \text{ min.} \\ 60\overline{)135} \\ 120 \\ \hline 15 \end{array}$
  b. $\begin{array}{r} \textbf{2 min. 15 sec.} \\ 60\overline{)135} \\ 120 \\ \hline 15 \end{array}$

3. a. $\begin{array}{r} 2\frac{2}{4} = 2\frac{1}{2} \text{ gal.} \\ 4\overline{)10} \\ 8 \\ \hline 2 \end{array}$
  b. $\begin{array}{r} \textbf{2 gal. 2 qt.} \\ 4\overline{)10} \\ 8 \\ \hline 2 \end{array}$

4. a. $\begin{array}{r} 1\frac{4}{16} = 1\frac{1}{4} \text{ lb.} \\ 16\overline{)20} \\ 16 \\ \hline 4 \end{array}$
  b. $\begin{array}{r} \textbf{1 lb. 4 oz.} \\ 16\overline{)20} \\ 16 \\ \hline 4 \end{array}$

5. a. $\begin{array}{r} 1\frac{8}{12} = 1\frac{2}{3} \text{ yr.} \\ 12\overline{)20} \\ 12 \\ \hline 8 \end{array}$
  b. $\begin{array}{r} \textbf{1 yr. 8 mo.} \\ 12\overline{)20} \\ 12 \\ \hline 8 \end{array}$

6. 10 mo. $= \frac{10}{12} = \frac{5}{6}$ yr.
45 min. $= \frac{45}{60} = \frac{3}{4}$ hr.
9 in. $= \frac{9}{12} = \frac{3}{4}$ ft.

7. 8 oz. $= \frac{8}{16} = \frac{1}{2}$ lb.
20 in. $= \frac{20}{36} = \frac{5}{9}$ yd.
2 qt. $= \frac{2}{4} = \frac{1}{2}$ gal.

8. 1800 lb. = $\frac{1800}{2000} = \frac{9}{10}$ **t.**   1320 ft. = $\frac{1320}{5280} = \frac{1}{4}$ **mi.**

   16 hr. = $\frac{16}{24} = \frac{2}{3}$ **da.**

9. $6 \times 4 =$ **24 qt.** $3\frac{1}{4} \times 16 =$

   $5 \times 24 =$ **120 hr.** $\frac{13}{4} \times \frac{\overset{4}{16}}{1} =$ **52 oz.**

10. $4 \times 36 =$ **144 in.** $2\frac{1}{2} \times 60 =$

    $1\frac{1}{2} \times 2000 =$ $\frac{5}{2} \times \frac{\overset{30}{60}}{1} =$ **150 sec.**

    $\frac{3}{2} \times \frac{\overset{1000}{2000}}{1} =$ **3000 lb.**

Lesson 6

1. a. $P = 2l + 2w$

   $P = \left(\frac{2}{1} \times \frac{3}{4}\right) + \left(\frac{2}{1} \times \frac{1}{2}\right)$

   $P = \frac{3}{2} + \frac{2}{2}$

   $P = \frac{5}{2} =$ **2½ in.**

   b. $A = lw$

   $A = \frac{3}{4} \times \frac{1}{2}$

   $A = \frac{3}{8}$ **in.²**

2. a. $P = 4s$

   $P = 4 \times 1\frac{1}{4}$

   $P = \frac{4}{1} \times \frac{5}{4}$

   $P =$ **5 in.**

   b. $A = s$

   $A = (1\frac{1}{4})^2$

   $A = \frac{5}{4} \times \frac{5}{4}$

   $A = \frac{25}{16} = 1\frac{9}{16}$ **in.²**

3. a. $P = a + b + c$

   $P = \frac{3}{16} + \frac{1}{4} + \frac{5}{16}$

   $P = \frac{3}{16} + \frac{4}{16} + \frac{5}{16}$

   $P = \frac{12}{16} = \frac{3}{4}$ **in.**

   b. $A = \frac{1}{2}bh$

   $A = \frac{1}{2} \times \frac{1}{4} \times \frac{3}{16}$

   $A = \frac{3}{128}$ **in.²**

4. $V = lwh$

   $V = 4 \times 1\frac{1}{2} \times 2\frac{1}{4}$

   $V = \frac{4}{1} \times \frac{3}{2} \times \frac{9}{4}$

   $V = \frac{27}{2} = 13\frac{1}{2}$ **ft.³**

5. $V = s^3$

   $V = (4\frac{1}{2})^3$

   $V = \frac{9}{2} \times \frac{9}{2} \times \frac{9}{2}$

   $V = \frac{729}{8} = 91\frac{1}{8}$ **ft.³**

Lesson 7

1. a. $\frac{8}{15} = \frac{16}{30}$ $\frac{1}{2} = \frac{15}{30}$ b. $\frac{2}{3} = \frac{6}{9}$ $\frac{5}{9} = \frac{5}{9}$

   $\frac{8}{15}$ **is larger.** $\frac{2}{3}$ **is larger.**

   c. $\frac{5}{8} = \frac{25}{40}$ $\frac{4}{5} = \frac{32}{40}$ d. $\frac{5}{12} = \frac{15}{36}$ $\frac{4}{9} = \frac{16}{36}$

   $\frac{4}{5}$ **is larger.** $\frac{4}{9}$ **is larger.**

2. $\frac{3}{5} = \frac{12}{20}$ $\frac{1}{2} = \frac{10}{20}$ $\frac{11}{20} = \frac{11}{20}$ $\frac{7}{10} = \frac{14}{20}$

   From smallest to largest: $\frac{1}{2}, \frac{11}{20}, \frac{3}{5}, \frac{7}{10}$

3. $\frac{5}{12} = \frac{20}{48}$ $\frac{1}{4} = \frac{12}{48}$ $\frac{5}{16} = \frac{15}{48}$ $\frac{1}{3} = \frac{16}{48}$

   From largest to smallest: $\frac{5}{12}, \frac{1}{3}, \frac{5}{16}, \frac{1}{4}$

4. $\frac{5}{8} = \frac{10}{16}$ $\frac{9}{16} = \frac{9}{16}$

   $\frac{5}{8}$ **in. is longer.**

5. **(4) C, E, D, B, A**

   A—$1\frac{1}{2} = 1\frac{8}{16}$ D—$2\frac{3}{16} = 2\frac{3}{16}$

   B—$1\frac{5}{8} = 1\frac{10}{16}$ E—$2\frac{1}{4} = 2\frac{4}{16}$

   C—$2\frac{3}{8} = 2\frac{6}{16}$

Lesson 8

(The numbers in each problem are in cents.)

1. $\frac{\text{lemons, imported}}{\text{domestic}} = \frac{30}{45} = \frac{2}{3}$

2. $\frac{\text{apples, imported}}{\text{domestic}} = \frac{48}{64} = \frac{3}{4}$

3. $\frac{\text{potatoes, imported}}{\text{domestic}} = \frac{12}{24} = \frac{1}{2}$

4. $\frac{\text{lamb, imported}}{\text{domestic}} = \frac{25}{150} = \frac{1}{6}$

5. $\frac{\text{tomatoes, imported}}{\text{domestic}} = \frac{50}{60} = \frac{5}{6}$

6. $\frac{\text{beef, imported}}{\text{domestic}} = \frac{45}{90} = \frac{1}{2}$

7. $\frac{\text{imported lemons}}{\text{imported tomatoes}} = \frac{30}{50} = \frac{3}{5}$

8. $\frac{\text{imported apples}}{\text{imported tomatoes}} = \frac{48}{50} = \frac{24}{25}$

9. $\frac{\text{imported potatoes}}{\text{imported apples}} = \frac{12}{48} = \frac{1}{4}$

10. $\frac{\text{domestic beef}}{\text{domestic lamb}} = \frac{90}{150} = \frac{3}{5}$

Lesson 9

1. 15:36 = **5:12**

2. width : length = 12:15 = **4:5**

3. town : country = 16:24 = **2:3**

4. 4:16 = **1:4**

5. $20 - 12 = 8$ women
   men : women = 12:8 = **3:2**

6. $\$450 + \$150 = \$600$ original price
   sale price : original price = 450:600 = **3:4**

7. $\$220 + \$880 = \$1100$ total
   rent : total = 220:1100 = **1:5**

8. $280 - 70 = 210$ lb., September weight
   March weight : September weight = 280:210 = **4:3**

9. $\$5000 - \$1500 = \$3500$ still owed
   amount paid : amount owed = 1500:3500 – **3:7**

10. $40 + 16 = 56$ total
    right : total = 40:56 = **5:7**

Lesson 10

1. $\frac{c}{6} = \frac{7}{30}$ $\frac{9}{a} = \frac{3}{2}$

   $30c = 42$ $3a = 18$

   $c = 1\frac{12}{30} = 1\frac{2}{5}$ $a = $ **6**

   $\frac{5}{8} = \frac{m}{20}$ $\frac{4}{w} = \frac{18}{6}$

   $8m = 100$ $18w = 24$

   $m = 12\frac{4}{8} = $ **12½** $w = 1\frac{6}{18} = 1\frac{1}{3}$

2. $\frac{9}{10} = \frac{27}{e}$ $\frac{x}{14} = \frac{4}{7}$

   $9e = 270$ $7x = 56$

   $e = $ **30** $x = $ **8**

   $\frac{1}{s} = \frac{8}{5}$ $\frac{8}{15} = \frac{r}{2}$

   $8s = 5$ $15r = 16$

   $s = \frac{5}{8}$ $r = 1\frac{1}{15}$

3. $3:n = 6:11$

$\frac{3}{n} = \frac{6}{11}$

$6n = 33$

$n = 5\frac{3}{6} = \mathbf{5\frac{1}{2}}$

$2:5 = t:60$

$\frac{2}{5} = \frac{t}{60}$

$5t = 120$

$t = \mathbf{24}$

$h:4 = 5:6$

$\frac{h}{4} = \frac{5}{6}$

$6h = 20$

$h = 3\frac{2}{6} = \mathbf{3\frac{1}{3}}$

$9:2 = 1:d$

$\frac{9}{2} = \frac{1}{d}$

$9d = 2$

$d = \mathbf{\frac{2}{9}}$

4. $6:n = 8:48$

$\frac{6}{n} = \frac{8}{48}$

$8n = 288$

$n = \mathbf{36}$

$3:5 = m:25$

$\frac{3}{5} = \frac{m}{25}$

$5m = 75$

$m = \mathbf{15}$

$k:3 = 9:27$

$\frac{k}{3} = \frac{9}{27}$

$27k = 27$

$k = \mathbf{1}$

$4:16 = 8:p$

$\frac{4}{16} = \frac{8}{p}$

$4p = 128$

$p = \mathbf{32}$

5. $7:35 = 2:r$

$\frac{7}{35} = \frac{2}{r}$

$7r = 70$

$r = \mathbf{10}$

$a:5 = 20:100$

$\frac{a}{5} = \frac{20}{100}$

$100a = 100$

$a = \mathbf{1}$

$9:x = 1:2$

$\frac{9}{x} = \frac{1}{2}$

$x = \mathbf{18}$

$8:24 = y:3$

$\frac{8}{24} = \frac{y}{3}$

$24y = 24$

$y = \mathbf{1}$

Lesson 11

1. length
   weight $\frac{6}{15} = \frac{16}{x}$

   $6x = 240$

   $x = \mathbf{40\ lb.}$

2. girls
   boys $\frac{5}{4} = \frac{120}{x}$

   $5x = 480$

   $x = \mathbf{96\ boys}$

3. hours
   wages $\frac{8}{52.80} = \frac{20}{x}$

   $8x = 1056$

   $x = \mathbf{\$132}$

4. acres
   bushels $\frac{12}{1440} = \frac{50}{x}$

   $12x = 72{,}000$

   $x = \mathbf{6{,}000\ bu.}$

5. width
   length $\frac{3}{5} = \frac{15}{x}$

   $3x = 75$

   $x = \mathbf{25\ in.}$

6. hours
   miles $\frac{1\frac{1}{2}}{72} = \frac{4}{x}$

   $1\frac{1}{2}x = 288$

   $x = 288 \div 1\frac{1}{2}$

   $x = \frac{288}{1} \div \frac{3}{2}$

   $x = \frac{\overset{96}{\cancel{288}}}{1} \times \frac{2}{\cancel{3}} = \mathbf{192\ mi.}$

7. inches
   miles $\frac{\frac{1}{2}}{15} = \frac{2}{x}$

   $\frac{1}{2}x = 30$

   $x = 30 \div \frac{1}{2}$

   $x = \frac{30}{1} \times \frac{2}{1} = \mathbf{60\ mi.}$

8. blue
   white $\frac{3}{2} = \frac{6}{x}$

   $3x = 12$

   $x = \mathbf{4\ gal.}$

9. wins
   losses $\frac{3}{5} = \frac{15}{x}$

   $3x = 75$

   $x = \mathbf{25\ games}$

10. divorces
    population $\frac{5}{1000} = \frac{x}{15{,}000}$

    $1000x = 75{,}000$

    $x = \mathbf{75\ divorces}$

Lesson 12

1. $\frac{15}{7500} = \mathbf{\frac{1}{500}}$

2. $12 + 20 + 18 + 30 = $
    80 total

    a. $\frac{20}{80} = \mathbf{\frac{1}{4}}$

    b. $\frac{30}{80} = \mathbf{\frac{3}{8}}$

    c. $12 + 18 = 30$

    $\frac{30}{80} = \mathbf{\frac{3}{8}}$

3. $10 + 15 + 8 = 33$ total

    $\frac{15}{33} = \mathbf{\frac{5}{11}}$

4. $5 + 7 + 3 = 15$ total

    a. $\frac{3}{15} = \mathbf{\frac{1}{5}}$

    b. There are now 4
    hardballs out of a total
    of 12 balls remaining.

    $\frac{4}{12} = \mathbf{\frac{1}{3}}$

5. $8 + 6 = 14$ total

    a. $\frac{8}{14} = \mathbf{\frac{4}{7}}$

    b. There are now 7 cans
    of tomato sauce and 5
    cans of green beans.

    $\mathbf{\frac{5}{12}}$

6. a. $\mathbf{\frac{1}{540}}$

    b. $\frac{3}{540} = \mathbf{\frac{1}{180}}$

7. $\frac{25}{100} = \mathbf{\frac{1}{4}}$

8. There are 25 red
    marbles out of a total of
    95 marbles remaining.

    $\frac{25}{95} = \mathbf{\frac{5}{19}}$

9. $\frac{20}{1000} = \mathbf{\frac{1}{50}}$

10. There are three dimes
    out of a total of eight
    coins remaining.

    $\mathbf{\frac{3}{8}}$

Level 2   Review

1. $\$240 + \$960 = \$1200$ total
    rent
    take-home $\frac{240}{1200} = \mathbf{\frac{1}{5}}$

2. $90 \div 2\frac{1}{4} = $

    $\frac{90}{1} \div \frac{9}{4} = $

    $\frac{\overset{10}{\cancel{90}}}{1} \times \frac{4}{\cancel{9}} = \mathbf{40\ mph}$

3. $c = nr$

    $c = 1\frac{3}{4} \times 1.84$

    $c = \frac{7}{\cancel{4}} \times \frac{\overset{46}{\cancel{1.84}}}{1} = \mathbf{\$3.22}$

4. $\left(\frac{3}{8}\right)^3 = \frac{3}{8} \times \frac{3}{8} \times \frac{3}{8} = \mathbf{\frac{27}{512}}$

5. $\frac{12}{16} = \frac{3}{4}$ **lb.**

6. $A = \frac{1}{2}bh$
$A = \frac{1}{2} \times 10 \times 5\frac{1}{4}$
$A = \frac{1}{2} \times \frac{\overset{5}{10}}{1} \times \frac{21}{4} = \frac{105}{4}$
$= \mathbf{26\frac{1}{4}}$ **in.²**

7. $\frac{1}{4} = \frac{9}{36}$ $\frac{4}{9} = \frac{16}{36}$ $\frac{1}{3} = \frac{12}{36}$ $\frac{5}{12} = \frac{15}{36}$
From smallest to largest: $\frac{1}{4}, \frac{1}{3}, \frac{5}{12}, \frac{4}{9}$

8. $65 + 25 + 10 = 100$ total
want : total = 65:100 = **13:20**

9. defective $\quad \frac{3}{100} = \frac{x}{20,000}$
total
$100x = 60,000$
$x = \mathbf{600}$ **defective bottles**

10. The bag now has 4 green + 5 blue + 3 black = 12 marbles.
$\frac{3}{12} = \frac{1}{4}$

Level 3 Fraction Problem Solving

Preview

1. won $\quad$ 8 $\quad$ lost $\quad \frac{5}{13} = \frac{15}{x}$
lost $\quad +5 \quad$ played
played $\quad 13 \quad\quad 5x = 195$
$\qquad\qquad\qquad x = \mathbf{39}$

2. total $\quad 7 \quad$ other $\quad \frac{3}{7} = \frac{x}{280}$
cash $\quad -4 \quad$ total
other $\quad 3 \quad\quad 7x = 840$
$\qquad\qquad\qquad x = \mathbf{120}$

3. (5) $\frac{(9 \times 20)}{17}$
$\frac{9}{x} = \frac{17}{20}$

4. 2 yd. = 2 × 3 = 6 ft.
8 in. = $\frac{8}{12} = \frac{2}{3}$ ft.
$V = lwh$
$V = 6 \times 1\frac{1}{2} \times \frac{2}{3}$
$V = \frac{6}{1} \times \frac{\overset{1}{3}}{\underset{1}{2}} \times \frac{\overset{1}{2}}{\underset{1}{3}} = \mathbf{6}$ **ft.³**

5. $c = nr$
$c = 10 \times 12.50 = \mathbf{\$125.00}$

Lesson 1

1. right $\quad 7 \quad$ right $\quad \frac{7}{9} = \frac{x}{36}$
wrong $\quad +2 \quad$ total
total $\quad 9 \quad\quad 9x = 252$
$\qquad\qquad\qquad x = \mathbf{28}$ **right**

2. total $\quad 5 \quad$ women $\quad \frac{2}{5} = \frac{x}{60}$
men $\quad -3 \quad$ total
women $\quad 2 \quad\quad 5x = 120$
$\qquad\qquad\qquad x = \mathbf{24}$ **women**

3. domestic $\quad 3 \quad$ domestic $\quad \frac{3}{5} = \frac{42}{x}$
imported $\quad +2 \quad$ total
total $\quad 5 \quad\quad 3x = 210$
$\qquad\qquad\qquad x = \mathbf{70}$ **cars**

4. total $\quad 500 \quad$ defective $\quad \frac{3}{500} = \frac{x}{15,000}$
good $\quad -497 \quad$ total
defective $\quad 3 \quad\quad 500x = 45,000$
$\qquad\qquad\qquad x = \mathbf{90}$

5. total $\quad 7 \quad$ car $\quad \frac{2}{5} = \frac{96}{x}$
car $\quad -2 \quad$ other
other $\quad 5 \quad\quad 2x = 480$
$\qquad\qquad\qquad x = \mathbf{\$240}$

6. total $\quad 8 \quad$ no $\quad \frac{3}{5} = \frac{120}{x}$
yes $\quad -5 \quad$ yes
no $\quad 3 \quad\quad 3x = 600$
$\qquad\qquad\qquad x = \mathbf{200}$ **people**

7. men $\quad 9 \quad$ women $\quad \frac{4}{13} = \frac{360}{x}$
women $\quad +4 \quad$ total
total $\quad 13 \quad\quad 4x = 4680$
$\qquad\qquad\qquad x = \mathbf{1170}$

8. total $\quad 10 \quad$ tea $\quad \frac{3}{10} = \frac{27}{x}$
coffee $\quad -7 \quad$ total
tea $\quad 3 \quad\quad 3x = 270$
$\qquad\qquad\qquad x = \mathbf{90}$

9. total $\quad 5 \quad$ no $\quad \frac{1}{4} = \frac{150}{x}$
yes $\quad -4 \quad$ yes
no $\quad 1 \quad\quad x = \mathbf{600}$

10. right $\quad 9 \quad$ right $\quad \frac{9}{10} = \frac{x}{120}$
wrong $\quad +1 \quad$ total
total $\quad 10 \quad\quad 10x = 1080$
$\qquad\qquad\qquad x = \mathbf{108}$

Lesson 2

1. (4) $\frac{20}{12}$

2. (3) $16 \times 4\frac{1}{2}$

3. (2) $\frac{(6 \times 9)}{3}$

4. (1) $12(5 \times 2\frac{1}{2})$

5. (1) $\frac{(12 \times 9)}{34}$

6. (3) $\frac{500}{90}$

7. (4) $\frac{(10 \times 275)}{75}$

8. (2) $\frac{(15 \times 12)}{32}$

9. (1) 85×60

10. (4) $\frac{(25,000)}{(85 \times 60)}$

Lesson 3

1. $l = \frac{30}{12} = 2\frac{6}{12} = 2\frac{1}{2}$ ft.
$A = lw$
$A = 2\frac{1}{2} \times 2$
$A = \frac{5}{2} \times \frac{\overset{1}{2}}{\underset{1}{1}} = \mathbf{5}$ **ft.²**

2. $l = 2$ yd. = 2 × 3 = 6 ft.
$w = 15$ in. = $\frac{15}{12} = 1\frac{3}{12} = 1\frac{1}{4}$ ft.
$P = 2l + 2w$
$P = 2(6) + 2(1\frac{1}{4})$
$P = 12 + \frac{2}{1} \times \frac{5}{\underset{2}{4}}$
$P = 12 + 2\frac{1}{2} = \mathbf{14\frac{1}{2}}$ **ft.**

3. $b = 1$ ft. = 1 × 12 = 12 in.
$A = bh$
$A = 12 \times 9 = \mathbf{108}$ **in.²**

4. 18 in. = $\frac{18}{12} = 1\frac{6}{12} = 1\frac{1}{2}$ in.
$A = s^2$
$A = (1\frac{1}{2})^2$
$A = 1\frac{1}{2} \times 1\frac{1}{2}$
$A = \frac{3}{2} \times \frac{3}{2} = \frac{9}{4} = \mathbf{2\frac{1}{4}}$ **ft.²**

5. $b = 28$ in. $= \frac{28}{12} = 2\frac{4}{12} = 2\frac{1}{3}$ ft.

  $h = 1$ yd. $= 1 \times 3 = 3$ ft.

  $A = \frac{1}{2} bh$

  $A = \frac{1}{2} \times 2\frac{1}{3} \times 3$

  $A = \frac{1}{2} \times \frac{7}{\overset{1}{\cancel{3}}} \times \frac{\overset{1}{\cancel{3}}}{1} = \frac{7}{2} = $ **$3\frac{1}{2}$ ft.²**

6. $h = 9$ in. $= \frac{9}{12} = \frac{3}{4}$ ft.

  $V = lwh$

  $V = \frac{20}{1} \times \frac{\overset{1}{\cancel{4}}}{1} \times \frac{3}{\cancel{4}} = $ **60 ft.³**

7. $l = 12$ ft. $= \frac{12}{3} = 4$ yd.

  $w = 6$ ft. $= \frac{6}{3} = 2$ yd.

  $h = 5$ ft. $= \frac{5}{3} = 1\frac{2}{3}$ yd.

  $V = lwh$

  $V = 4 \times 2 \times 1\frac{2}{3}$

  $V = \frac{4}{1} \times \frac{2}{1} \times \frac{5}{3} = \frac{40}{3} = $ **$13\frac{1}{3}$ yd.²**

8. $l = 80$ in. $= \frac{80}{12} = 6\frac{8}{12} = 6\frac{2}{3}$ ft.

  $w = 2$ ft.

  $h = 3$ yd. $= 3 \times 3 = 9$ ft.

  $V = lwh$

  $V = 6\frac{2}{3} \times 2 \times 9$

  $V = \frac{20}{\underset{1}{\cancel{3}}} \times 2 \times \overset{3}{\cancel{9}} = $ **120 ft.³**

9. $s = 18$ in. $= \frac{18}{12} = 1\frac{1}{2}$ ft.

  $V = s^3$

  $V = 1\frac{1}{2} \times 1\frac{1}{2} \times 1\frac{1}{2}$

  $V = \frac{3}{2} \times \frac{3}{2} \times \frac{3}{2} = \frac{27}{8} = $ **$3\frac{3}{8}$ ft.³**

10. $l = 24$ in. $= \frac{24}{36} = \frac{2}{3}$ yd.

  $w = 1$ ft. $= \frac{1}{3}$ yd.

  $h = 1$ yd.

  $V = lwh$

  $V = \frac{2}{3} \times \frac{1}{3} \times 1 = $ **$\frac{2}{9}$ yd.³**

Lesson 4

1. 24 gallons is extraneous.

  $d = rt$

  $d = 48 \times 6\frac{1}{2}$

  $d = \frac{\overset{24}{\cancel{48}}}{1} \times \frac{13}{\underset{1}{\cancel{2}}} = $ **312 mi.**

2. 48 with vaccinations is extraneous.

  boys 32   girls $\frac{28}{60} = \frac{7}{15}$

  girls $\underline{+28}$ total

  total 60

3. 10 inches wide is extraneous.

  $3\frac{1}{2} \times 12 =$

  $\frac{7}{\cancel{2}} \times \frac{\overset{6}{\cancel{12}}}{1} = $ **42 lb.**

4. $3.80 and $1.89 a pound are extraneous.

  $4\frac{1}{4}$

  $\underline{+5\frac{3}{4}}$

  $9\frac{4}{4} = $ **10 lb.**

---

5. 8 is extraneous.

  right 54   right $\frac{54}{60} = \frac{9}{10}$

  wrong $\underline{+6}$ total

  total 60

6. $1.12 is extraneous.

  $15\frac{8}{16} = $ **$15\frac{1}{2}$ mi.**

  $16\overline{)248}$
  $\quad\underline{16}$
  $\quad 88$
  $\quad\underline{80}$
  $\quad\ \ 8$

7. $7\frac{1}{2}$ feet high is extraneous.

  $A = lw$

  $A = 60 \times 20 = $ **1200 ft.²**

8. 3 gallons of thinner is extraneous.

  gray $\frac{2}{5} = \frac{x}{10}$

  green

  $5x = 20$

  $x = $ **4 gal. of gray paint**

9. 50 mph is extraneous.

  $\begin{array}{r}7 \text{ gal.}\\25\overline{)175}\end{array}$ $\begin{array}{r}\$1.30\\\times\quad 7\\\hline \mathbf{\$9.10}\end{array}$

10. $210 is extraneous.

  $\begin{array}{r}\$1800\\-\underline{1200}\\\$\ 600\end{array}$ $\frac{600}{1800} = \frac{1}{3}$

Level 3   Review

1. mgmt 3   mgmt $\frac{3}{17} = \frac{x}{340}$

  labor $\underline{+14}$ total

  total 17 $17x = 1020$

  $x = $ **60 people**

2. girls 5   boys $\frac{4}{9} = \frac{84}{x}$

  boys $\underline{+4}$ total

  total 9 $4x = 756$

  $x = $ **189 children**

3. total 9   rail $\frac{2}{7} = \frac{x}{140}$

  rail $\underline{-2}$ truck

  truck 7 $7x = 280$

  $x = $ **40 lb.**

4. (1) $\frac{(4 \times 11)}{16}$

  $\frac{11}{16} \times \frac{x}{4}$

5. (3) $\frac{(310 + 165)}{450}$

6. (4) $20 \times 8 \times 5$

7. 15 in. $= \frac{15}{12} = 1\frac{3}{12}$

  $\quad\quad = 1\frac{1}{4}$ ft.

  $A = \frac{1}{2} bh$

  $A = \frac{1}{2} \times 2 \times 1\frac{1}{4}$

  $A = \frac{1}{\cancel{2}} \times \frac{\cancel{2}}{1} \times \frac{5}{4} = \frac{5}{4}$

  $\quad = 1\frac{1}{4}$ ft.²

8. 18 in. $= \frac{18}{12} = 1\frac{6}{12}$

  $\quad\quad = 1\frac{1}{2}$ ft.

  $V = s^3$

  $V = (1\frac{1}{2})^3$

  $V = \frac{3}{2} \times \frac{3}{2} \times \frac{3}{2} = \frac{27}{8}$

  $\quad = 3\frac{3}{8}$ ft.³

9. food $\frac{360}{960} = \frac{3}{8}$

  total

10. $\frac{1}{\cancel{5}} \times \frac{\overset{480}{\cancel{2400}}}{1} = $ **$480**

Chapter 2 Quiz

1. $\frac{3}{4} \times \frac{\overset{3550}{14,200}}{1} = $10,650$

2. Donna $\frac{120}{180} = \frac{2}{3}$ Steve

3. $c = nr$
$c = 2\frac{1}{4} \times 384$
$c = \frac{9}{4} \times \frac{\overset{96}{384}}{1} = 864

4. $d = rt$
$d = \frac{3}{4} \times \frac{\overset{115}{460}}{1} = 345\text{ mi.}$

5. $\sqrt{\frac{49}{100}} = \frac{7}{10}$

6. $16\overline{)40} = 2\frac{1}{2}\text{ lb.}$ $\frac{2\frac{8}{16}}{\frac{32}{8}}$

7. $A = lw$
$A = 6 \times 3\frac{3}{4}$
$A = \frac{\overset{3}{6}}{1} \times \frac{15}{\underset{2}{4}} = \frac{45}{2} = 22\frac{1}{2}\text{ in.}^2$

8. (2) E, A, C, D, B
A—2 ft. 9 in. = 33 in.
B—37 in. = 37 in.
C—33$\frac{1}{4}$ in. = 33$\frac{1}{4}$ in.
D—1 yd. = 36 in.
E—2$\frac{1}{2}$ ft. = 30 in.

9. total 128 whole:other
whole −80 80:48
other 48 **5:3**

10. flour $\frac{5}{2} = \frac{8}{x}$ sugar
$5x = 16$
$x = 3\frac{1}{5}\text{ c.}$

11. (4) $\frac{5}{12}$
$15 + 12 + 9 = 36\text{ total}$
$\frac{15}{36} = \frac{5}{12}$

12. (3) $\frac{3}{11}$
$13 + 11 + 9 = 33\text{ total now}$
$\frac{9}{33} = \frac{3}{11}$

13. (3) $80,000
total 5
raised −3
need 2
$\frac{\text{raised}}{\text{need}} \quad \frac{3}{2} = \frac{120,000}{x}$
$3x = 240,00$
$x = $80,000$

14. (1) 882
children 5
adults +2
total 7
children $\frac{5}{7} = \frac{630}{x}$
total
$5x = 4410$
$x = 882$

15. (5) $\frac{(18 \times 25)}{3}$
$\frac{3}{18} = \frac{25}{x}$

16. (2) $\frac{(6 \times 380)}{50}$

17. (4) $6\frac{1}{4}$
$30\text{ in.} = \frac{30}{12} = 2\frac{6}{12} = 2\frac{1}{2}\text{ ft.}$
$A = s^2$
$A = (2\frac{1}{2})^2$
$A = \frac{5}{2} \times \frac{5}{2} = \frac{25}{4} = 6\frac{1}{4}\text{ ft.}^2$

18. (4) 6
1 yd. = 1 × 3 = 3 ft.
3 in. = $\frac{3}{12} = \frac{1}{4}$ ft.
$V = lwh$
$V = \frac{\overset{2}{8}}{1} \times \frac{3}{1} \times \frac{1}{\underset{1}{4}} = 6\text{ ft.}^3$

19. (2) $\frac{3}{20}$
downpymt $\frac{6,000}{40,000} = \frac{3}{20}$
total

20. (3) 32
$304 \div 9\frac{1}{2} =$
$\frac{304}{1} \div \frac{19}{2} =$
$\frac{\overset{16}{304}}{1} \times \frac{2}{\underset{1}{19}} = 32\text{ mpg}$

Chapter 3 Decimals

Level 1 Decimal Skills

Preview
1. **18.023**
2. .409 .28 +.7 **1.389**
3. .820 −.197 **.623**
4. 4.5 ×.26 270 90 1.170 = **1.17**
5. $.26\frac{10}{15} = .26\frac{2}{3}$ 15$\overline{)4.00}$ 3 0 1 00 90 10

Lesson 1
1. a. **two** b. **three** c. **three** d. **five** e. **two** f. **one**
2. a. **two** b. **four** c. **four** d. **two** e. **one** f. **five**
3. a. **20.067** b. **.409** c. **28.7** d. **1.208** e. **3.6** f. **4.5**
4. a. **3.04** b. **.4** c. **.607** d. **.003** e. **.107** f. **.59**
5. a. **100.0107** b. **.39** c. **74,000.1** d. **6080.01** e. **150.2** f. **19.2**

Lesson 2
1. **d** 2. **c** 3. **e** 4. **b** 5. **a**
6. **.7** 7. **.036** 8. **.0519**
9. **.00854** 10. **.003768**

Lesson 3
1. .36 .5 +.607 **1.467** | .38 .619 +.2 **1.199** | .3 .9 +.7 **1.9** | .006 .05 +.8 **.856**
2. 2.5 18. +1.07 **21.57** | .506 3.1 +9. **12.606** | 38. 4.078 +.0195 **42.0975** | 9.1 .87 +.143 **10.113**
3. 6.0 −2.5 **3.5** | 8.00 −.19 **7.81** | .300 −.258 **.042** | 5.900 −2.114 **3.786**

4.
```
   .015        1.0000         9.00          .60
  −.009       − .0865       − .32         −.24
   .006         .9135         8.68          .36
```

5.
```
   .0174       4.8300          .62          .206
  −.0162      −1.7123         3.18         3.605
   .0012       3.1177        + .0132      +2.7
                              3.8132       6.511
```

Lesson 4

1.
```
   3.5            .56
  × 7           × 8
  24.5          4.48

    29            .06
  ×.04          ×.5
  1.16          .030 = .03
```

2.
```
   .47            .185
  ×16            × .4
  2 82           .0740 = .074
  4 7
  7.52

   .82            .59
  ×3.6          ×.004
  492            .00236
  2 46
  2.952
```

3.
```
   2.09           .088
  × 30          ×.52
  62.70 = 62.7   176
                 440
                 .04576

   .0065          215
  × .6          ×.04
  .00390 = .0039  8.60 = 8.6
```

4.
```
      2.7            .48
  6)16.2         9)4.32
     12             3 6
      4 2            72
      4 2            72

      8.065          9.7
  3)24.195       15)145.5
     24             135
      0 19           10 5
        18           10 5
        15
        15
```

5.
```
      .053           1.9
  14).742        25)47.5
     70             25
      42             22 5
      42             22 5

      .018           6.6
  64)1.152       31)204.6
     64             186
      512            18 6
      512            18 6
```

6.
```
       4 00               6 5
  .32)128.00        2.6)169.0
     128                156
       0 00              13 0
                         13 0

       7 00               9 0
  .08)56.00         1.2)108.0
                         108
                           0 0
```

7.
```
       300               43 5
  .008)2.400         .6)261.0
                        24
                        21
                        18
                         3 0
                         3 0

       6 2               12
  7.5)465.0         .026).312
     450                 26
      15 0               52
      15 0               52
```

8.
```
      .7                .36
  .9).6 3           .7).2 52
                        2 1
                        42
                        42

      4.8               .53
  .03).14 4         .06).03 18
      12                 3 0
       2 4               18
       2 4               18
```

9.
```
      1.06               .04
  3.6)3.8 16         .48).01 92
     3 6                 1 92
      2 16
      2 16

       6                1.5
  .052).312         .125).187 5
      312                125
                          62 5
                          62 5
```

10.
```
      .38               47 2
  .8).3 04          1.5)708.0
     2 4                60
      64                108
      64                105
                          3 0
                          3 0

       .9               1.16
  .703).632 7        .8).9 28
                         8
                         1 2
                          8
                          4 8
                          4 8
```

Lesson 5

1. $.8 = \frac{8}{10} = \frac{4}{5}$ $.04 = \frac{4}{100} = \frac{1}{25}$

   $.35 = \frac{35}{100} = \frac{7}{20}$ $.005 = \frac{5}{1000} = \frac{1}{200}$

2. $.065 = \frac{65}{1000} = \frac{13}{200}$ $.0075 = \frac{75}{10,000} = \frac{3}{400}$

   $.002 = \frac{2}{1000} = \frac{1}{500}$ $.875 = \frac{875}{1000} = \frac{7}{8}$

3. $.6 = \frac{6}{10} = \frac{3}{5}$ $.075 = \frac{75}{1000} = \frac{3}{40}$

   $.006 = \frac{6}{1000} = \frac{3}{500}$ $.925 = \frac{925}{1000} = \frac{37}{40}$

4. $8.4 = 8\frac{4}{10} = 8\frac{2}{5}$ $2.85 = 2\frac{85}{100} = 2\frac{17}{20}$

   $1\frac{4}{1000} = 1\frac{1}{250}$ $6.3 = 6\frac{3}{10}$

5. $10.125 = 10\frac{125}{1000} = 10\frac{1}{8}$ $3.009 = 3\frac{9}{1000}$

   $4.80 = 4\frac{80}{100} = 4\frac{4}{5}$ $12.16 = 12\frac{16}{100} = 12\frac{4}{25}$

6. $\frac{9}{10} = 10\overline{)9.0} \rightarrow .9$ $\frac{4}{5} = 5\overline{)4.0} \rightarrow .8$

   $\frac{9}{50} = 50\overline{)9.00} \rightarrow .18$ $\frac{3}{8} = 8\overline{)3.00} \rightarrow .37\frac{4}{8} = .37\frac{1}{2}$ or $.375$
   $\underline{5\ 0}$ $\underline{2\ 4}$
   $4\ 00$ 60
   $\underline{4\ 00}$ $\underline{56}$
    4

7. $\frac{1}{2} = 2\overline{)1.0} \rightarrow .5$ $\frac{2}{3} = 3\overline{)2.00} \rightarrow .66\frac{2}{3}$
    $\underline{1\ 8}$
    20
    $\underline{18}$
    2

   $\frac{8}{25} = 25\overline{)8.00} \rightarrow .32$ $\frac{1}{6} = 6\overline{)1.00} \rightarrow .16\frac{4}{6} = .16\frac{2}{3}$
   $\underline{7\ 5}$ $\underline{6}$
   50 40
   $\underline{50}$ $\underline{36}$
    4

8. $\frac{2}{9} = 9\overline{)2.00} \rightarrow .22\frac{2}{9}$ $\frac{3}{4} = 4\overline{)3.00} \rightarrow .75$
   $\underline{1\ 8}$ $\underline{2\ 8}$
   20 20
   $\underline{18}$ $\underline{20}$
   2

   $\frac{5}{12} = 12\overline{)5.00} \rightarrow .41\frac{8}{12} = .41\frac{2}{3}$ $\frac{7}{8} = 8\overline{)7.00} \rightarrow .87\frac{4}{8} = .87\frac{1}{2}$ or $.875$
   $\underline{4\ 8}$ $\underline{6\ 4}$
   20 60
   $\underline{12}$ $\underline{56}$
   8 4

9. $\frac{7}{10} = 10\overline{)7.0} \rightarrow .7$ $\frac{5}{8} = 8\overline{)5.00} \rightarrow .62\frac{4}{8} = .62\frac{1}{2}$ or $.625$
    $\underline{4\ 8}$
    20
    $\underline{16}$
    4

   $\frac{7}{12} = 12\overline{)7.00} \rightarrow .58\frac{4}{12} = .58\frac{1}{3}$ $\frac{1}{8} = 8\overline{)1.00} \rightarrow .12\frac{4}{8} = .12\frac{1}{2}$ or $.125$
   $\underline{6\ 0}$ $\underline{8}$
   $1\ 00$ 20
   $\underline{96}$ $\underline{16}$
   4 4

10. $\frac{4}{9} = 9\overline{)4.00} \rightarrow .44\frac{4}{9}$ $\frac{3}{5} = 5\overline{)3.00} \rightarrow .6$
   $\underline{3\ 6}$
   40
   $\underline{36}$
   4

   $\frac{5}{6} = 6\overline{)5.00} \rightarrow .83\frac{2}{6} = .83\frac{1}{3}$ $\frac{2}{50} = 50\overline{)2.00} \rightarrow .04$
   $\underline{4\ 8}$
   20
   $\underline{18}$
   2

Level 1   Review

1. **(3) seven and six hundredths**

2. **60.0012**   3. $.385$ 4. $.0580$
   $.6$ $\underline{-.0496}$
   $\underline{+.09}$ $\mathbf{.0084}$
   $\mathbf{1.075}$

5. 12.8 6. $\mathbf{.032}$
   $\times.35$ $76\overline{)2.432}$
   640 $\underline{2\ 28}$
   $\underline{3\ 84}$ 152
   $4.480 = \mathbf{4.48}$ $\underline{152}$

7. $\mathbf{9.4}$ 8. $\frac{625}{1000} = \frac{5}{8}$
   $.07\overline{)\,.65\ 8}$
   $\underline{63}$
   $2\ 8$
   $\underline{2\ 8}$

9. $4.32 = 4\frac{32}{100} = \mathbf{4\frac{8}{25}}$ 10. $.31\frac{4}{16} = .31\frac{1}{4}$
   $\frac{5}{16} = 16\overline{)5.00}$
   $\underline{4\ 8}$
   20
   $\underline{16}$
   4

Level 2   Decimal Applications

Preview

1. **0.284**
2. $\$3.48$
   $\underline{\times3.6}$
   $2\ 088$
   $\underline{10\ 44}$
   $\$12.528$ to the nearest cent = **$12.53**
3. $(.5)^3 = .5 \times .5 \times .5 = \mathbf{.125}$

4. $\underset{1000\overline{)655.000}}{\text{.655 m}}$

5. $A = \pi r^2$
$A = 3.14 \times (0.4)^2$
$A = 3.14 \times 0.4 \times 0.4$
$A = 0.5024$ to the nearest tenth $= \textbf{0.5 m}^2$

Lesson 1

1. **.6**   **.4**   **.1**   **5.2**   **0.3**
2. **.2**   **.1**   **4.8**   **.2**   **.7**
3. **.53**   **.48**   **2.02**   **8.30**   **0.91**
4. **.80**   **.40**   **1.03**   **6.92**   **.09**
5. **.139** **.058** **1.781** **0.105** **6.433**

Lesson 2

1. $12.5 million
   $+ \quad 6.75$
   $19.25 million

2.   8   **5.85 lb.**
     4.5 $4\overline{)23.40}$
     7.65 $\underline{20}$
   $+3.25$ 3 4
     23.40 $\underline{3\,2}$
                    2 0
                    $\underline{2\,0}$

3.   4534.2 mi.
   -3789.6
   744.6 mi.

4.   .268
   $\times\ \ 45$
   1 340
   $\underline{10\,72}$
   12.060 to the nearest whole number = **12 hits**

5. $8.40
   $\times\ \ 6.5$
   4 200
   $\underline{50\,40}$
   $54.600 = **$54.60**

6. $\underset{8.4\overline{)200.0\,0}}{2\ 3.8}$ to the nearest gallon = **24 mpg**
   168
   32 0
   $\underline{25\,2}$
    6 8 0
    $\underline{6\,7\,2}$

7. $0.036
   $\times\ \ 45$
   180
   $\underline{1\,44}$
   $1.620 = **$1.62**

8.   98.6°   104.4°
   $+\ 5.8°$ $-\ 3.9°$
   104.4°   **100.5°**

9.   6.25
   $\times\ \ 55$
   31 25
   $\underline{312\,5}$
   343.75 to the nearest mile = **344 mi.**

10. $\underset{6.80\overline{)238.00}}{\textbf{35 hr.}}$
    204 0
    34 00
    $\underline{34\,00}$

Lesson 3

1. $(.5)^2 = .5 \times .5 = \textbf{.25}$ $(.02)^3 = .02 \times .02 \times .02$
   $(.4)^2 = .4 \times .4 = \textbf{.16}$ $= \textbf{.000008}$
                                        $(.12)^2 = .12 \times .12 = \textbf{.0144}$

2. $(.07)^2 = .07 \times .07$ $(.009)^2 = .009 \times .009$
           $= \textbf{.0049}$ $= \textbf{.000081}$
   $(.1)^4 = .1 \times .1 \times .1 \times .1 = \textbf{.0001}$
   $(1.5)^2 = 1.5 \times 1.5 = \textbf{2.25}$

3. $\sqrt{.16} = \textbf{.4}$ $\sqrt{.81} = \textbf{.9}$
   $\sqrt{.0036} = \textbf{.06}$ $\sqrt{.0001} = \textbf{.01}$

4. $\sqrt{.0004} = \textbf{.02}$ $\sqrt{.0121} = \textbf{.11}$
   $\sqrt{.000009} = \textbf{.003}$ $\sqrt{.0625} = \textbf{.25}$

5. $\sqrt{.0025} = \textbf{.05}$ $\sqrt{.0049} = \textbf{.07}$
   $\sqrt{.64} = \textbf{.8}$ $\sqrt{.000144} = \textbf{.012}$

Lesson 4

1. $2.5 \times 100 = \textbf{250 cm}$ 2. $3 \times 1000 = \textbf{3000 g}$
3. $4.8 \times 1000 = \textbf{4800 ml}$ 4. $6.5 \times 1000 = \textbf{6500 m}$
5. $\underset{1000\overline{)1250.00}}{\textbf{1.25 km}}$ 6. $\underset{10\overline{)8.0}}{\textbf{0.8 l}}$
7. $\underset{1000\overline{)385.000}}{\textbf{0.385 kg}}$ 8. $\underset{100\overline{)195.00}}{\textbf{1.95 m}}$
9. $\underset{1000\overline{)1680.00}}{\textbf{1.68 g}}$ 10. $1.6 \times 100 = \textbf{160 cl}$

Lesson 5

1. a. $P = 2l + 2w$ b. $A = lw$
      $P = 2(8.5) + 2(4.2)$ $A = 8.5 \times 4.2$
      $P = 17 + 8.4 = \textbf{25.4 m}$ $A = \textbf{35.7 m}^2$

2. a. $P = 4s$ b. $A = s^2$
      $P = 4(3.4) = \textbf{13.6 cm}$ $A = (3.4)^2$
                                        $A = 3.4 \times 3.4 = \textbf{11.56 cm}^2$

3. a. $P = a + b + c$ b. $A = \frac{1}{2}bh$
      $P = 6 + 3.6 + 4.8$ $A = \frac{1}{2} \times \frac{8}{1} \times \frac{3.6}{1}$
      $P = \textbf{14.4 ft.}$
                                        $A = \textbf{10.8 ft.}^2$

4. $A = bh$ 5. $V = lwh$
   $A = 4.1 \times 2.2$ $V = 20 \times 4.5 \times 3.6$
   $A = \textbf{9.02 m}^2$ $V = \textbf{32.4 cm}^3$

Lesson 6

1. a. $r = \frac{d}{2}$

   $r = \frac{8}{2}$

   $r = $ **4 in.**

   c. $A = \pi r^2$

   $A = 3.14 \times 4^2$

   $A = 3.14 \times 16$

   $A = $ **50.24 in.²**

   b. $C = \pi d$

   $C = 3.14 \times 8$

   $C = $ **25.12 in.**

2. a. $d = 2r$

   $d = 2 \times 30$

   $d = $ **60 ft.**

   c. $A = \pi r^2$

   $A = 3.14 \times 30^2$

   $A = 3.14 \times 900$

   $A = $ **2826 ft.²**

   b. $C = \pi d$

   $C = 3.14 \times 60$

   $C = $ **188.4 ft.**

3. a. $d = 2r$

   $d = 2 \times 1.2$

   $d = $ **2.4 m**

   c. $A = \pi r^2$

   $A = 3.14 \times (1.2)^2$

   $A = 3.14 \times 1.2 \times 1.2$

   $A = 4.5216$ to the
   nearest tenth =
   4.5 m²

   b. $C = \pi d$

   $C = 3.14 \times 2.4$

   $C = 7.536$ to the nearest
   tenth = **7.5 m**

4. a. $r = \frac{d}{2}$

   $r = \frac{40}{2}$

   $r = $ **20 in.**

   c. $A = \pi r^2$

   $A = 3.14 \times 20^2$

   $A = 3.14 \times 400$

   $A = $ **1256 in.²**

   b. $C = \pi d$

   $C = 3.14 \times 40$

   $C = $ **125.6 in.**

5. a. $C = \pi d$

   $C = \frac{22}{7} \times \frac{2.8}{1}$

   $C = $ **8.8 in.**

   b. $A = \pi r^2$

   $A = \frac{22}{7} \times (1.4)^2$

   $A = \frac{22}{7} \times \frac{1.4}{1} \times \frac{1.4}{1}$

   $A = $ **6.16 m²**

Lesson 7

1. $V = \pi r^2 h$

   $V = 3.14 \times 5^2 \times 20$

   $V = 3.14 \times 5 \times 5 \times 20$

   $V = $ **1570 ft.³**

2. $V = \pi r^2 h$

   $V = 3.14 \times (.2)^2 \times 1.5$

   $V = 3.14 \times .2 \times .2 \times 1.5$

   $V = 0.1884$ to the
   nearest tenth =
   0.2 m³

3. $V = \pi r^2 h$

   $V = 3.14 \times 30^2 \times 1$

   $V = 3.14 \times 30 \times 30 \times 1$

   $V = $ **2826 ft.³**

4. $V = \pi r^2 h$

   $V = 3.14 \times 20^2 \times 0.5$

   $V = 3.14 \times 20 \times 20 \times 0.5$

   $V = $ **628 m³**

5. $V = \pi r^2 h$

   $V = 3.14 \times 1^2 \times 25$

   $V = 3.14 \times 1 \times 1 \times 25$

   $V = $ **78.5 ft.³**

Lesson 8

1. a. $0.056 = 0.056$
   $0.05 = 0.050$
   0.056 is larger.

   b. $0.19 = 0.19$
   $0.2 = 0.20$
   0.2 is larger.

   c. $1.08 = 1.080$
   $1.082 = 1.082$
   1.082 is larger.

   d. $0.075 = 0.075$
   $0.57 = 0.570$
   0.57 is larger.

2. $0.021 = 0.021$, $0.012 = 0.012$, $0.21 = 0.210$, $0.201 = 0.201$
   From smallest to largest: **0.012, 0.021, 0.201, 0.21**

3. $0.38 = 0.380$, $0.8 = 0.800$, $0.083 = 0.083$, $0.308 = 0.308$
   From largest to smallest: **0.8, 0.38, 0.308, 0.083**

4. $0.705 = 0.705$, $0.75 = 0.750$, $0.075 = 0.075$
   0.75 m is largest.

5. **(2) D, E, B, A, C**
   A 0.65   = 0.650 kg
   B 1.05   = 1.050 kg
   C 0.065 = 0.065 kg
   D 1.65   = 1.650 kg
   E 1.5   = 1.500 kg

Lesson 9

1. **2.4 cm (24 mm)** 2. **3.9 cm (39 mm)** 3. **4.4 cm (44 mm)**

4. **7.1 cm (71 mm)** 5. **8.5 cm (85 mm)** 6. **7.1 − 3.9 = 3.2 cm (32 mm)**

7. **3.9 − 2.4 = 1.5 cm (15 mm)** 8. **4.4 − 3.9 = .5 cm (5 mm)**
9. **7.1 − 4.4 = 2.7 (27 mm)**   10. **8.5 − 7.1 = 1.4 (14 mm)**

Level 2   Review

1. **6.30**

2. $\quad .2833$ to the nearest thousandth = **.283**

   $60\overline{)17.0000}$
   $\quad \underline{12\ 0}$
   $\quad \ 5\ 00$
   $\quad \ \underline{4\ 80}$
   $\quad \quad \ 200$
   $\quad \quad \ \underline{180}$
   $\quad \quad \ 200$
   $\quad \quad \ \underline{180}$

3. $\quad 22.5$
   $\underline{\times 8.4}$
   $\quad 9\ 00$
   $\underline{180\ 0}$
   $189.00 = $ **189 mi.**

4. $\sqrt{.0049} = $ **.07**

5. $2.4 \times 1000 = $ **2400 g**

6. $A = \frac{1}{2} bh$

   $A = \frac{1}{2} \times \frac{5.4}{1} \times \frac{2.8}{1} = 7.56$
   to the nearest tenth
   = **7.6 m²**

7. $A = \pi r^2$
   $A = 3.14 \times 8^2$
   $A = 3.14 \times 8 \times 8$
   $A = 200.96$ to the
   nearest square
   inch = **201 in.²**

8. $V = \pi r^2 h$
   $V = 3.14 \times 3^2 \times 5$
   $V = 3.14 \times 3 \times 3 \times 5$
   $V = $ **141.3 ft.³**

9. $0.06 = 0.060$, $0.066 = 0.066$, $0.6 = 0.600$, $0.065 = 0.065$

In order: **0.06, 0.065, 0.066, 0.6**

10. **2.7 cm (27 mm)**

Level 3   Decimal Problem Solving

Preview

1. **(2) 1.7 million**

$$\begin{array}{r} 27.6 \\ -25.9 \\ \hline 1.7 \text{ million} \end{array}$$

2. **(1) more than $1\frac{1}{2}$ times**

$$\begin{array}{r} 1.58 \text{, which is more than } 1\frac{1}{2} \\ 14\overline{)22.20} \\ \underline{14} \\ 8\,2 \\ \underline{7\,0} \\ 1\,20 \\ 1\,12 \end{array}$$

3.

$d = rt = 42 \times 2.5 = 105$ mi.

$d = rt = 15 \times 2.5 = 37.5$ mi.

$$\begin{array}{r} 105.0 \text{ mi} \\ -37.5 \\ \hline \textbf{67.5 mi.} \end{array}$$

4.

$d = rt = 350 \times 1.5 = 525$ mi.

$d = rt = 420 \times 1.5 = 630$ mi.

$$\begin{array}{r} 525 \text{ mi} \\ +630 \\ \hline \textbf{1155 mi.} \end{array}$$

5. **(5) Insufficient data is given to solve the problem.**

Lesson 1

1. $\begin{array}{r} 9.3 \\ -5.6 \\ \hline 3.7 \end{array}$ 2. $\begin{array}{r} 8.9 \\ -7.9 \\ \hline 1.0 \end{array}$

3. $\begin{array}{r} 15 \\ 1000\overline{)15{,}000} \end{array}$ $\begin{array}{r} 5.6 \\ \times\ 15 \\ \hline 28\,0 \\ 56 \\ \hline \textbf{84.0 beds} \end{array}$

4. $\begin{array}{r} 280 \\ 1000\overline{)280{,}000} \end{array}$ $\begin{array}{r} 280 \\ \times\ 6 \\ \hline \textbf{1680 beds} \end{array}$

5. **(4) The rate gradually decreased.**

6. **1980**

7. $\begin{array}{r} 14.1 \\ 11.4 \\ +11.1 \\ \hline 36.6 \end{array}$ $\begin{array}{r} 45.2 \\ -36.6 \\ \hline \textbf{8.6 million} \end{array}$

8. $\begin{array}{r} 43.7 \\ -30.4 \\ \hline \textbf{13.3 million} \end{array}$

9. $\begin{array}{r} 10.5 \\ -9.5 \\ \hline \textbf{1.0 million} \end{array}$

10. **(2) about $\frac{1}{3}$**

14.1 is close to 15.

45.2 is close to 45.

$\frac{15}{45} = \frac{1}{3}$

Lesson 2

1.

$d = rt = 55 \times 0.5 = 27.5$ mi.

$d = rt = 45 \times 0.5 = 22.5$ mi.

$$\begin{array}{r} 27.5 \\ +22.5 \\ \hline \textbf{50.0 mi.} \end{array}$$

2.

$d = rt = 40 \times 2.5 = 100$ mi.

$d = rt = 30 \times 2.5 = 75$ mi.

$$\begin{array}{r} 100 \\ -75 \\ \hline \textbf{25 mi.} \end{array}$$

3.

$d = rt = 60 \times 4.5 = 270$ mi.

$d = rt = 50 \times 4.5 = 225$ mi.

$$\begin{array}{r} 270 \\ -225 \\ \hline \textbf{45 mi.} \end{array}$$

4.

$d = rt = 12 \times 1.25 = 15$ mi.

$d = rt = 10 \times 1.25 = 12.5$ mi.

$$\begin{array}{r} 15 \\ +12.5 \\ \hline \textbf{27.5 mi.} \end{array}$$

5.

Express travels from 10:30 to 1:00, or 2.5 hr.

$d = rt = 50 \times 2.5 = 125$ mi.

Local travels from 11:30 to 1:00, or 1.5 hr.

$d = rt = 30 \times 1.5 = 45$ mi.

$$\begin{array}{r} 125 \\ -45 \\ \hline \textbf{80 mi.} \end{array}$$

6.

$d = rt = 24 \times 0.75 = 18$ mi.

$d = rt = 16 \times 0.75 = 12$ mi.

$$\begin{array}{r} 18 \\ -12 \\ \hline \textbf{6 mi.} \end{array}$$

7. $d = rt = 360 \times 2.25 = 810$ mi.

$d = rt = 480 \times 2.25 = 1080$ mi.

$$\begin{array}{r} 810 \\ +1080 \\ \hline \textbf{1890 mi.} \end{array}$$

8.

$d = rt = 20 \times 3.5 = 70$ mi.

$d = rt = 4 \times 3.5 = 14$ mi.

$$\begin{array}{r} 70 \\ -14 \\ \hline \textbf{56 mi.} \end{array}$$

9. $d = rt = 55 \times 1.5 = 82.5$ mi.

$d = rt = 30 \times 1.5 = 45$ mi.

$$\begin{array}{r} 82.5 \text{ mi.} \\ +45. \\ \hline \textbf{127.5 mi.} \end{array}$$

10.

$d = rt = 40 \times 4.5 = 180$ mi.

$d = rt = 45 \times 3 = 135$ mi.

$d = rt = 55 \times 1.5 = 82.5$ mi.

$$\begin{array}{rr} 135.0 & 180 \\ -82.5 & +52.5 \\ \hline 52.5 \text{ mi.} & \textbf{232.5 mi.} \end{array}$$

Lesson 3

1. The weight of the fourth package is missing.
2. The height is missing.
3. The price of the beef is missing.
4. The breakdown of males and females is missing.
5. The average number of beds per hospital is missing.
6. Their total monthly income is missing.
7. The overtime pay rate is missing.
8. The total weight of the nails is missing.
9. The cost of parking is missing.
10. The height is missing.

Level 3   Review

1. pork

2. (2) more than $2\frac{1}{2}$ times

$$\begin{array}{r} 2.6, \text{ which is more than } 2\frac{1}{2} \\ 8.3\overline{)21.7\,0} \\ \underline{16\ 6} \\ 5\ 1\ 0 \\ \underline{4\ 9\ 8} \end{array}$$

3. $\begin{array}{r} 72.6 \\ -55.4 \\ \hline \textbf{17.2 lb.} \end{array}$ 4. **1975**

5. $d = rt = 3.5 \times 1.75 = 6.125$ mi.

$d = rt = 4.5 \times 1.75 = 7.875$ mi.

$$\begin{array}{r} 6.125 \\ +7.875 \\ \hline \textbf{14.000 mi.} \end{array}$$

6.

$d = rt = 60 \times 2.25 = 135$ mi.

$d = rt = 40 \times 2.25 = \ \ 90$ mi.

$$\begin{array}{r} 135 \\ -\ 90 \\ \hline \textbf{45 mi.} \end{array}$$

7.

$d = rt = 38 \times 3.5 = 133$ mi.

$d = rt = 42 \times 2.5 = 105$ mi.

$d = rt = 54 \times 1 = 54$ mi.

$$\begin{array}{rr} 105 & 133 \\ -\ 54 & -\ 51 \\ \hline 51 \text{ mi.} & \textbf{82 mi.} \end{array}$$

8. **(5) Insufficient data is given to solve the problem.**

9. **(3) $19.60**

$$\begin{array}{ccc} \$\ 2.75 & 6.5 & \$13.75 \\ \times\ \ 5 & \times\$.90 & +5.85 \\ \hline \$13.75 & \$5.85 & \$19.60 \end{array}$$

10. **(5) Insufficient data is given to solve the problem.**

Chapter 3   Quiz

1. **206.7**   2. **0.40**

3. $c = nr$

   $= 6.5 \times \$.895$

   $= \$5.8175$ to the nearest penny $= \textbf{\$5.82}$

4. $\begin{array}{r} 5.8 \\ 2.75 \\ 6.95 \\ +5.9 \\ \hline 21.40 \end{array}$ $\begin{array}{r} \textbf{5.35 lb.} \\ 4\overline{)21.40} \\ \underline{20} \\ 1\ 4 \\ \underline{1\ 2} \\ 20 \\ \underline{20} \end{array}$

5.   4.20
   -1.95
   2.25 kg

6. $11.50
   $\times\ 7.5$
   5 750
   80 50
   $86.250 = **$86.25**

7.   .3066 to the nearest thousandth = **.307**
   75)23.0000
      22 5
       50
        0
       500
       450
       500
       450

8. $(2.8)^2 = 2.8 \times 2.8 =$ **7.84**

9. $6.2 \times 1000 =$ **6200 ml**

10. $A = lw$
    $= 12.4 \times 8.5 =$ **105.4 m²**

11. $d = 2r = 2 \times 0.75 = 1.5$ m

    $C = \pi d$
    $= 3.14 \times 1.5$
    $=$ **4.71 m**

12. $A = \pi r^2$
    $= 3.14 \times 1.6 \times 1.6$
    $= 8.0384$ to the nearest tenth $=$ **8.0 m²**

13. $V = \pi r^2 h$
    $= 3.14 \times 3^2 \times 10$
    $= 3.14 \times 9 \times 10$
    $=$ **282.6 m³**

14. **(2) A, C, D, B, E**

    A = 1.400 m
    B = 0.430 m
    C = 1.300 m
    D = 1.034 m
    E = 0.340 m

15. **(1) California**

16. **(4) over $\frac{1}{2}$**   29.4
                    27.0
                    20.2
                  $+13.0$
                    89.6

    $\frac{89.6}{170.7}$ is more than $\frac{1}{2}$ because the numerator is more than half the denominator.

17. **(5) 1 million**   6.8   13.9
                     $+7.1$ -12.9
                      13.9   1.0 million

18. **(4) more than 4 times**
              4.3 is more than 4 times
       6.8)29.4 0
          27 2
           2 20
           2 04

19. **(2) 247.5**

    $d = rt = 55 \times 3.5 = 192.5$ mi.

    $d = rt = 25 \times 3 = 75$ mi.

    $d = rt = 40 \times .5 = 20$ mi.

         75   192.5
       -20 $+55.0$
         55 mi.   247.5 mi.

20. **(5) Insufficient data is given to solve the problem.**

Chapter 4 **Percents**

Level 1 **Percent Skills**

1. $.035 = .03\ 5 =$ **3.5%**   2. $84\% = \frac{84}{100} = \frac{21}{25}$

3. $\frac{7}{\underset{3}{12}} \times \frac{\overset{25}{100\%}}{1} = \frac{175}{3} = 58\frac{1}{3}\%$ 4. $\frac{120}{48} = \frac{5}{2}$

                                    $\frac{5}{\underset{1}{2}} \times \frac{\overset{50}{100\%}}{1} =$ **250%**

5. $75\% = \frac{3}{4}$

   $72 \div \frac{3}{4} =$

   $\frac{\overset{24}{72}}{1} \times \frac{4}{\underset{1}{3}} =$ **96**

Lesson 1

1. $75\% = 75\ \% = .75$ $4\% = 04\ \% = .04$

   $62.5\% = 62.5\% = .625$ $7\% = 07\ \% = .07$

2. $60\% = 60\ \% = .6$ $150\% = 1\ 50\ \% = 1.5$

   $1\% = 01\ \% = .01$ $300\% = 3\ 00\ \% = 3.$

3. $8\frac{3}{4}\% = 08\ \frac{3}{4}\% = .08\frac{3}{4}$ $12.6\% = 12.6\% = .126$

   $.6\% = 00.6\% = .006$ $15\% = 15\ \% = .15$

4. $.1\% = 00.1\% = .001$ $.25\% = 00.25\% = .0025$

   $2.5\% = 02.5\% = .025$ $200\% = 2\ 00\ \% = 2.$

5. $12\% = 12\ \% = .12$ $105\% = 1\ 05\ \% = 1.05$

   $9.7\% = 09.7\% = .097$ $.2\% = 00.2\% = .002$

6. $.46 = .46 = 46\%$ $.08 = .08 = 8\%$

   $.045 = .04\ 5 = 4.5\%$ $.08\frac{1}{3} = .08\ \frac{1}{3} = 8\frac{1}{3}\%$

7. $.9 = .90 = 90\%$ $.25 = .25 = 25\%$

   $.005 = .00\ 5 = .5\%$ $.05 = .05 = 5\%$

8. $.0825 = .08\ 25 = 8.25\%$ $.4 = .40 = 40\%$

   $.675 = .67\ 5 = 67.5\%$ $1.2 = 1.20 = 120\%$

9. $4.75 = 4.75 = 475\%$ $.625 = .62\ 5 = 62.5\%$

   $8.0 = 8.00 = 800\%$ $.66\frac{2}{3} = .66\frac{2}{3} = 66\frac{2}{3}\%$

10. $.003 = .00\ 3 = .3\%$ $.80 = .80 = 80\%$

    $6.35 = 6.35 = 635\%$ $.3\frac{2}{3} = .30\ \frac{2}{3} = 30\frac{2}{3}\%$

Lesson 2

1. $15\% = \frac{15}{100} = \frac{3}{20}$ $85\% = \frac{85}{100} = \frac{17}{20}$

$96\% = \frac{96}{100} = \frac{24}{25}$ $60\% = \frac{60}{100} = \frac{3}{5}$

2. $275\% = \frac{275}{100} = \frac{11}{4} = 2\frac{3}{4}$ $8\% = \frac{8}{100} = \frac{2}{25}$

$450\% = \frac{450}{100} = \frac{9}{2} = 4\frac{1}{2}$ $42\% = \frac{42}{100} = \frac{21}{50}$

3. $1.5\% = 01.5\% = .015$ $4.8\% = 04.8\% = .048$

$\frac{15}{1000} = \frac{3}{200}$ $\frac{48}{1000} = \frac{6}{125}$

$12.5\% = .125$ $6.25\% = 06.25\% = .0625$

$\frac{125}{1000} = \frac{1}{8}$ $\frac{625}{10,000} = \frac{1}{16}$

4. $6\frac{1}{4}\% = \frac{6\frac{1}{4}}{100}$ $18\frac{3}{4}\% = \frac{18\frac{3}{4}}{100}$

$6\frac{1}{4} \div 100 =$ $18\frac{3}{4} \div 100 =$

$\frac{25}{4} \div \frac{100}{1} =$ $\frac{75}{4} \div \frac{100}{1} =$

$\frac{\overset{1}{\cancel{25}}}{4} \times \frac{1}{\underset{4}{\cancel{100}}} = \frac{1}{16}$ $\frac{\overset{3}{\cancel{75}}}{4} \times \frac{1}{\underset{4}{\cancel{100}}} = \frac{3}{16}$

$13\frac{1}{3}\% = \frac{13\frac{1}{3}}{100}$ $56\frac{1}{4}\% = \frac{56\frac{1}{4}}{100}$

$13\frac{1}{3} \div 100 =$ $56\frac{1}{4} \div 100 =$

$\frac{40}{3} \div \frac{100}{1} =$ $\frac{225}{4} \div \frac{100}{1} =$

$\frac{\overset{2}{\cancel{40}}}{3} \times \frac{1}{\underset{5}{\cancel{100}}} = \frac{2}{15}$ $\frac{\overset{9}{\cancel{225}}}{4} \times \frac{1}{\underset{4}{\cancel{100}}} = \frac{9}{16}$

5. $5\frac{5}{8}\% = \frac{5\frac{5}{8}}{100}$ $2.75\% = 02.75\% = .0275$

$5\frac{5}{8} \div 100 =$ $\frac{275}{10,000} = \frac{11}{400}$

$\frac{45}{8} \div \frac{100}{1} =$

$\frac{\overset{9}{\cancel{45}}}{8} \times \frac{1}{\underset{20}{\cancel{100}}} = \frac{9}{160}$

$80.5\% = 80.5\% = .805$ $61\frac{1}{4}\% = \frac{61\frac{1}{4}}{100}$

$\frac{805}{1000} = \frac{161}{200}$ $61\frac{1}{4} \div 100$

                 $\frac{245}{4} \div \frac{100}{1}$

                 $\frac{\overset{49}{\cancel{245}}}{4} \times \frac{1}{\underset{20}{\cancel{100}}} = \frac{49}{80}$

6. $\frac{3}{\cancel{5}} \times \frac{\overset{20}{\cancel{100}}\%}{1} = 60\%$ $\frac{12}{\cancel{25}} \times \frac{\overset{4}{\cancel{100}}\%}{1} = 48\%$

$\frac{1}{\cancel{2}} \times \frac{\overset{50}{\cancel{100}}\%}{1} = 50\%$ $\frac{2}{3} \times \frac{100\%}{1} = \frac{200}{3} = 66\frac{2}{3}\%$

7. $\frac{23}{\cancel{100}} \times \frac{\overset{1}{\cancel{100}}\%}{1} = 23\%$ $\frac{3}{\cancel{10}} \times \frac{\overset{10}{\cancel{100}}\%}{1} = 30\%$

$\frac{5}{\cancel{8}} \times \frac{\overset{25}{\cancel{100}}\%}{1} = \frac{125}{2}$ $\frac{3}{\cancel{16}} \times \frac{\overset{25}{\cancel{100}}\%}{1} = \frac{75}{4}$

       $= 62\frac{1}{2}\% \ (62.5\%)$ $= 18\frac{3}{4}\% \ (18.75\%)$

8. $\frac{2}{7} \times \frac{100\%}{1} = \frac{200}{7} = 28\frac{4}{7}\%$ $\frac{5}{\cancel{6}} \times \frac{\overset{50}{\cancel{100}}\%}{1} = \frac{250}{3} = 83\frac{1}{3}\%$

$\frac{19}{\cancel{20}} \times \frac{\overset{5}{\cancel{100}}\%}{1} = 95\%$ $\frac{4}{\cancel{5}} \times \frac{\overset{20}{\cancel{100}}\%}{1} = 80\%$

9. $\frac{4}{\cancel{5}} \times \frac{\overset{20}{\cancel{100}}\%}{1} = 80\%$ $\frac{3}{\cancel{2}} \times \frac{\overset{25}{\cancel{100}}}{1} = \frac{75}{2}$

$\frac{5}{7} \times \frac{100\%}{1} = \frac{500}{7} = 71\frac{3}{7}\%$ $= 37\frac{1}{2}\% \ (37.5\%)$

                 $\frac{8}{9} \times \frac{100\%}{1} = \frac{800}{9} = 88\frac{8}{9}\%$

10. $\frac{1}{\cancel{6}} \times \frac{\overset{50}{\cancel{100}}\%}{1} = \frac{50}{3} = 16\frac{2}{3}\%$ $\frac{3}{7} \times \frac{100\%}{1} = \frac{300}{7} = 42\frac{6}{7}\%$

$\frac{4}{\cancel{15}} \times \frac{\overset{20}{\cancel{100}}\%}{1} = \frac{80}{3} = 26\frac{2}{3}\%$ $\frac{1}{\cancel{20}} \times \frac{\overset{5}{\cancel{100}}\%}{1} = 5\%$

Lesson 3

1. $6\% = .06$ $40\% = .4$

$\begin{array}{r} 150 \\ \times .06 \\ \hline 9.00 = 9 \end{array}$ $\begin{array}{r} 160 \\ \times \ .4 \\ \hline 64.0 = 64 \end{array}$

$5.4\% = .054$ $0.8\% = .008$

$\begin{array}{r} .054 \\ \times 80 \\ \hline 4.320 = 4.32 \end{array}$ $\begin{array}{r} .008 \\ \times 50 \\ \hline .400 = .4 \end{array}$

2. $9\% = .09$ $3\% = .03$

$\begin{array}{r} 90 \\ \times .09 \\ \hline 8.10 = 8.1 \end{array}$ $\begin{array}{r} 300 \\ \times .03 \\ \hline 9.00 = 9 \end{array}$

$125\% = 1.25$ $1.9\% = .019$

$\begin{array}{r} 1.25 \\ \times \ 36 \\ \hline 7 \ 50 \\ 37 \ 5 \\ \hline 45.00 = 45 \end{array}$ $\begin{array}{r} .019 \\ \times 200 \\ \hline 3.800 = 3.8 \end{array}$

3. $6.25\% = .0625$ $10.4\% = .104$

$\begin{array}{r} .0625 \\ \times 300 \\ \hline 18.7500 = 18.75 \end{array}$ $\begin{array}{r} .104 \\ \times 500 \\ \hline 52.000 = 52 \end{array}$

$12.8\% = .128$ $.2\% = .002$

$\begin{array}{r} .128 \\ \times 80 \\ \hline 10.240 = 10.24 \end{array}$ $\begin{array}{r} 100 \\ \times .002 \\ \hline .200 = .2 \end{array}$

4. $66\frac{2}{3}\% = \frac{2}{3}$ $12\frac{1}{2}\% = \frac{1}{8}$

$\frac{2}{3} \times \frac{\overset{80}{\cancel{240}}}{1} = 160$ $\frac{1}{8} \times \frac{\overset{50}{\cancel{400}}}{1} = 50$

$33\frac{1}{3}\% = \frac{1}{3}$ $16\frac{2}{3}\% = \frac{1}{6}$

$\frac{1}{3} \times \frac{\overset{6}{\cancel{18}}}{1} = 6$ $\frac{1}{6} \times \frac{\overset{16}{\cancel{96}}}{1} = 16$

5. $12\frac{1}{2}\% = \frac{1}{8}$ $37\frac{1}{2}\% = \frac{3}{8}$

$\frac{1}{8} \times \frac{\overset{3}{\cancel{24}}}{1} = 3$ $\frac{3}{8} \times \frac{\overset{125}{\cancel{1000}}}{1} = 375$

$62\frac{1}{2}\% = \frac{5}{8}$ $83\frac{1}{3}\% = \frac{5}{6}$

$\frac{5}{8} \times \frac{\overset{6}{\cancel{48}}}{1} = 30$ $\frac{5}{6} \times \frac{\overset{20}{\cancel{120}}}{1} = 100$

6. $37\frac{1}{2}\% = \frac{3}{8}$ $87\frac{1}{2}\% = \frac{7}{8}$

$\frac{3}{8} \times \frac{\overset{3}{\cancel{24}}}{1} = 9$ $\frac{7}{8} \times \frac{\overset{150}{\cancel{1200}}}{1} = 1050$

$33\frac{1}{3}\% = \frac{1}{3}$ $87\frac{1}{2}\% = \frac{7}{8}$

$\frac{1}{3} \times \frac{\overset{200}{\cancel{600}}}{1} = 200$ $\frac{7}{8} \times \frac{\overset{750}{\cancel{6000}}}{1} = 5250$

7. $75\% = \frac{3}{4}$ $8.5\% = .085$

$\frac{3}{4} \times \frac{\overset{21}{\cancel{84}}}{1} = 63$ $\begin{array}{r} .085 \\ \times 400 \\ \hline 34.000 = 34 \end{array}$

$250\% = 2.5$ $60\% = \frac{3}{5}$

$\begin{array}{r} 36 \\ \times 2.5 \\ \hline 18 \ 0 \\ 72 \\ \hline 90.0 = 90 \end{array}$ $\frac{3}{5} \times \frac{\overset{40}{\cancel{200}}}{1} = 120$

8. $20\% = \frac{1}{5}$

$\frac{1}{\cancel{5}} \times \frac{\cancel{200}^{40}}{1} = 40$

$300\% = 3$

$\begin{array}{r} 21 \\ \times\ 3 \\ \hline 63 \end{array}$

$150\% = 1.5$

$\begin{array}{r} 1.5 \\ \times 1000 \\ \hline 1500.0 \end{array}$

$62\frac{1}{2}\% = \frac{5}{8}$

$\frac{5}{\cancel{8}} \times \frac{\cancel{800}^{100}}{1} = 500$

9. $90\% = .9$

$\begin{array}{r} 130 \\ \times\ .9 \\ \hline 117.0 \end{array} = 117$

$25\% = \frac{1}{4}$

$\frac{1}{\cancel{4}} \times \frac{\cancel{116}^{29}}{1} = 29$

$50\% = \frac{1}{2}$

$\frac{1}{\cancel{2}} \times \frac{\cancel{28}^{14}}{1} = 14$

$35\% = .35$

$\begin{array}{r} 260 \\ \times\ .35 \\ \hline 13\ 00 \\ 78\ 0 \\ \hline 91.00 \end{array} = 91$

10. $76\% = .76$

$\begin{array}{r} .76 \\ \times\ 80 \\ \hline 60.80 \end{array} = 60.8$

$70\% = \frac{7}{10}$

$\frac{7}{\cancel{10}} \times \frac{\cancel{30}^{3}}{1} = 21$

$40\% = \frac{2}{5}$

$\frac{2}{\cancel{5}} \times \frac{\cancel{35}^{7}}{1} = 14$

$22\% = .22$

$\begin{array}{r} 190 \\ \times\ .22 \\ \hline 380 \\ 380 \\ \hline 41.80 \end{array} = 41.8$

Lesson 4

1. $\frac{28}{70} = \frac{2}{5}$ $\frac{104}{160} = \frac{13}{20}$ $\frac{16}{20} = \frac{4}{5}$

$\frac{2}{\cancel{5}} \times \frac{\cancel{100}^{20}\%}{1} =$ $\frac{13}{\cancel{20}} \times \frac{\cancel{100}^{5}\%}{1} =$ $\frac{4}{\cancel{5}} \times \frac{\cancel{100}^{20}\%}{1} = 80\%$

       40% 65%

2. $\frac{45}{135} = \frac{1}{3}$ $\frac{24}{32} = \frac{3}{4}$ $\frac{30}{48} = \frac{5}{8}$

$\frac{1}{3} \times \frac{100\%}{1} =$ $\frac{3}{\cancel{4}} \times \frac{\cancel{100}^{25}\%}{1} = 75\%$ $\frac{5}{\cancel{8}} \times \frac{\cancel{100}^{25}\%}{1} = \frac{125}{2}$

    $\frac{100}{3} = 33\frac{1}{3}\%$ $= 62\frac{1}{2}\%$

3. $\frac{225}{375} = \frac{3}{5}$ $\frac{27}{72} = \frac{3}{8}$ $\frac{36}{24} = \frac{3}{2}$

$\frac{3}{\cancel{5}} \times \frac{\cancel{100}^{20}\%}{1} =$ $\frac{3}{\cancel{8}} \times \frac{\cancel{100}^{25}\%}{1} = \frac{75}{2} =$ $\frac{3}{\cancel{2}} \times \frac{\cancel{100}^{50}\%}{1} =$

       60% $37\frac{1}{2}\%$ $= 150\%$

4. $\frac{15}{120} = \frac{1}{8}$ $\frac{36}{40} = \frac{9}{10}$ $\frac{150}{75} = \frac{2}{1}$

$\frac{1}{\cancel{8}} \times \frac{\cancel{100}^{25}\%}{1} = \frac{25}{2}$ $\frac{9}{\cancel{10}} \times \frac{\cancel{100}^{10}\%}{1} =$ $\frac{2}{1} \times \frac{100\%}{1} =$

   $= 12\frac{1}{2}\%$ 90% 200%

5. $\frac{15}{90} = \frac{1}{6}$ $\frac{14}{20} = \frac{7}{10}$ $\frac{55}{220} = \frac{1}{4}$

$\frac{1}{\cancel{6}} \times \frac{\cancel{100}^{50}\%}{1} = \frac{50}{3}$ $\frac{7}{\cancel{10}} \times \frac{\cancel{100}^{10}\%}{1} =$ $\frac{1}{\cancel{4}} \times \frac{\cancel{100}^{25}\%}{1} = 25\%$

   $= 16\frac{2}{3}\%$ 70%

Lesson 5

1. $60\% = .6$

$\begin{array}{r} 4\ 0 \\ .6\overline{)24.0} \end{array}$

$25\% = \frac{1}{4}$

$18 \div \frac{1}{4} =$

$\frac{18}{1} \times \frac{4}{1} = 72$

2. $12\frac{1}{2}\% = \frac{1}{8}$

$15 \div \frac{1}{8} =$

$\frac{15}{1} \times \frac{8}{1} = 120$

$4\% = .04$

$\begin{array}{r} 2\ 40 \\ .04\overline{)9.60} \\ 8 \\ \hline 1\ 6 \\ 1\ 6 \\ \hline 00 \end{array}$

3. $66\frac{2}{3}\% = \frac{2}{3}$

$52 \div \frac{2}{3} =$

$\frac{\cancel{52}^{26}}{1} \times \frac{3}{\cancel{2}} = 78$

$87\frac{1}{2}\% = \frac{7}{8}$

$112 \div \frac{7}{8} =$

$\frac{\cancel{112}^{16}}{1} \times \frac{8}{\cancel{7}} = 128$

4. $2.5\% = .025$

$\begin{array}{r} 320 \\ .025\overline{)8.000} \\ 7\ 5 \\ \hline 50 \\ 50 \\ \hline 00 \end{array}$

$35\% = .35$

$\begin{array}{r} 4\ 00 \\ .35\overline{)140.00} \\ 140 \\ \hline 0\ 00 \end{array}$

5. $90\% = .9$

$\begin{array}{r} 2\ 6 \\ .9\overline{)23.4} \\ 18 \\ \hline 5\ 4 \\ 5\ 4 \\ \hline 0\ 0 \end{array}$

$62\frac{1}{2}\% = \frac{5}{8}$

$45 \div \frac{5}{8} =$

$\frac{\cancel{45}^{9}}{1} \times \frac{8}{\cancel{5}} = 72$

6. $8.5\% = .085$

$\begin{array}{r} 200 \\ .085\overline{)17.000} \end{array}$

$16\frac{2}{3}\% = \frac{1}{6}$

$36 \div \frac{1}{6} =$

$\frac{36}{1} \times \frac{6}{1} = 216$

7. $37\frac{1}{2}\% = \frac{3}{8}$

$24 \div \frac{3}{8} =$

$\frac{\cancel{24}^{8}}{1} \times \frac{8}{\cancel{3}} = 64$

$40\% = .4$

$\begin{array}{r} 12\ 0 \\ .4\overline{)48.0} \end{array}$

8. $33\frac{1}{3}\% = \frac{1}{3}$

$112 \div \frac{1}{3} =$

$\frac{112}{1} \times \frac{3}{1} = 336$

$18\% = .18$

$\begin{array}{r} 2\ 00 \\ .18\overline{)36.00} \end{array}$

9. $83\frac{1}{3}\% = \frac{5}{6}$

$50 \div \frac{5}{6} =$

$\frac{\cancel{50}^{10}}{1} \times \frac{6}{\cancel{5}} = 60$

$22\% = .22$

$\begin{array}{r} 3\ 50 \\ .22\overline{)77.00} \\ 66 \\ \hline 11\ 0 \\ 11\ 0 \\ \hline 00 \end{array}$

10. $32\% = .32$

$\begin{array}{r} 3\ 50 \\ .32\overline{)112.00} \\ 96 \\ \hline 16\ 0 \\ 16\ 0 \\ \hline 00 \end{array}$

$16\% = .16$

$\begin{array}{r} 2\ 00 \\ .16\overline{)32.00} \end{array}$

Level 1   Review

1. $9.6\% = 09.6\% = .096$

2. $.0145 = .01\,45 = 1.45\%$

3. $8\frac{1}{3}\% = \dfrac{8\frac{1}{3}}{100}$

   $8\frac{1}{3} \div 100 =$

   $\dfrac{25}{3} \div \dfrac{100}{1} =$

   $\overset{1}{\cancel{\dfrac{25}{3}}} \times \dfrac{1}{\underset{4}{\cancel{100}}} = \dfrac{1}{12}$

4. $\dfrac{9}{\underset{4}{\cancel{16}}} \times \dfrac{\overset{25}{\cancel{100\%}}}{1} = \dfrac{225}{4} = 56\frac{1}{4}\%$

5. $4.5\% = .045$

     $2\,400$
     $\times .045$
    $\overline{12\,000}$
    $96\,00$
   $\overline{108.000}$

6. $83\frac{1}{3}\% = \dfrac{5}{6}$

   $\dfrac{5}{\cancel{6}} \times \dfrac{\overset{55}{\cancel{330}}}{1} = 275$

7. $\dfrac{24}{72} = \dfrac{1}{3}$

   $\dfrac{1}{3} \times \dfrac{100\%}{1} = \dfrac{100}{3} = 33\frac{1}{3}\%$

8. $\dfrac{105}{60} = \dfrac{7}{4}$

   $\dfrac{7}{\cancel{4}} \times \dfrac{\overset{25}{\cancel{100\%}}}{1} = 175\%$

9. $65\% = .65$

       80
  $.65\overline{)52.00}$
     $\underline{52\,0}$
      00

10. $16\frac{2}{3}\% = \dfrac{1}{6}$

   $25 \div \dfrac{1}{6} =$

   $\dfrac{25}{1} \times \dfrac{6}{1} = 150$

Level 2   Percent Applications

1. $80\% = .8$

     60
   $\times .8$
  $\overline{48.0} = 48$ right

2. by car $\dfrac{20}{24} = \dfrac{5}{6}$
    total

   $\dfrac{5}{\underset{3}{\cancel{6}}} \times \dfrac{\overset{50}{\cancel{100\%}}}{1} = \dfrac{250}{3} = 83\frac{1}{3}\%$

3. agreed $\dfrac{65}{500} = \dfrac{13}{100}$
   total

   $\dfrac{13}{100} \times \dfrac{\overset{1}{\cancel{100\%}}}{1} = 13\%$

4. $75\% = \dfrac{3}{4}$

   $\$1350 \div \dfrac{3}{4} =$

   $\dfrac{\overset{450}{\cancel{\$1350}}}{1} \times \dfrac{4}{3} = \$1800$

5. $i = \dfrac{\overset{14}{\cancel{\$1400}}}{1} \times \dfrac{6.5}{\underset{1}{\cancel{100}}} \times 1 = \$91$

Lesson 1

1. $6\% = .06$

   $\$48,000$
   $\times \quad .06$
  $\overline{\$2880.00}$

2. $65\% = .65$

      320
     $\times .65$
    $\overline{16\,00}$
   $192\,0$
  $\overline{208.00} = 208$ women

3. a. $15\% = .15$ b. $\quad\$69.00$
                  $- \quad 10.35$
                  $\overline{\$58.65}$

      $\$69$
    $\times .15$
    $\overline{3\,45}$
    $6\,9$
   $\overline{\$10.35}$

4. $5\% = .05$

   $\$36,400$
   $\times \quad .05$
  $\overline{\$1820.00}$

5. $1.5\% = .015$

   $\$430$
   $\times .015$
  $\overline{2\,150}$
  $4\,30$
  $\overline{\$6.450} = \$6.45$

6. $37\frac{1}{2}\% = \dfrac{3}{8}$

   $\dfrac{3}{8} \times \dfrac{\overset{4,200}{\cancel{33,600}}}{1} = \$12,600$

     $\$33,600$
    $+ 12,600$
   $\overline{\$46,200}$

7. $\$49 + \$12 = \$61$

   $7\% = .07$ $\$61$ $\$61.00$
             $\underline{\times .07}$ $\underline{+ 4.27}$
             $\$4.27$ $\$65.27$

8. $40\% = .4$

   $\$28$ $\$28.00$
  $\underline{\times .4}$ $\underline{+ 11.20}$
  $\$11.2$ $\$39.20$ to the
             nearest dol-
             lar $= \$39$

9. $1.8\% = .018$

   $\$1650$
   $\underline{\times .018}$
  $13\,200$
  $16\,50$
  $\overline{\$\,29.700} = \$29.70$

10. $16\frac{2}{3}\% = \dfrac{1}{6}$

   $\dfrac{1}{6} \times \overset{15,000}{\cancel{90,000}} = \$15,000$

     $\$90,000$
    $+ \; 15,000$
   $\overline{\$105,000}$

Lesson 2

1. right $\dfrac{34}{40} = \dfrac{17}{20}$
  total

   $\dfrac{17}{\underset{1}{\cancel{20}}} \times \dfrac{\overset{5}{\cancel{100\%}}}{1} = 85\%$

2. late $\dfrac{6}{48} = \dfrac{1}{8}$
  total

   $\dfrac{1}{\underset{2}{\cancel{8}}} \times \dfrac{\overset{25}{\cancel{100\%}}}{1} = \dfrac{25}{2} = 12\frac{1}{2}\%$

3. tax $\dfrac{1.80}{22.50} = \dfrac{6}{75} = \dfrac{2}{25}$
  total

   $\dfrac{2}{\cancel{25}} \times \dfrac{\overset{4}{\cancel{100\%}}}{1} = 8\%$

4. cash $\dfrac{280}{320} = \dfrac{7}{8}$
  total

   $\dfrac{7}{\underset{2}{\cancel{8}}} \times \dfrac{\overset{25}{\cancel{100\%}}}{1} = \dfrac{175}{2} = 87\frac{1}{2}\%$

5. 1986 value $\dfrac{2560}{6400} = \dfrac{64}{160} = \dfrac{2}{5}$
   1980 value

   $\dfrac{2}{\cancel{5}} \times \dfrac{\overset{4}{\cancel{100\%}}}{1} = 40\%$

6. rent $\dfrac{252}{1200} = \dfrac{63}{300} = \dfrac{21}{100}$
  total

   $\dfrac{21}{100} \times \dfrac{\overset{1}{\cancel{100\%}}}{1} = 21\%$

7. sale price $\dfrac{42,000}{24,000} = \dfrac{7}{4}$
   purchase price

   $\dfrac{7}{\cancel{4}} \times \dfrac{\overset{25}{\cancel{100\%}}}{1} = 175\%$

8. amount lost $\dfrac{75}{250} = \dfrac{3}{10}$
   original weight

   $\dfrac{3}{\underset{1}{\cancel{10}}} \times \dfrac{\overset{10}{\cancel{100\%}}}{1} = 30\%$

9. agreed $\dfrac{60}{72} = \dfrac{5}{6}$
  total

   $\dfrac{5}{\underset{3}{\cancel{6}}} \times \dfrac{\overset{50}{\cancel{100\%}}}{1} = \dfrac{250}{3} = 83\frac{1}{3}\%$

10. tax $\dfrac{5}{60} = \dfrac{1}{12}$
   total

   $\dfrac{1}{\underset{3}{\cancel{12}}} \times \dfrac{\overset{25}{\cancel{100\%}}}{1} = \dfrac{25}{3} = 8\frac{1}{3}\%$

Lesson 3

1. $80\% = .8$

     $2\,0$ problems
  $.8\overline{)16.0}$

2. $9\% = .09$

     $\$\,70\,00$
  $.09\overline{)\$630.00}$

3. $75\% = .75$

       $5\,60$ mi.
  $.75\overline{)420.00}$
     $\underline{375}$
     $45\,0$
     $\underline{45\,0}$
      00

4. $15\% = .15$

      $1\,60$ employees
  $.15\overline{)24.00}$
     $\underline{15}$
     $9\,0$
     $\underline{9\,0}$
      00

5. $60\% = .6$

     $\$\,240\,0$
  $.6\overline{)\$1440.0}$
    $\underline{12}$
    24
    $\underline{24}$
    000

6. $22\% = .22$

      $\$\,24,0\,00$
  $.22\overline{)\$528\,0.00}$
     $\underline{44}$
     88
     $\underline{88}$
     $0\,0\,00$

7. $66\frac{2}{3}\% = \dfrac{2}{3}$

   $180 \div \dfrac{2}{3} =$

   $\dfrac{\overset{90}{\cancel{180}}}{1} \times \dfrac{3}{2} = 270$ lb.

8. $37\frac{1}{2}\% = \dfrac{3}{8}$

   $450 \div \dfrac{3}{8} =$

   $\dfrac{\overset{150}{\cancel{450}}}{1} \times \dfrac{8}{3} = 1200$ people

9. $40\% = .4$

$$\begin{array}{r} 21\ 0\ \text{mi.} \\ .4\overline{)84.0} \end{array}$$

10. $18\% = .18$

$$\begin{array}{r} 220\ 00 \\ .18\overline{)3960.00} \\ \underline{36}\ \ \ \\ 36\ \ \\ \underline{36}\ \\ 0000 \end{array}$$

Lesson 4

1. $i = \dfrac{\overset{6}{\cancel{600}}}{1} \times \dfrac{4}{\cancel{100}} \times 1 = \$24$

2. $i = \dfrac{\cancel{400}}{1} \times \dfrac{8.5}{\cancel{100}} \times 1 = \$34$

3. $i = \dfrac{\overset{40}{\cancel{4000}}}{1} \times \dfrac{12}{\cancel{100}} \times 1 = \$480$

4. $i = \dfrac{\overset{12}{\cancel{1200}}}{1} \times \dfrac{21}{\cancel{4}} \times 1 = \$63$

5. $6\ \text{mo.} = \dfrac{6}{12} = \dfrac{1}{2}\ \text{yr.}$
$i = \dfrac{\overset{5}{\cancel{500}}}{1} \times \dfrac{4}{\cancel{100}} \times \dfrac{1}{2} = \$20$

6. $8\ \text{mo.} = \dfrac{8}{12} = \dfrac{2}{3}\ \text{yr.}$
$i = \dfrac{\overset{10}{\cancel{1000}}}{1} \times \dfrac{9}{\cancel{100}} \times \dfrac{2}{3} = \$60$

7. $1\ \text{yr. }3\ \text{mo.} = 1\dfrac{3}{12} = 1\dfrac{1}{4}\ \text{yr.}$
$i = \dfrac{\overset{1}{\cancel{400}}}{\cancel{100}} \times \dfrac{11}{2} \times \dfrac{5}{\cancel{4}} = \dfrac{55}{2} = \$27.50$

8. $9\ \text{mo.} = \dfrac{9}{12} = \dfrac{3}{4}\ \text{yr.}$
$i = \dfrac{\overset{2}{\cancel{800}}}{1} \times \dfrac{4.5}{\cancel{100}} \times \dfrac{3}{\cancel{4}} = \$27$

9. $i = \dfrac{\overset{25}{\cancel{2500}}}{1} \times \dfrac{15}{\cancel{100}} \times \dfrac{3}{1} = \$1125$

10. $10\ \text{mo.} = \dfrac{10}{12} = \dfrac{5}{6}\ \text{yr.}$
$i = \dfrac{\overset{18}{\cancel{1800}}}{\cancel{100}} \times \dfrac{15}{4} \times \dfrac{5}{\cancel{6}} = \dfrac{225}{4} = \$56.25$

Level 2 Review

1. $2\% = .02$
$$\begin{array}{r} 1250 \\ \times\ .02 \\ \hline 25.00 = 25\ \text{buckets} \end{array}$$

2. $6\% = .06$
$$\begin{array}{r} \$6850 \\ \times\ .06 \\ \hline \$411.00 \end{array}$$

3. $18\% = .18$
$$\begin{array}{r} \$6.50 \\ \times\ .18 \\ \hline 52\ 00 \\ 65\ 0 \\ \hline \$1.17\ 00 \end{array} \qquad \begin{array}{r} \$6.50 \\ +\ 1.17 \\ \hline \$7.67 \end{array}$$

4. empty $\dfrac{65}{520} = \dfrac{1}{8}$
total
$\dfrac{1}{\underset{2}{\cancel{8}}} \times \dfrac{\overset{25}{\cancel{100\%}}}{1} = \dfrac{25}{2} = 12\dfrac{1}{2}\%$

5. $\dfrac{\text{tax}}{\text{total}}\ \dfrac{.75}{12.50} = \dfrac{3}{50}$
$\dfrac{3}{\underset{1}{\cancel{50}}} \times \dfrac{\overset{2}{\cancel{100\%}}}{1} = 6\%$

6. $\dfrac{\text{food}}{\text{total}}\ \dfrac{115}{345} = \dfrac{1}{3}$
$\dfrac{1}{3} \times \dfrac{100\%}{1} = \dfrac{100}{3} = 33\dfrac{1}{3}\%$

7. $60\% = .6$
$$\begin{array}{r} \$\ 2\ 8,00\ 0 \\ .6\overline{)\$16,8\ 00.0} \\ \underline{12}\ \ \ \ \ \ \\ 48\ \ \ \ \\ \underline{48}\ \ \ \\ 000 \end{array}$$

8. $4\% = .04$
$$\begin{array}{r} \$1\ 24 \\ .04\overline{)\$4.96} \\ \underline{4}\ \ \ \\ 0\ 9 \\ \underline{8}\ \\ 16 \\ \underline{16} \end{array}$$

9. $i = \dfrac{\overset{24}{\cancel{2400}}}{1} \times \dfrac{14.5}{\cancel{100}} \times 1 = \$348$

10. $1\ \text{yr. }8\ \text{mo.} = {}_1 1\dfrac{8}{12} = 1\dfrac{2}{3}\ \text{yr.}$
$i = \dfrac{360}{1} \times \dfrac{7\dfrac{2}{}}{100} \times 1\dfrac{2}{3}$
$i = \dfrac{\overset{9}{\cancel{360}}}{\cancel{5}} \times \dfrac{15}{2} \times \dfrac{5}{3} = \$45$

Level 3 Percent Problem Solving

1. $80\% = .8$
$$\begin{array}{r} \$\ 24\ 0 \\ .8\overline{)\$192.0} \end{array}$$
$$\begin{array}{r} \$240 \\ -\ 192 \\ \hline \$\ 48\ \text{saved} \end{array}$$

2. $75\% = \dfrac{3}{4}$
$54 \div \dfrac{3}{4} =$
$\dfrac{\overset{18}{\cancel{54}}}{1} \times \dfrac{4}{\cancel{3}} = 72$
$$\begin{array}{r} 72 \\ -\ 54 \\ \hline 18\ \text{mi.} \end{array}$$

3. $25\% = .25$
$$\begin{array}{r} .25 \\ \times\ \$700 \\ \hline \$175.00 \end{array} \qquad \begin{array}{r} \$700 \\ -\ 175 \\ \hline \$525 \end{array}$$

4. $50\% = .5$
$$\begin{array}{r} \$700 \\ \times\ .5 \\ \hline \$350.0 \end{array} \quad \begin{array}{r} \$700 \\ -\ 350 \\ \hline \$350 \end{array} \quad \begin{array}{r} \$525 \\ -\ 350 \\ \hline \$175 \end{array}$$

5.
$$\begin{array}{r} \$700 \\ +\ 400 \\ \hline \$1100 \end{array} \quad \begin{array}{r} \$1100 \\ \times\ .8 \\ \hline \$880.0 \end{array} \quad \begin{array}{r} \$1100 \\ \times\ 880 \\ \hline \$\ 220 \end{array} \quad \begin{array}{r} \$220 \\ \times\ .05 \\ \hline \$11.00 \end{array} \quad \begin{array}{r} \$220 \\ +\ 11 \\ \hline \$231 \end{array}$$

Lesson 1

1.
$$\begin{array}{r} \$.96 \\ -\ .88 \\ \hline \$.08 \end{array}$$
$\dfrac{\text{change}}{\text{original}}\ \dfrac{8}{96} = \dfrac{1}{12}$
$\dfrac{1}{\underset{3}{\cancel{12}}} \times \dfrac{\overset{25}{\cancel{100\%}}}{1} = \dfrac{25}{3} = 8\dfrac{1}{3}\%$

2.
$$\begin{array}{r} \$360 \\ 306 \\ \hline \$\ 54 \end{array}$$
$\dfrac{\text{change}}{\text{original}}\ \dfrac{54}{360} = \dfrac{6}{40} = \dfrac{3}{20}$
$\dfrac{3}{\underset{1}{\cancel{20}}} \times \dfrac{\overset{5}{\cancel{100\%}}}{1} = 15\%$

3.
$$\begin{array}{r} 15,000 \\ 12,000 \\ \hline 3,000 \end{array}$$
$\dfrac{\text{change}}{\text{original}}\ \dfrac{3,000}{12,000} = \dfrac{1}{4}$
$\dfrac{1}{4} \times \dfrac{\overset{25}{\cancel{100\%}}}{1} = 25\%$

4.
$$\begin{array}{r} \$24 \\ -\ 16 \\ \hline \$\ 8 \end{array}$$
$\dfrac{\text{change}}{\text{original}}\ \dfrac{8}{16} = \dfrac{1}{2}$
$\dfrac{1}{2} \times \dfrac{\overset{50}{\cancel{100\%}}}{1} = 50\%$

5.
$$\begin{array}{r} 84 \\ +126 \\ \hline 210 \end{array}$$
$\dfrac{\text{women}}{\text{total}}\ \dfrac{84}{210} = \dfrac{12}{30} = \dfrac{2}{5}$
$\dfrac{2}{\underset{1}{\cancel{5}}} \times \dfrac{\overset{20}{\cancel{100\%}}}{1} = 40\%$

6.
$$\begin{array}{r} \$270 \\ +\ 630 \\ \hline \$900 \end{array}$$
$\dfrac{\text{food}}{\text{total}}\ \dfrac{270}{900} = \dfrac{3}{10}$
$\dfrac{3}{\underset{1}{\cancel{10}}} \times \dfrac{\overset{10}{\cancel{100\%}}}{1} = 30\%$

7.
$$\begin{array}{r} \$4800 \\ -\ 1800 \\ \hline \$3000 \end{array}$$
$\dfrac{\text{change}}{\text{original}}\ \dfrac{3000}{4800} = \dfrac{5}{8}$
$\dfrac{5}{\underset{2}{\cancel{8}}} = \dfrac{\overset{25}{\cancel{100\%}}}{1} = \dfrac{125}{2} = 62\dfrac{1}{2}\%$

8.
$$\begin{array}{r} 792 \\ -720 \\ \hline 72 \end{array}$$
$\dfrac{\text{change}}{\text{original}}\ \dfrac{72}{720} = \dfrac{1}{10}$
$\dfrac{1}{\underset{1}{\cancel{10}}} \times \dfrac{\overset{10}{\cancel{100\%}}}{1} = 10\%$

9.
$$\begin{array}{r} \$45 \\ -\ 40 \\ \hline \$\ 5 \end{array}$$
$\dfrac{\text{change}}{\text{original}}\ \dfrac{5}{40} = \dfrac{1}{8}$
$\dfrac{1}{\underset{2}{\cancel{8}}} \times \dfrac{\overset{25}{\cancel{100\%}}}{1} = \dfrac{25}{2} = 12\dfrac{1}{2}\%$

10.
$$\begin{array}{r} \$200 \\ -\ 120 \\ \hline \$\ 80 \end{array}$$
$\dfrac{\text{change}}{\text{original}}\ \dfrac{80}{120} = \dfrac{2}{3}$
$\dfrac{2}{3} \times \dfrac{100\%}{1} = \dfrac{200}{3} = 66\dfrac{2}{3}\%$

Lesson 2

1. 60% = .6

```
        $ 45 0
.6)$270.0
   24
   30
   30
    0 0
```

```
  $450
− 270
 $180
```

2. 75% = $\frac{3}{4}$

900 ÷ $\frac{3}{4}$

$\frac{\overset{300}{\cancel{900}}}{1} \times \frac{4}{\overset{3}{\cancel{3}}_1} = 1200$

```
  1200
−  900
   300 signatures
```

3. 40% = .4

```
       28 0
.4)112.0
    8
    32
    32
     0 0
```

```
  280
  112
  168 needed
```

4. 85% = .85

```
        1 20
.85)102.00
    85
    17 0
    17 0
      00
```

```
  $120
− 102
 $ 18 saved
```

5. 70% = .7

```
       60 0
.7)420.0
   600
  −420
   180 failed
```

6. 90% = .9

```
       24 0
.9)216.0
   18
   36
   36
    0 0
```

```
  240
− 216
  24 pounds lost
```

7. 65% = .65

```
         5 20
65)338 00
   325
   13 0
   13 0
     00
```

```
  520
− 338
  182 mi. to go
```

8. 21% = .21

```
        $ 16,5 00
.21)$346 5.00
    21
    136
    126
     10 5
     10 5
       0 00
```

```
  $16,500
−  3,465
  $13,035
```

9. 40% = .4

```
       57 5
.4)230.0
   20
   30
   28
    2 0
    2 0
     0 0
```

```
  575
− 230
  345 mi. to go
```

10. 75% = .75

```
        6 00
.75)450.00
```

```
  $600
− 450
 $150 saved
```

Lesson 3

1. B. 15% of the yearly rent of $6000

  A. $35 × 30 = $1050
  B. 15% = .15   .15 × $6000 = $900
  C. $25 × 30 = $750 $750 + $250 = $1000
  D. $285 × 4 = $1140

2. D. $250 a week for half a year

  A. 50% = $\frac{1}{2}$ $\frac{1}{2}$ × $10,000 = $5000
  B. $800 × 6 = $4800
  C. 3 × $1500 = $4500
  D. $\frac{1}{2}$ yr. = $\frac{52}{2}$ = 26 wk. $250 × 26 = $6500

3. B. 1300 more people per year for three years

  A. 15% = .15   .15 × 25,000 = 3750
  B. 1300 × 3 = 3900
  C. 5% = .05   .05 × 25,000 = 1250
     1250 + 1000 + 1000 = 3250
  D. 3 yr. = 3 × 12 = 36 mo.   36 × 80 = 2880

4. C. $150 for each 100 pounds

```
A.   500        40      $12.50     $300
   −100    10)400     ×   40     + 500
   400 lb.              $500.00    $800
```

```
B.      10        $65      $650      5% = .05
   50)500       × 10     + 100
                 $650      $750

                          $750       $750.00
                        × .05      + 37.50
                          $37.50     $787.50
```

```
C.       5       $150
   100)500      × 5
                 $750
```

```
D.  $1.25         $625
  ×  500        + 150
  $625.00         $775
```

5. C. $295 and $.05 a mile

```
A. $125      $500        $500
 ×   4     × .15       −  75
 $500        $75.00       $425
```

```
B.   $50      2000        $200
   ×  4     ×  .1       + 200
   $200       $200.0       $400
```

```
C.   2000      $295
   × $.05     + 100
   $100.00      $395
```

```
D.  $450       $450
  ×  .10     −  45
  $45.00       $405
```

Level 3   Review

1.
```
  $24.00
−  20.40
 $ 3.60
```

$\frac{3.60}{24.00} = \frac{3}{20}$

$\frac{3}{\underset{1}{\cancel{20}}} \times \frac{\overset{5}{\cancel{100\%}}}{1} = 15\%$

2.
```
  $390
+ 260
 $650
```

$\frac{390}{650} = \frac{3}{5}$

$\frac{3}{\underset{1}{\cancel{5}}} \times \frac{\overset{20}{\cancel{100\%}}}{1} = 60\%$

3. 85% = .85

$$\begin{array}{r} 1\ 60 \text{ lb.} \\ .85\overline{)136.00} \\ \underline{85} \\ 51\ 0 \\ \underline{51\ 0} \\ 00 \end{array}$$

$$\begin{array}{r} 160 \\ -136 \\ \hline 24 \text{ lb. lost} \end{array}$$

4. 75% = .75

$$\begin{array}{r} 9\ 00 \\ .75\overline{)675.00} \\ \underline{675} \\ 0\ 00 \end{array}$$

$$\begin{array}{r} 900 \\ -675 \\ \hline 225 \text{ empty seats} \end{array}$$

5. D. $550 each quarter

    A. $40 × 52 = $2080
    B. $2000
    C. 10% = .1   .1 × $1200 = $120/mo.
       $120 × 12 = $1440
    D. $550 × 4 = $2200

6. 9% = .09

$$\begin{array}{r} 100,000 \\ \times .09 \\ \hline \$9000.00 \end{array}$$

7. 6% = .06

$$\begin{array}{r} \$100,000 \\ \times .06 \\ \hline \$6,000.00 \end{array} \qquad \begin{array}{r} \$6,000 \\ +\ 6,000 \\ \hline \$12,000 \end{array}$$

8. 20% = .2

$$\begin{array}{r} \$14,000 \\ \times .2 \\ \hline \$2,800.0 \end{array} \qquad \begin{array}{r} \$14,000 \\ -\ 2,800 \\ \hline \$11,200 \end{array}$$

9.

$$\begin{array}{r} \$250,000 \\ \times .09 \\ \hline \$22,500.00 \end{array}$$

10.

$$\begin{array}{r} \$180,000 \\ \times .06 \\ \hline \$10,800.00 \end{array} \qquad \begin{array}{r} \$10,800 \\ +\ 6,000 \\ \hline \$16,800 \end{array}$$

Chapter 4 **Quiz**

1. .09 5 = 9.5%

2. $41\frac{2}{3}\% = \dfrac{41\frac{2}{3}}{100}$
$= 41\frac{2}{3} \div 100$
$= \dfrac{125}{3} \div \dfrac{100}{1}$
$= \dfrac{\overset{5}{\cancel{125}}}{3} \times \dfrac{1}{\cancel{100}} = \dfrac{5}{12}$

3. 7.6% = .076

$$\begin{array}{r} 230 \\ \times .076 \\ \hline 1\ 380 \\ 16\ 10 \\ \hline 17.480 = 17.48 \end{array}$$

4. $16\frac{2}{3}\% = \frac{1}{6}$

$\dfrac{1}{\cancel{6}} \times \dfrac{\overset{8}{\cancel{48}}}{1} = 8$ employees

5. 6.5% = .065

$$\begin{array}{r} \$128 \\ \times .065 \\ \hline 640 \\ 7\ 68 \\ \hline \$8.320 \end{array} \qquad \begin{array}{r} \$128.00 \\ +\ 8.32 \\ \hline \$136.32 \end{array}$$

6. $\dfrac{42}{120} = \dfrac{7}{20}$

$\dfrac{7}{\cancel{20}} \times \dfrac{\overset{5}{\cancel{100}\%}}{1} = 35\%$

7. $\dfrac{1,680}{11,200} = \dfrac{42}{280} = \dfrac{6}{40} = \dfrac{3}{20}$

$\dfrac{3}{\cancel{20}} \times \dfrac{\overset{5}{\cancel{100}\%}}{1} = 15\%$

8. $\dfrac{2.40}{40.00} = \dfrac{3}{50}$

$\dfrac{3}{\cancel{50}} \times \dfrac{\overset{2}{\cancel{100}\%}}{1} = 6\%$

9. 45% = .45

$$\begin{array}{r} 1\ 60 \\ .45\overline{)72.00} \\ \underline{45} \\ 27\ 0 \\ \underline{27\ 0} \\ 00 \end{array}$$

10. 35% = .35

$$\begin{array}{r} 6\ 00 \text{ mi.} \\ .35\overline{)210.00} \\ \underline{210} \\ 0\ 00 \end{array}$$

11. 1 yr. 6 mo. $= 1\frac{6}{12} = 1\frac{1}{2}$ yr.

$i = \dfrac{\overset{3}{\cancel{600}}}{1} \times \dfrac{8.5}{\cancel{100}} \times \dfrac{3}{\cancel{2}} = \$76.50$

12.

$$\begin{array}{r} \$29 \\ -20 \\ \hline \$\ 9 \end{array} \qquad \begin{array}{l} \text{change} \\ \text{original} \end{array} \dfrac{9}{20}$$

$\dfrac{9}{\cancel{20}} \times \dfrac{\overset{5}{\cancel{100}\%}}{1} = 45\%$

13.

$$\begin{array}{r} 165 \\ +135 \\ \hline 300 \end{array} \begin{array}{l} \text{women} \\ \text{total} \end{array} \dfrac{165}{300} = \dfrac{55}{100}$$

$\dfrac{55}{\cancel{100}} \times \dfrac{\overset{1}{\cancel{100}\%}}{1} = 55\%$

14. 60% = .6

$$\begin{array}{r} 12,00\ 0 \\ .6\overline{)72\ 00.0} \\ \underline{6} \\ 12 \\ \underline{12} \\ 0\ 00\ 0 \end{array} \qquad \begin{array}{r} 12,000 \\ -\ 7,200 \\ \hline 4,800 \end{array}$$

15. 8% = .08

$$\begin{array}{r} \$\ 12,7\ 50 \\ .08\overline{)\$102\ 0.00} \\ \underline{8} \\ 22 \\ \underline{16} \\ 6\ 0 \\ \underline{5\ 6} \\ 4\ 0 \\ \underline{4\ 0} \\ 00 \end{array} \qquad \begin{array}{r} \$12,750 \\ -\ 1,020 \\ \hline \$11,730 \end{array}$$

16. C. 30% of the amount Jeff paid for the car

    A. 24 × $50 = $1200
    B. 9 × $60 = $540 $540 + $500 = $1040
    C. 30% = .3   .3 × $4500 = $1350
    D. 44 × $30 = $1320

17. **(2) $3710**

6% = .06

$$\begin{array}{r} \$3500 \\ \times\ \ .06 \\ \hline \$210.00 \end{array} \qquad \begin{array}{r} \$3500 \\ +\ 210 \\ \hline \$3710 \end{array}$$

18. **(5) Insufficient data is given to solve the problem.**

19. **(2) $1600**

10% = .1

$$\begin{array}{r} \$16,000 \\ \times .1 \\ \hline \$1600.0 \end{array}$$

20. **(1) $3776**

Assistant makes

$$\begin{array}{r} \$\ 40 \\ \times\ 4 \\ \hline 160 \\ \times\ 8 \\ \hline \$1280 \end{array}$$

$$\begin{array}{ll} \text{materials} & \$16,000 \\ \text{expenses} & 1,600 \\ \text{assistant} & +1,280 \\ \text{total} & \$18,880 \end{array}$$

20% = .2

$$\begin{array}{r} \$18,880 \\ \times .2 \\ \hline \$3776.0 \end{array}$$

Chapter 5 **Graphs**

Lesson 1

1. a. **5.5 million** b. **4.5 million** c. **14 million** d. **4.5 million**

2. **Iowa**

3.   4.5 million
      + 3.5
      ‾‾‾‾‾‾‾‾‾‾
       8.0 million

4. **(2) slightly more than the production in Iowa**

    5.5 million
    4.5
    + 4.5
    ‾‾‾‾‾‾‾‾‾
    14.5 million

5. $\frac{3.5}{14} =$

$$\begin{array}{r} .25 = 25\% \\ 14\overline{)3.50} \\ \underline{2\,8} \\ 70 \\ \underline{70} \end{array}$$

6. **1 billion bushels of corn**

7.   6.5 bil. bu.
      − 6.
      ‾‾‾‾‾‾‾‾‾‾
       0.5 bil. bu.

8. $\frac{4}{6} = \frac{2}{3}$ 9. **1980**   10. **1985**

Lesson 2

1. a. **26%**   b. **13%**   c. **27%**   d. **3%**

2. **education**

3. $\frac{4}{100} = \frac{1}{25}$

4.   4%
       3
      + 13
      ‾‾‾‾‾‾
      $20\% = \frac{1}{5}$

5. **(2) a little over $\frac{1}{2}$**
       26%
      + 27
      ‾‾‾‾‾
      53% is a little over
      50%.

6. **divorced**

7.   7.3%
      − 3.2
      ‾‾‾‾‾‾
       4.1%

8. **(2) $\frac{1}{4}$**
      8.9%
     + 16.2
     ‾‾‾‾‾‾‾
     25.1%, which is
     close to
     $25\% = \frac{1}{4}$

9.   21.8%
      − 16.2
      ‾‾‾‾‾‾‾
       5.6%

10. **(1) While the percent of adults who were married or widowed decreased, the percent who were single or divorced increased.**

Lesson 3

1. **(2) $83**

2. **(4) $178**

3. **(4) 1984**

4. **(4) $80**
      1985 is about $215
      1980 is about − 135
      ‾‾‾‾‾‾‾‾‾‾‾‾‾‾‾‾‾‾‾‾‾‾‾‾
      The difference is $ 80

5. **(1) $\frac{1}{2}$**
      1978 is about $110.
      1985 is about $215.
      $\frac{110}{215}$ is close to $\frac{100}{200} = \frac{1}{2}$.

6. **(2) $2\frac{1}{2}$ times**
      1976 is about $80.
      1984 is about $200.

$$\begin{array}{r} 2\frac{40}{80} = 2\frac{1}{2} \\ 80\overline{)200} \\ \underline{160} \\ 40 \end{array}$$

7. **(3) 1980 to 1982**

8. **(2) $1400**
    The 1982 price is
    about $180. $180 × 8
    = $1440, which is
    close to $1400.

9. **(4) $800**
    The 1976 price is about $80.
    $80 × 10 = $800

10. **(2) $800**
    The 1980 price is about $135.
    $135 × 10 = $1350
    The 1985 price is about $215.
    $215 × 10 = $2150

          $2150
        − 1350
        ‾‾‾‾‾‾‾
        $ 800

Lesson 4

1. a. **145,000**   b. **267,000**   c. **47,000**   d. **160,000**

2.   47
       48
      219
       99
      ‾‾‾‾‾‾‾‾‾‾‾‾
      413 thousand

3.   94
       76
      309
     + 160
     ‾‾‾‾‾‾‾‾‾‾‾‾
     639 thousand

4. **1978**   5. **1984**   6. **1982**

7. **(4) $\frac{1}{2}$**
    413 is about 400.
    816 is about 800.
    $\frac{400}{800} = \frac{1}{2}$

8. **1984**

9. **(2) 50%**
    309 is about 300.
    639 is about 600.
    $\frac{300}{600} = \frac{1}{2} = 50\%$

10. **the South**

Lesson 5

1. a. **9%**   b. **14%**   c. **8%**

2. **1978**

3. **1982**

4.   14%
      − 6
      ‾‾‾‾‾
      8%

5. $\frac{6\%}{12\%} = \frac{1}{2}$

6. **(2) $135 billion**

7. **(2) $25 billion**
      $101 billion
      − 75
      ‾‾‾‾‾‾‾‾‾‾‾
      $ 26 billion

8. **(4) $30 billion**
      $166 billion
      − 136
      ‾‾‾‾‾‾‾‾‾‾‾
      $ 30 billion

9. **(1) 1979**

10. **(4) 1982–1983**

Chapter 5 **Quiz**

1. **(1) $22\frac{1}{2}$**
    $4\frac{1}{2} \times 5 =$
    $\frac{9}{2} \times \frac{5}{1} = \frac{45}{2} = 22\frac{1}{2}$

2. **(4) 2 times**
$$\begin{array}{r} 2 \\ 20\overline{)40} \end{array}$$

3. **(4) 10**
    $7\frac{1}{2} \times 5 = \quad 37\frac{1}{2}$
    $5\frac{1}{2} \times 5 = \underline{-27\frac{1}{2}}$
    ‾‾‾‾‾‾‾‾‾‾‾‾‾‾‾‾‾‾
         10 gal.

4. **(2) 20**
    $8 \times 5 = \quad 40$
    $4 \times 5 = \underline{-20}$
    ‾‾‾‾‾‾‾‾‾‾‾‾‾‾‾
       20 gal.

5. **(2) 13.9%**
      26.3%
     − 12.4
     ‾‾‾‾‾‾‾
     13.9%

6. **(4) 7.6%**
      22.3%
     − 14.7
     ‾‾‾‾‾‾‾
      7.6%

7. **(3) a little over $\frac{1}{4}$**
      14.7%
     + 12.4
     ‾‾‾‾‾‾‾
     27.1% is a little
     more than
     $25\% = \frac{1}{4}$.

8. **(5) 50%**
      26.3%
     + 24.3
     ‾‾‾‾‾‾‾
     50.6% is about
     50%.

9. **(1) $372**
   12.4% = .124   .124
   $$\times\ 3000$$
   $372.000

10. **(2) $441**
   14.7% = .147   .147
   $$\times\ 3000$$
   $441.000

11. **(5) $800 billion**

12. **(2)** $\frac{1}{4}$
   $\frac{200}{800} = \frac{1}{4}$

13. **(4)** $\frac{2}{3}$
   $\frac{200}{300} = \frac{2}{3}$

14. **(2) 1979–1980**

15. **(5) 1976–1978**

16. **(3) millions of students**

17. **(1) 1978**

18. **(2) 500,000**
   a little over 500,000

19. **(1) 1 million**

20. **(3) The number of female students increased steadily, while the number of male students gradually decreased.**

Chapter 6 **Algebra**

Level 1 **Algebra Skills**

1. point **C**

2. $(12) + (-9) + (-15) =$
   $12 - 9 - 15 =$
   $12 - 24 = $ **-12**

3. $(3ab)(-4a) = $ **$-12a^2b$**

4. $\frac{-4x^2y^2}{-8xy} = \frac{xy}{2}$

5. $3m - mn =$
   $3(-5) - (-5)(-2) =$
   $-15 - (+10) =$
   $-15 - 10 = $ **-25**

Lesson 1

1. *I*   2. *C*   3. F   4. *A*   5. *G*

6. *B*   7. *E*   8. *H*   9. *D*   10. *J*

Lesson 2

1. $+7 - 6 = $ **1** $-13 + 13 = $ **0** $-14 + 18 = $ **4**

2. $(+9) + (-15)$ $(-8) + (-11) =$ $(-12) + (+12)$
   $= +9 - 15$ $-8 - 11 = $ **-19** $= -12 + 12$
   $= -6$ $= 0$

3. $(-24) + (+7)$ $(-3) + (+14) =$ $(-19) + (-19)$
   $= -24 + 7$ $-3 + 14 = $ **11** $= -19 - 19$
   $= -17$ $= -38$

4. $+3 - 9 + 7$ $-4 - 6 - 3 = $ $+2 + 8 - 10$
   $= +10 - 9$ **-13** $=$
   $= 1$ $+10 - 10 = $ **0**

5. $(+3) + (+11) + (+5) =$ $(-9) + (-1) + (+4) =$
   $+3 + 11 + 5 = $ **$+19$** $-10 + 4 = $ **-6**

   $-8 - 1 - 6 + 10 =$
   $-15 + 10 = $ **-5**

6. $(-8) + (+12) + (+16) + (-11) =$
   $-8 + 12 + 16 - 11 =$
   $-19 + 28 = $ **9**

   $(+7) + (-12) + (+3) + (+2) =$
   $+7 - 12 + 3 + 2 =$
   $+12 - 12 = $ **0**

7. $a + b = (-6) + (-4) = -6 - 4 = $ **-10**

8. $m + n + p = (3) + (9) + (-7) = 12 - 7 = $ **5**

9. $x + y = (15) + (-15) = $ **0**

10. $r + s + t = (-1) + (-8) + (7) = -9 + 7 = $ **-2**

Lesson 3

1. $(+8) - (+9)$ $(+7) - (-6) =$ $(-9) - (+3) =$
   $= +8 - 9 =$ $+7 + 6 = $ **$+13$** $-9 - 3 = $ **-12**
   -1

2. $(-3) - (-14)$ $(-10) - (+15)$ $(+6) - (-11) =$
   $= -3 + 14$ $= -10 - 15$ $+6 + 11 = $ **$+17$**
   $= 11$ $= -25$

3. $(+20) - (-4)$ $(+2) - (+18)$ $(-16) - (-5) =$
   $= +20 + 4$ $= +2 - 18$ $-16 + 5 = $ **-11**
   $= 24$ $= -16$

4. $(-11) -$ $(+7) - (-8)$ $(-9) - (+9) =$
   $(+15)$ $= +7 + 8$ $-9 - 9 = $ **-18**
   $= -11 - 15$ $= +15$
   $= -26$

5. $(+8) - (+7) + (6) =$ $(-4) + (-5) - (+3) =$
   $+8 - 7 + 6 =$ $-4 - 5 - 3 = $ **-12**
   $+14 - 7 = $ **7**

6. $(-10) - (-7) - (-1) =$ $(+11) - (-14) + (-3) =$
   $-10 + 7 + 1 =$ $+11 + 14 - 3 =$
   $-10 + 8 = $ **-2** $+25 - 3 = $ **22**

7. $a - b = (-6) - (-7) = -6 + 7 = $ **1**

8. $m - n = (14) - (-4) = 14 + 4 = $ **18**

9. $w - y = (-6) - (+7) = -6 - 7 = $ **-13**

10. $c - d = (15) - (-9) = 15 + 9 = $ **24**

Lesson 4

1. $(+6)(-10) = $ **-60** $(+12)(-1) = $ **-12**
   $(-5)(-13) = $ **65**

2. $(+8)(+\frac{1}{2}) = $ **4** $(-\frac{1}{3})(-15) = $ **5**
   $(-9)(+8) = $ **-72**

3. $(-\frac{3}{4})(+24) = $ **-18** $(+20)(-3) = $ **-60**
   $(-15)(-2) = $ **30**

4. $(+4)(-10)(+8) = $ **-320** $(+5)(-9)(-\frac{2}{3}) = $ **30**

5. $(-\frac{1}{2})(-12)(-5) = $ **-30** $(-8)(+12)(-1) = $ **96**

6. $(-4)(+10)(+\frac{3}{4})(+2) = $ **-60**
   $(+5)(-1)(+9)(-6) = $ **270**

7. $ab = (-7)(-6) = $ **42**

8. $pqr = (5)(-3)(-10) = $ **150**

9. $mn = (-9)(+9) = $ **-81**

10. $xyz = (-1)(-\frac{1}{2})(-\frac{1}{4}) = $ **$-\frac{1}{8}$**

Lesson 5

1. $\frac{-24}{-8} = $ **$+3$** $\frac{-30}{+10} = -3$ $\frac{+5}{-10} = -\frac{1}{2}$ $\frac{+8}{+2} = +4$

2. $\frac{-12}{+12} = -1$ $\frac{-15}{-20} = +\frac{3}{4}$ $\frac{+100}{-20} = -5$ $\frac{-20}{-40} = \frac{1}{2}$

3. $\frac{63}{-9} = -7$ $\frac{-48}{-12} = +4$ $\frac{-80}{+100} = \frac{-4}{5}$ $\frac{+18}{-6} = -3$

4. $\frac{150}{-25} = -6$ $\frac{-35}{49} = \frac{-5}{7}$ $\frac{-20}{-24} = \frac{5}{6}$ $\frac{-96}{+12} = -8$

5. $\frac{a}{b} = \frac{-14}{-2} = 7$

6. $\frac{x}{y} = \frac{-12}{15} = \frac{-4}{5}$

7. $\frac{m}{n} = \frac{23}{-1} = -23$

8. $\frac{c}{d} = \frac{+18}{-24} = \frac{-3}{4}$

9. $\frac{e}{f} = \frac{22}{-11} = -2$

10. $\frac{u}{v} = \frac{-8}{32} = \frac{-1}{4}$

Lesson 6

1. $5m + 3m = $ **$8m$** $2c - 6c = $ **$-4c$**
   $8x - x = $ **$7x$**

2. $-4p + 5p = $ **p** $18xy - 19xy = $ **$-xy$**
   $-6st - 5st = $ **$-11st$**

3. $9w - 3w + 4w = $ $-8y - 3y - y = $ **$-12y$**
   $13w - 3w = $ **$10w$**

   $-d + 5d - 4d = $
   $-5d + 5d = $ **0**

4. $(-8e) + (-6e) + (11e) = $ $(7n) + (-12n) + (-4n) = $
   $-14e + 11e = $ **$-3e$** $+7n - 16n = $ **$-9n$**

5. $(-a) + (5a) + (-9a) + (+3a) = -10a + 8a = $ **$-2a$**

   $(-7w) + (-2w) + (-3w) + (11w) = -12w + 11w = $ **$-w$**

Lesson 7

1. $(+5p) - (+3p) = 5p - 3p = $ **$2p$**
   $(+5p) - (-3p) = 5p + 3p = $ **$8p$**
   $(-5p) - (-3p) = -5p + 3p = $ **$-2p$**

2. $(-6a) - (-3a) = -6a + 3a = $ **$-3a$**
   $(+4c) - (-4c) = +4c + 4c = $ **$8c$**
   $(-7mn) - (+3mn) = -7mn - 3mn = $ **$10mn$**

3. $(-9f) - (-f) + (-2f) = $ $(10y) - (3y) - (-7y) = $
   $-9f + f - 2f = $ $10y - 3y + 7y = $
   $-11f + f = $ **$-10f$** $17y - 3y = $ **$14y$**

4. $(-8cd) + (-2cd) - (3cd) = -8cd - 2cd - 3cd = $ **$-13cd$**
   $(-12t) - (-3t) + (9t) = -12t + 3t + 9t = -12t + 12t = $ **0**

5. $(4bc) - (-8bc) - (2bc) = $ $(-16x) - (2x) + (-5x) = $
   $4bc + 8bc - 2bc = $ $-16x - 2x - 5x = $ **$-23x$**
   $12bc - 2bc = $ **$10bc$**

Lesson 8

1. $c^2 \cdot c^3 = $ **c^5** $m^4 \cdot m = $ **m^5**
   $x^3 \cdot x^5 = $ **x^8**

2. $a \cdot a = $ **a^2** $a \cdot b = $ **ab**
   $(a^2b^2)(a^3b^4) = $ **a^5b^6**

3. $(2x)(-3x) = $ **$-6x^2$** $(-4y^2)(-2y^5) = $ **$8y^7$**
   $(-5p)(2p^3) = $ **$-10p^4$**

4. $(5yz)(-6yz) = $ **$-30y^2z^2$** $(-mn^2)(-4m^2n^3) = $ **$4m^3n^5$**
   $(-2rs)(-9r^3s) = $ **$18r^4s^2$**

5. $(-ef^2)(2e^2f) = $ **$-2e^3f^3$** $(-4kt^3)(-k^3t) = $ **$4k^4t^4$**
   $(-5d^2h^3)(6dh) = $ **$-30d^3h^4$**

Lesson 9

1. $\frac{m^6}{m^2} = $ **m^4** $\frac{st}{s} = $ **t** $\frac{c^5}{c^2} = $ **c^3**

2. $\frac{a^5}{a^4} = $ **a** $\frac{x^3y^2}{xy} = $ **x^2y** $\frac{x^6}{x^6} = $ **1**

3. $\frac{-12x^2}{3x} = $ **$-4x$** $\frac{20m^3n}{-4m} = $ **$-5m^2n$** $\frac{-36a^5}{-9a^4} = $ **$4a$**

4. $\frac{-4m^2n^2}{16mn} = $ **$\frac{-mn}{4}$** $\frac{-18c^3d^4}{-12c^3d} = $ **$\frac{3d^3}{2}$** $\frac{+24x}{+30x} = $ **$\frac{4}{5}$**

5. $\frac{2x^3y^2}{-18y} = $ **$\frac{x^3y}{-9}$** $\frac{-4a^3b}{20ab} = $ **$\frac{-a^2}{5}$** $\frac{-20m^2n}{-4mn} = $ **$5m$**

Lesson 10

1. $ab - c = $
   $(-4)(-3) - (-5) = $
   $+12 + 5 = $ **$+17$**

2. $m(m - n) = $
   $-2(-2 - 7) = $
   $-2(-9) = $ **$+18$**

3. $x^2y = $
   $(-3)^2 \cdot (-4) = $
   $(+9)(-4) = $ **-36**

4. $s(s + t) - t = $
   $-5(-5 + 1) - 1 = $
   $-5(-4) - 1 = $
   $+20 - 1 = $ **$+19$**

5. $e + ef = $
   $6 + (6)(-4) = $
   $6 + (-24) = $
   $6 - 24 = $ **-18**

6. $ab^2 = $
   $(-2)(-5)^2 = $
   $(-2)(+25) = $ **-50**

7. $x(x - y) = $
   $-8(-8 - 2) = $
   $-8(-10) = $ **$+80$**

8. $(j - k)^2 = $
   $(-3 - 1)^2 = $
   $(-4)^2 = $ **$+16$**

9. $\frac{a + b}{2} = $
   $\frac{-6 + (-2)}{2} = $
   $\frac{-6 - 2}{2} = $
   $\frac{-8}{2} = $ **-4**

10. $\frac{m - n}{n} = $
    $\frac{8 - (-4)}{-4} = $
    $\frac{8 + 4}{-4} = $
    $\frac{+12}{-4} = $ **-3**

Level 1   Review

1. **point B**

2. $(-13) + (8) + (-7) = $
   $-13 + 8 - 7 = $
   $-20 + 8 = $ **-12**

3. $(-9) - (-11) - (+4) = $
   $-9 + 11 - 4 = -13 + $
   $11 = $ **-2**

4. $(-6)\left(-\frac{2}{3}\right)(-5) = $ **-20**

5. $\frac{+24}{-36} = $ **$-\frac{2}{3}$**

6. $(4m) + (-2m) + (-7m)$
   $= 4m - 2m - 7m = $
   $4m - 9m = $ **$-5m$**

7. $(-n) - (-6n) + (-4n)$
   $= -n + 6n - 4n = $
   $-5n + 6n = $ **n**

8. $(-9x^2y)(-2xy) = $ **$18x^3y^2$**

9. $\frac{12a^2c}{-3a} = $ **$-4ac$**

10. $ab - 4b = $
    $(-7)(-3) - 4(-3) = $
    $+21 - (-12) = $
    $+21 + 12 = $ **$+33$**

Level 2   Algebra Applications

1. **$4(x - 7)$**

2. **$x - 12$**

3. $m^2 + 12m = $ **$m(m + 12)$**

4. $\sqrt{12} = \sqrt{4} \cdot \sqrt{3} = $ **$2\sqrt{3}$**

5. $x^2 - 100 = 0$
   $(x + 10)(x - 10) = 0$

$x + 10 = $	0	$x - 10 = $	0
-10	-10	$+10$	$+10$

   x = **-10** and x = **$+10$**

Lesson 1

1.
$$\begin{aligned} m + 11 &= 30 \\ -11 & \quad -11 \\ \hline m &= 19 \end{aligned}$$
$$\frac{8w}{8} = \frac{56}{8}$$
$$w = 7$$
$$\begin{aligned} c - 12 &= 5 \\ +12 & \quad +12 \\ \hline c &= 17 \end{aligned}$$

ANSWER CHECK   ANSWER CHECK   ANSWER CHECK
$$\begin{aligned} 19 + 11 &= 30 \\ 30 &= 30 \end{aligned}$$
$$\begin{aligned} 8 \times 7 &= 56 \\ 56 &= 56 \end{aligned}$$
$$\begin{aligned} 17 - 12 &= 5 \\ 5 &= 5 \end{aligned}$$

2.
$$\begin{aligned} 16 &= f - 4 \\ +4 & \quad +4 \\ \hline 20 &= f \end{aligned}$$
$$4 \cdot \frac{c}{4} = 5 \cdot 4$$
$$c = 20$$
$$\frac{6n}{6} = \frac{9}{6}$$
$$n = 1\tfrac{3}{6} = 1\tfrac{1}{2}$$

ANSWER CHECK   ANSWER CHECK   ANSWER CHECK
$$\begin{aligned} 16 &= 20 - 4 \\ 16 &= 16 \end{aligned}$$
$$\frac{\overset{5}{\cancel{20}}}{\underset{1}{\cancel{4}}} = 5$$
$$5 = 5$$
$$6 \cdot 1\tfrac{1}{2} = 9$$
$$\overset{3}{\cancel{6}} \cdot \frac{3}{\underset{1}{\cancel{2}}} = 9$$
$$9 = 9$$

3.
$$9 \cdot 2 = \frac{y}{9} \cdot 9$$
$$18 = y$$
$$\frac{12}{18} = \frac{18p}{18}$$
$$\frac{2}{3} = p$$
$$\begin{aligned} 14 &= a + 3 \\ -3 & \quad -3 \\ \hline 11 &= a \end{aligned}$$

ANSWER CHECK   ANSWER CHECK   ANSWER CHECK
$$2 = \frac{\overset{2}{\cancel{18}}}{\underset{1}{\cancel{9}}}$$
$$2 = 2$$
$$12 = \overset{6}{\cancel{18}} \cdot \frac{2}{\underset{1}{\cancel{3}}}$$
$$12 = 12$$
$$\begin{aligned} 14 &= 11 + 3 \\ 14 &= 14 \end{aligned}$$

4.
$$\begin{aligned} g - 9 &= 41 \\ +9 & \quad +9 \\ \hline g &= 50 \end{aligned}$$
$$\begin{aligned} e + 6 &= -8 \\ -6 & \quad -6 \\ \hline e &= -14 \end{aligned}$$
$$12 \cdot \frac{n}{12} = 1 \cdot 12$$
$$n = 12$$

ANSWER CHECK   ANSWER CHECK   ANSWER CHECK
$$\begin{aligned} 50 - 9 &= 41 \\ 41 &= 41 \end{aligned}$$
$$\begin{aligned} -14 + 6 &= -8 \\ -8 &= -8 \end{aligned}$$
$$\frac{\overset{1}{\cancel{12}}}{\underset{1}{\cancel{12}}} = 1$$
$$1 = 1$$

5.
$$\begin{aligned} 15 &= i - 8 \\ +8 & \quad +8 \\ \hline 23 &= i \end{aligned}$$
$$\frac{200}{25} = \frac{25r}{25}$$
$$8 = r$$
$$\frac{4}{3} \cdot \frac{3}{4}s = 24 \cdot \frac{4}{3}$$
$$s = 32$$

ANSWER CHECK   ANSWER CHECK   ANSWER CHECK
$$\begin{aligned} 15 &= 23 - 8 \\ 15 &= 15 \end{aligned}$$
$$\begin{aligned} 200 &= 25 \cdot 8 \\ 200 &= 200 \end{aligned}$$
$$\frac{3}{4} \cdot \overset{8}{\underset{1}{\cancel{32}}} = 24$$
$$24 = 24$$

6.
$$2 \cdot 10 = \tfrac{1}{2}w \cdot 2$$
$$20 = w$$
$$\begin{aligned} 21 &= d + 16 \\ -16 & \quad -16 \\ \hline 5 &= d \end{aligned}$$
$$5 \cdot 10 = \frac{z}{5} \cdot 5$$
$$50 = z$$

ANSWER CHECK   ANSWER CHECK   ANSWER CHECK
$$10 = \tfrac{1}{2} \cdot \overset{10}{\underset{1}{\cancel{20}}}$$
$$10 = 10$$
$$\begin{aligned} 21 &= 5 + 16 \\ 21 &= 21 \end{aligned}$$
$$10 = \frac{\overset{10}{\cancel{50}}}{\underset{1}{\cancel{5}}}$$
$$10 = 10$$

7.
$$\frac{24f}{24} = \frac{12}{24}$$
$$f = \tfrac{1}{2}$$
$$\begin{aligned} p + 14 &= 4 \\ -14 & \quad -14 \\ \hline p &= -10 \end{aligned}$$
$$\frac{8}{3} \cdot \frac{3}{8}x = 15 \cdot \frac{8}{3}$$
$$x = 40$$

ANSWER CHECK   ANSWER CHECK   ANSWER CHECK
$$\overset{12}{\underset{1}{\cancel{24}}} \cdot \tfrac{1}{2} = 12$$
$$12 = 12$$
$$\begin{aligned} -10 + 14 &= 4 \\ 4 &= 4 \end{aligned}$$
$$\frac{3}{8} \cdot \overset{5}{\underset{1}{\cancel{40}}} = 15$$
$$15 = 15$$

8.
$$\frac{16y}{16} = \frac{4}{16}$$
$$y = \tfrac{1}{4}$$
$$\begin{aligned} a + 3 &= 10 \\ -3 & \quad -3 \\ \hline a &= 7 \end{aligned}$$
$$4 \cdot \frac{p}{4} = 3 \cdot 4$$
$$p = 12$$

ANSWER CHECK   ANSWER CHECK   ANSWER CHECK
$$\overset{4}{\cancel{16}} \cdot \frac{1}{\underset{1}{\cancel{4}}} = 4$$
$$4 = 4$$
$$\begin{aligned} 7 + 3 &= 10 \\ 10 &= 10 \end{aligned}$$
$$\frac{\overset{3}{\cancel{12}}}{\underset{1}{\cancel{4}}} = 3$$
$$3 = 3$$

9.
$$\frac{20}{2} = \frac{2k}{2}$$
$$10 = k$$
$$\begin{aligned} b - 3 &= 17 \\ +3 & \quad +3 \\ \hline b &= 20 \end{aligned}$$
$$2 \cdot 40 = \frac{x}{2} \cdot 2$$
$$80 = x$$

ANSWER CHECK   ANSWER CHECK   ANSWER CHECK
$$\begin{aligned} 20 &= 2 \cdot 10 \\ 20 &= 20 \end{aligned}$$
$$\begin{aligned} 20 - 3 &= 17 \\ 17 &= 17 \end{aligned}$$
$$40 = \frac{80}{2}$$
$$40 = 40$$

10.
$$18 \cdot \frac{z}{18} = 2 \cdot 18$$
$$z = 36$$
$$\begin{aligned} 7 &= a + 1 \\ -1 & \quad -1 \\ \hline 6 &= a \end{aligned}$$
$$\frac{5}{4} \cdot \frac{4c}{5} = 40 \cdot \frac{5}{4}$$
$$c = 50$$

ANSWER CHECK   ANSWER CHECK   ANSWER CHECK
$$\frac{\overset{2}{\cancel{36}}}{\underset{1}{\cancel{18}}} = 2$$
$$2 = 2$$
$$\begin{aligned} 7 &= 6 + 1 \\ 7 &= 7 \end{aligned}$$
$$\frac{4}{8} \cdot \overset{10}{\cancel{50}} = 40$$
$$40 = 40$$

Lesson 2

1.
$$\begin{aligned} 6m + 5 &= 47 \\ -5 & \quad -5 \\ \hline \frac{6m}{6} &= \frac{42}{6} \\ m &= 7 \end{aligned}$$
$$\begin{aligned} 3x - 2 &= 28 \\ +2 & \quad +2 \\ \hline \frac{3x}{3} &= \frac{30}{3} \\ x &= 10 \end{aligned}$$
$$\begin{aligned} \frac{c}{4} + 1 &= 8 \\ -1 & \quad -1 \\ \hline 4 \cdot \frac{c}{4} &= 7 \cdot 4 \\ c &= 28 \end{aligned}$$

ANSWER CHECK   ANSWER CHECK   ANSWER CHECK
$$\begin{aligned} 6 \cdot 7 + 5 &= 47 \\ 42 + 5 &= 47 \\ 47 &= 47 \end{aligned}$$
$$\begin{aligned} 3 \cdot 10 - 2 &= 28 \\ 30 - 2 &= 28 \\ 28 &= 28 \end{aligned}$$
$$\frac{\overset{7}{\cancel{28}}}{\underset{1}{\cancel{4}}} + 1 = 8$$
$$7 + 1 = 8$$
$$8 = 8$$

2.
$$\begin{aligned} 17 &= 7a + 3 \\ -3 & \quad -3 \\ \hline \frac{14}{7} &= \frac{7a}{7} \\ 2 &= a \end{aligned}$$
$$\begin{aligned} 50 &= 9d - 4 \\ +4 & \quad +4 \\ \hline \frac{54}{9} &= \frac{9d}{9} \\ 6 &= d \end{aligned}$$
$$\begin{aligned} 8 &= \frac{x}{7} - 2 \\ +2 & \quad +2 \\ \hline 7 \cdot 10 &= \frac{x}{7} \cdot 7 \\ 70 &= x \end{aligned}$$

ANSWER CHECK   ANSWER CHECK   ANSWER CHECK
$$\begin{aligned} 17 &= 7 \cdot 2 + 3 \\ 17 &= 14 + 3 \\ 17 &= 17 \end{aligned}$$
$$\begin{aligned} 50 &= 9 \cdot 6 - 4 \\ 50 &= 54 - 4 \\ 50 &= 50 \end{aligned}$$
$$8 = \frac{\overset{10}{\cancel{70}}}{\underset{1}{\cancel{7}}} - 2$$
$$8 = 10 - 2$$
$$8 = 8$$

3.
$$\begin{aligned} 2n - 11 &= 3 \\ +11 & \quad +11 \\ \hline \frac{2n}{2} &= \frac{14}{2} \\ n &= 7 \end{aligned}$$
$$\begin{aligned} \tfrac{3}{4}a + 5 &= 17 \\ -5 & \quad -5 \\ \hline \tfrac{4}{3} \cdot \tfrac{3}{4}a &= 12 \cdot \tfrac{4}{3} \\ a &= 16 \end{aligned}$$
$$\begin{aligned} \frac{s}{10} - 6 &= 3 \\ +6 & \quad +6 \\ \hline 10 \cdot \frac{s}{10} &= 9 \cdot 10 \\ s &= 90 \end{aligned}$$

ANSWER CHECK

$2 \cdot 7 - 11 = 3$
$14 - 11 = 3$
$3 = 3$

ANSWER CHECK

$\frac{3}{4} \cdot \overset{4}{16} + 5 = 17$
$12 + 5 = 17$
$17 = 17$

ANSWER CHECK

$\frac{\overset{9}{90}}{10} - 6 = 3$
$9 - 6 = 3$
$3 = 3$

4.
$$5y + 7 = -3$$
$$\underline{-7 \quad -7}$$
$$\frac{5y}{5} = \frac{-10}{5}$$
$$y = -2$$

$$2 = \frac{w}{9} + 11$$
$$\underline{-11 \quad -11}$$
$$9 \cdot -9 = \frac{w}{9} \cdot 9$$
$$-81 = w$$

$$\frac{4}{5}f - 7 = 1$$
$$\underline{+7 \quad +7}$$
$$\frac{5}{4} \cdot \frac{4}{5}f = 8 \cdot \frac{5}{4}$$
$$f = 10$$

ANSWER CHECK

$5 \cdot -2 + 7 =$
-3
$-10 + 7 =$
-3
$-3 =$
-3

ANSWER CHECK

$2 = \frac{\overset{9}{-81}}{\underset{1}{9}} + 11$
$2 = -9 + 11$
$2 = 2$

ANSWER CHECK

$\frac{4}{\underset{1}{5}} \cdot \overset{2}{10} - 7 = 1$
$8 - 7 = 1$
$1 = 1$

5.
$$8z + 3 = 9$$
$$\underline{-3 \quad -3}$$
$$\frac{8z}{8} = \frac{6}{8}$$
$$z = \frac{3}{4}$$

$$2 = 9p - 4$$
$$\underline{+4 \quad\quad +4}$$
$$\frac{6}{9} = \frac{9p}{9}$$
$$\frac{2}{3} = p$$

$$-4 = 7t + 3$$
$$\underline{-3 \quad\quad -3}$$
$$\frac{-7}{7} = \frac{7t}{7}$$
$$-1 = t$$

ANSWER CHECK

$\overset{2}{8} \cdot \frac{3}{\underset{1}{4}} + 3 = 9$
$6 + 3 = 9$
$9 = 9$

ANSWER CHECK

$2 = \overset{3}{9} \cdot \frac{2}{\underset{1}{3}} - 4$
$2 = 6 - 4$
$2 = 2$

ANSWER CHECK

$-4 = 7 \cdot -1 + 3$
$-4 = -7 + 3$
$-4 = -4$

Lesson 3

1.
$$9a - 2a = 21$$
$$\frac{7a}{7} = \frac{21}{7}$$
$$a = 3$$

$$5m = 18 + 2m$$
$$\underline{-2m \quad\quad -2m}$$
$$\frac{3m}{3} = \frac{18}{3}$$
$$m = 6$$

$$8r = 15 + 3r$$
$$\underline{-3r \quad\quad -3r}$$
$$\frac{5r}{5} = \frac{15}{5}$$
$$r = 3$$

ANSWER CHECK

$9 \cdot 3 - 2 \cdot 3 = 21$
$27 - 6 = 21$
$21 = 21$

ANSWER CHECK

$5 \cdot 6 = 18 + 2 \cdot 6$
$30 = 18 + 12$
$30 = 30$

ANSWER CHECK

$8 \cdot 3 = 15 + 3 \cdot 3$
$24 = 15 + 9$
$24 = 24$

2.
$$12 - 5x = 7x$$
$$\underline{+5x \quad\quad +5x}$$
$$\frac{12}{12} = \frac{12x}{12}$$
$$1 = x$$

$$16 = 13c + 7c$$
$$\frac{16}{20} = \frac{20c}{20}$$
$$\frac{4}{5} = c$$

$$3p + 7 = 10p$$
$$\underline{-3p \quad\quad -3p}$$
$$\frac{7}{7} = \frac{7p}{7}$$
$$1 = p$$

ANSWER CHECK

$12 - 5 \cdot 1 = 7 \cdot 1$
$12 - 5 = 7$
$7 = 7$

ANSWER CHECK

$16 = 13 \cdot \frac{4}{5} + 7 \cdot \frac{4}{5}$
$16 = \frac{52}{5} + \frac{28}{5}$
$16 = \frac{\overset{16}{80}}{\underset{1}{5}}$
$16 = 16$

ANSWER CHECK

$3 \cdot 1 + 7 = 10 \cdot 1$
$3 + 7 = 10$
$10 = 10$

3.
$$6y - y = 10$$
$$\frac{5y}{5} = \frac{10}{5}$$
$$y = 2$$

$$3t = 9 + 2t$$
$$\underline{-2t \quad\quad -2t}$$
$$t = 9$$

$$4 - 2n = 6n$$
$$\underline{+2n \quad\quad +2n}$$
$$\frac{4}{8} = \frac{8n}{8}$$
$$\frac{1}{2} = n$$

ANSWER CHECK

$6 \cdot 2 - 2 = 10$
$12 - 2 = 10$
$10 = 10$

ANSWER CHECK

$3 \cdot 9 = 9 + 2 \cdot 9$
$27 = 9 + 18$
$27 = 27$

ANSWER CHECK

$4 - 2 \cdot \frac{1}{\underset{1}{2}} = \overset{3}{6} \cdot \frac{1}{\underset{1}{2}}$
$4 - 1 = 3$
$3 = 3$

4.
$$5x + 4 = 3x + 20$$
$$\underline{-3x \quad\quad -3x}$$
$$2x + 4 = 20$$
$$\underline{-4 \quad\quad\quad -4}$$
$$\frac{2x}{2} = \frac{16}{2}$$
$$x = 8$$

$$8w - 5 = 7w + 13$$
$$\underline{-7w \quad\quad -7}$$
$$w - 5 = 13$$
$$\underline{+5 \quad\quad +5}$$
$$w = 18$$

$$3p + 12 = 8p - 23$$
$$\underline{-3p \quad\quad -3p}$$
$$12 = 5p - 23$$
$$\underline{+23 \quad\quad +23}$$
$$\frac{35}{5} = \frac{5p}{5}$$
$$7 = p$$

ANSWER CHECK

$5 \cdot 8 + 4 = 3 \cdot 8 + 20$
$40 + 4 = 24 + 20$
$44 = 44$

ANSWER CHECK

$8 \cdot 18 - 5 = 7 \cdot 18 + 13$
$144 - 5 = 126 + 13$
$139 = 139$

ANSWER CHECK

$3 \cdot 7 + 12 = 8 \cdot 7 - 23$
$21 + 12 = 56 - 23$
$33 = 33$

5.
$$7c - 3c = c + 27$$
$$4c = c + 27$$
$$\underline{-c \qquad\qquad -c}$$
$$\frac{3c}{3} = \frac{27}{3}$$
$$c = \mathbf{9}$$

$$9m - 12 = m + 20$$
$$\underline{-m \qquad\qquad -m}$$
$$8m - 12 = 20$$
$$\underline{+12 \qquad\qquad +12}$$
$$\frac{8m}{8} = \frac{32}{8}$$
$$m = \mathbf{4}$$

$$2d - 8 = 7d + 12$$
$$\underline{-2d \qquad\qquad -2d}$$
$$-8 = 5d + 12$$
$$\underline{-12 \qquad\qquad -12}$$
$$\frac{-20}{5} = \frac{5d}{5}$$
$$\mathbf{-4} = d$$

ANSWER CHECK
$$7 \cdot 9 - 3 \cdot 9 = 9 + 27$$
$$63 - 27 = 36$$
$$36 = 36$$

ANSWER CHECK
$$9 \cdot 4 - 12 = 4 + 20$$
$$36 - 12 = 24$$
$$24 = 24$$

ANSWER CHECK
$$2 \cdot -4 - 8 = 7 \cdot -4 + 12$$
$$-8 - 8 = -28 + 12$$
$$-16 = -16$$

Lesson 4

1.
$$4(m - 3) = 20$$
$$4m - 12 = 20$$
$$\underline{+12 \qquad\qquad +12}$$
$$\frac{4m}{4} = \frac{32}{4}$$
$$m = \mathbf{8}$$

$$5(a + 2) = 15$$
$$5a + 10 = 15$$
$$\underline{-10 \qquad\qquad -10}$$
$$\frac{5a}{5} = \frac{5}{5}$$
$$a = \mathbf{1}$$

$$9 = 2(x - 3)$$
$$9 = 2x - 6$$
$$\underline{+6 \qquad\qquad +6}$$
$$\frac{15}{2} = \frac{2x}{2}$$
$$\mathbf{7\tfrac{1}{2}} = x$$

ANSWER CHECK
$$4(8 - 3) = 20$$
$$4(5) = 20$$
$$20 = 20$$

ANSWER CHECK
$$5(1 + 2) = 15$$
$$5(3) = 15$$
$$15 = 15$$

ANSWER CHECK
$$9 = 2(7\tfrac{1}{2} - 3)$$
$$9 = 2(4\tfrac{1}{2})$$
$$9 = \overset{1}{2} \cdot \underset{1}{\tfrac{9}{2}}$$
$$9 = 9$$

2.
$$3(c + 4) = 2c + 17$$
$$3c + 12 = 2c + 17$$
$$\underline{-2c \qquad\qquad -2c}$$
$$c + 12 = 17$$
$$-12 \qquad\qquad -12$$
$$c = \mathbf{5}$$

$$8n - 7 = 6(n - 1)$$
$$8n - 7 = 6n - 6$$
$$\underline{-6n \qquad\qquad -6n}$$
$$2n - 7 = -6$$
$$\underline{+7 \qquad\qquad +7}$$
$$\frac{2n}{2} = \frac{1}{2}$$
$$n = \mathbf{\tfrac{1}{2}}$$

$$9(p + 2) = p + 20$$
$$9p + 18 = p + 20$$
$$\underline{-p \qquad\qquad -p}$$
$$8p + 18 = 20$$
$$\underline{-18 \qquad\qquad -18}$$
$$\frac{8p}{8} = \frac{2}{8}$$
$$p = \mathbf{\tfrac{1}{4}}$$

ANSWER CHECK
$$3(5 + 4) = 2 \cdot 5 + 17$$
$$3(9) = 10 + 17$$
$$27 = 27$$

ANSWER CHECK
$$\overset{4}{8} \cdot \underset{1}{\tfrac{1}{2}} - 7 = 6(\tfrac{1}{2} - 1)$$
$$4 - 7 = \overset{3}{6}(\underset{1}{-\tfrac{1}{2}})$$
$$-3 = -3$$

ANSWER CHECK
$$9(\tfrac{1}{4} + 2) = \tfrac{1}{4} + 20$$
$$9(2\tfrac{1}{4}) = 20\tfrac{1}{4}$$
$$9 \cdot \tfrac{9}{4} = 20\tfrac{1}{4}$$
$$\tfrac{81}{4} = 20\tfrac{1}{4}$$
$$20\tfrac{1}{4} = 20\tfrac{1}{4}$$

3.
$$4(a - 5) = 3(a + 2)$$
$$4a - 20 = 3a + 6$$
$$\underline{-3a \qquad\qquad -3a}$$
$$a - 20 = 6$$
$$\underline{+20 \qquad\qquad +20}$$
$$a = \mathbf{26}$$

$$6(d - 1) = 3(d + 2)$$
$$6d - 6 = 3d + 6$$
$$\underline{-3d \qquad\qquad -3d}$$
$$3d - 6 = 6$$
$$\underline{+6 \qquad\qquad +6}$$
$$\frac{3d}{3} = \frac{12}{3}$$
$$d = \mathbf{4}$$

$$5(y + 2) = 3(y - 8)$$
$$5y + 10 = 3y - 24$$
$$\underline{-3y \qquad\qquad -3y}$$
$$2y - 10 = -24$$
$$\underline{-10 \qquad\qquad -10}$$
$$\frac{2y}{2} = \frac{-34}{2}$$
$$y = \mathbf{-17}$$

ANSWER CHECK
$$4(26 - 5) = 3(26 + 2)$$
$$4(21) = 3(28)$$
$$84 = 84$$

ANSWER CHECK
$$6(4 - 1) = 3(4 + 2)$$
$$6(3) = 3(6)$$
$$18 = 18$$

ANSWER CHECK
$$5(-17 + 2) = 3(-17 - 8)$$
$$5(-15) = 3(-25)$$
$$-75 = -75$$

4. $20 = 4(15 - x)$

$20 = 60 - 4x$

$\underline{-60 \quad\quad -60}$

$\dfrac{-40 =}{4} \quad \dfrac{-4x}{4}$

$\mathbf{10} = x$

$4a - 8 = 2(a + 4)$

$4a - 8 = 2a + 8$

$\underline{-2a \quad\quad -2a}$

$2a - 8 = \quad\quad 8$

$\underline{\quad + 8 \quad\quad +8}$

$\dfrac{2a}{2} = \dfrac{16}{2}$

$a = \mathbf{8}$

$4(y - 8) = 2(y + 8)$

$4y - 32 = 2y + 16$

$\underline{-2y \quad\quad -2y}$

$2y - 32 = \quad\quad 16$

$\underline{\quad + 32 \quad\quad +32}$

$\dfrac{2y}{2} = \dfrac{48}{2}$

$y = \mathbf{24}$

ANSWER CHECK

$20 = 4(15 - 10)$

$20 = 4(5)$

$20 = 20$

ANSWER CHECK

$4 \cdot 8 - 8 = 2(8 + 4)$

$32 - 8 = 2(12)$

$24 = 24$

ANSWER CHECK

$4(24 - 8) = 2(24 + 8)$

$4(16) = 2(32)$

$64 = 64$

5. $4(c - 5) = 2(c + 5)$

$4c - 20 = 2c + 10$

$\underline{-2c \quad\quad -2c}$

$2c - 20 = \quad\quad 10$

$\underline{\quad + 20 \quad\quad +20}$

$\dfrac{2c}{2} = \dfrac{30}{2}$

$c = \mathbf{15}$

$12(p + 6) = 8(p + 10)$

$12p + 72 = 8p + 80$

$\underline{- 8p \quad\quad -8p}$

$4p + 72 = \quad\quad 80$

$\underline{\quad - 72 \quad\quad -72}$

$\dfrac{4p}{4} = \dfrac{8}{4}$

$p = \mathbf{2}$

$6(d - 3) = 3(d + 4)$

$6d - 18 = 3d + 12$

$\underline{-3d \quad\quad -3d}$

$3d - 18 = \quad\quad 12$

$\underline{\quad + 18 \quad\quad +18}$

$\dfrac{3d}{3} = \dfrac{30}{3}$

$d = \mathbf{10}$

ANSWER CHECK

$4(15 - 5) = 2(15 + 5)$

$4(10) = 2(20)$

$40 = 40$

ANSWER CHECK

$12(2 + 6) = 8(2 + 10)$

$12(8) = 8(12)$

$96 = 96$

ANSWER CHECK

$6(10 - 3) = 3(10 + 4)$

$6(7) = 3(14)$

$42 = 42$

Lesson 5

1. $a + 6 > 9$

$\underline{\quad - 6 \quad -6}$

$a > \mathbf{3}$

$c - 12 \le 3$

$\underline{\quad + 12 \quad +12}$

$c \le \mathbf{15}$

$2 \cdot \dfrac{n}{2} < 7 \cdot 2$

$n < \mathbf{14}$

2. $\dfrac{16r}{16} \ge \dfrac{20}{16}$

$r \ge \mathbf{1\tfrac{1}{4}}$

$6m - 2 < 22$

$\underline{\quad\quad + 2 \quad +2}$

$\dfrac{6m}{6} < \dfrac{24}{6}$

$m < \mathbf{4}$

$\dfrac{5}{3} \cdot \dfrac{3}{5}x \le 18 \cdot \dfrac{5}{3}$

$x \le \mathbf{30}$

3. $3p - 4 > p + 6$

$\underline{\quad - p \quad\quad -p}$

$2p - 4 > \quad\quad 6$

$\underline{\quad + 4 \quad\quad +4}$

$\dfrac{2p}{2} > \dfrac{10}{2}$

$p > \mathbf{5}$

$9w + 2 \ge w + 10$

$\underline{\quad - w \quad\quad -w}$

$8w + 2 \ge \quad\quad 10$

$\underline{\quad - 2 \quad\quad -2}$

$\dfrac{8w}{8} \ge \dfrac{8}{8}$

$w > \mathbf{1}$

$3(m - 2) < 9$

$3m - 6 < 9$

$\underline{\quad + 6 \quad +6}$

$\dfrac{3m}{3} < \dfrac{15}{3}$

$m < \mathbf{5}$

4. $\dfrac{1}{2}y - 4 \le 1$

$\underline{\quad + 4 \quad +4}$

$2 \cdot \dfrac{1}{2}y \le 5 \cdot 2$

$y \le \mathbf{10}$

$4 > 2(n - 9)$

$4 > 2n - 18$

$\underline{+18 \quad\quad + 18}$

$\dfrac{22}{2} > \dfrac{2n}{2}$

$\mathbf{11} > n$

$8t - 5 \le 2t + 1$

$\underline{-2t \quad\quad -2t}$

$6t - 5 \le \quad\quad 1$

$\underline{\quad + 5 \quad\quad +5}$

$\dfrac{6t}{6} \le \dfrac{6}{6}$

$t \le \mathbf{1}$

5. $6(x - 12) \le 18$

$6x - 72 \le 18$

$\underline{\quad + 72 \quad +72}$

$\dfrac{6x}{6} \le \dfrac{90}{6}$

$x \le \mathbf{15}$

$5a - 3 \ge 3a + 3$

$\underline{-3a \quad\quad -3a}$

$2a - 3 \ge \quad\quad 3$

$\underline{\quad + 3 \quad\quad +3}$

$\dfrac{2a}{2} \ge \dfrac{6}{2}$

$a \ge \mathbf{3}$

$c - 1 \le \dfrac{3}{4}c + 2$

$\underline{-\tfrac{3}{4}c \quad\quad -\tfrac{3}{4}c}$

$\dfrac{1}{4}c - 1 \le \quad\quad 2$

$\underline{\quad + 1 \quad\quad +1}$

$\dfrac{4}{1} \cdot \dfrac{1}{4}c \le 3 \cdot 4$

$c \le \mathbf{12}$

Lesson 6

1. $x + 12$
2. $\frac{x}{4}$
3. $x + 5x$
4. $6(x + 8)$
5. $(x + 7)/3$, or $\frac{(x + 7)}{3}$
6. $\frac{1}{4}p$
7. a. $x - 5$
   b. $x + 3$
8. $4r$
9. $.22g$
10. $x - 32$

Lesson 7

1.
$$\begin{array}{r} x + 5 \\ x + 2 \\ \hline 2x + 10 \\ x^2 + 5x \\ \hline x^2 + 7x + 10 \end{array} \qquad \begin{array}{r} x + 3 \\ x + 1 \\ \hline x + 3 \\ x^2 + 3x \\ \hline x^2 + 4x + 3 \end{array} \qquad \begin{array}{r} x + 2 \\ x + 6 \\ \hline 6x + 12 \\ x^2 + 2x \\ \hline x^2 + 8x + 12 \end{array}$$

2.
$$\begin{array}{r} x - 4 \\ x - 3 \\ \hline -3x + 12 \\ x^2 - 4x \\ \hline x^2 - 7x + 12 \end{array} \qquad \begin{array}{r} x - 1 \\ x - 8 \\ \hline -8x + 8 \\ x^2 - x \\ \hline x^2 - 9x + 8 \end{array} \qquad \begin{array}{r} x - 4 \\ x - 5 \\ \hline -5x + 20 \\ x^2 - 4x \\ \hline x^2 - 9x + 20 \end{array}$$

3.
$$\begin{array}{r} x + 5 \\ x - 2 \\ \hline -2x - 10 \\ x^2 + 5x \\ \hline x^2 + 3x - 10 \end{array} \qquad \begin{array}{r} x - 6 \\ x + 7 \\ \hline +7x - 42 \\ x^2 - 6x \\ \hline x^2 + x - 42 \end{array} \qquad \begin{array}{r} x + 12 \\ x - 10 \\ \hline -10x - 120 \\ x^2 + 12x \\ \hline x^2 + 2x - 120 \end{array}$$

4.
$$\begin{array}{r} x + 8 \\ x - 8 \\ \hline -8x - 64 \\ x^2 + 8x \\ \hline x^2 \quad - 64 \end{array} \qquad \begin{array}{r} x + 10 \\ x - 10 \\ \hline -10x - 100 \\ x^2 + 10x \\ \hline x^2 \quad - 100 \end{array} \qquad \begin{array}{r} x - 3 \\ x + 3 \\ \hline +3x - 9 \\ x^2 - 3x \\ \hline x^2 \quad - 9 \end{array}$$

5.
$$\begin{array}{r} x - 7 \\ x + 3 \\ \hline +3x - 21 \\ x^2 - 7x \\ \hline x^2 - 4x - 21 \end{array} \qquad \begin{array}{r} x + 6 \\ x - 9 \\ \hline -9x - 54 \\ x^2 + 6x \\ \hline x^2 - 3x - 54 \end{array} \qquad \begin{array}{r} x - 11 \\ x + 6 \\ \hline +6x - 66 \\ x^2 - 11x \\ \hline x^2 - 5x - 66 \end{array}$$

Lesson 8

1. $3x + 9 = 3(x + 3)$ $8w - 12 = 4(2w - 3)$
   $10c - 5d = 5(2c - d)$
2. $4x - 16 = 4(x - 4)$ $7a + 21 = 7(a + 3)$
   $5y - 20z = 5(y - 4z)$
3. $36f - 12 = 12(3f - 1)$ $9m + 21 = 3(3m + 7)$
   $6w - 9 = 3(2w - 3)$
4. $c^2 + 8c = c(c + 8)$ $m^2 + 6m = m(m + 6)$
   $x^2 - 5x = x(x - 5)$
5. $p^2 + 10p = p(p + 10)$ $a^2 - a = a(a - 1)$
   $n^2 - 2n = n(n - 2)$

6. $x^2 + 6x + 8 =$ $x^2 + 8x + 7 =$
   $(x + 4)(x + 2)$ $(x + 7)(x + 1)$

ANSWER CHECK   ANSWER CHECK

$$\begin{array}{r} x + 4 \\ x + 2 \\ \hline + 2x + 8 \\ x^2 + 4x \\ \hline x^2 + 6x + 8 \end{array} \qquad \begin{array}{r} x + 7 \\ x + 1 \\ \hline + x + 7 \\ x^2 + 7x \\ \hline x^2 + 8x + 7 \end{array}$$

$x^2 + 13x + 40 =$
$(x + 8)(x + 5)$

ANSWER CHECK

$$\begin{array}{r} x + 8 \\ x + 5 \\ \hline + 5x + 40 \\ x^2 + 8x \\ \hline x^2 + 13x + 40 \end{array}$$

7. $x^2 - 13x + 36 =$ $x^2 - 12x + 36 =$
   $(x - 4)(x - 9)$ $(x - 6)(x - 6)$

ANSWER CHECK   ANSWER CHECK

$$\begin{array}{r} x - 4 \\ x - 9 \\ \hline - 9x + 36 \\ x^2 - 4x \\ \hline x^2 - 13x + 36 \end{array} \qquad \begin{array}{r} x - 6 \\ x - 6 \\ \hline - 6x + 36 \\ x^2 - 6x \\ \hline x^2 - 12x + 36 \end{array}$$

$x^2 - 10x + 9 =$
$(x - 1)(x - 9)$

ANSWER CHECK

$$\begin{array}{r} x - 1 \\ x - 9 \\ \hline - 9x + 9 \\ x^2 - x \\ \hline x^2 - 10x + 9 \end{array}$$

8. $x^2 + 5x - 14 =$ $x^2 + 2x - 8 =$
   $(x + 7)(x - 2)$ $(x - 2)(x + 4)$

ANSWER CHECK   ANSWER CHECK

$$\begin{array}{r} x + 7 \\ x - 2 \\ \hline - 2x - 14 \\ x^2 + 7x \\ \hline x^2 + 5x - 14 \end{array} \qquad \begin{array}{r} x - 2 \\ x + 4 \\ \hline + 4x - 8 \\ x^2 - 2x \\ \hline x^2 + 2x - 8 \end{array}$$

$x^2 + x - 56 =$
$(x + 8)(x - 7)$

ANSWER CHECK

$$\begin{array}{r} x + 8 \\ x - 7 \\ \hline - 7x - 56 \\ x^2 + 8x \\ \hline x^2 + x - 56 \end{array}$$

9. $x^2 - 36 =$
$(x + 6)(x - 6)$

ANSWER CHECK

$$\begin{array}{r} x + 6 \\ x - 6 \\ \hline -6x - 36 \\ x^2 + 6x \\ \hline x^2 \quad\quad - 36 \end{array}$$

$x^2 - 4 =$
$(x + 2)(x - 2)$

ANSWER CHECK

$$\begin{array}{r} x + 2 \\ x - 2 \\ \hline -2x - 4 \\ x^2 + 2x \\ \hline x^2 \quad\quad - 4 \end{array}$$

$x^2 - 81 =$
$(x + 9)(x - 9)$

ANSWER CHECK

$$\begin{array}{r} x + 9 \\ x - 9 \\ \hline -9x - 81 \\ x^2 + 9x \\ \hline x^2 \quad\quad - 81 \end{array}$$

10. $x^2 - 3x - 10 =$
$(x - 5)(x + 2)$

ANSWER CHECK

$$\begin{array}{r} x - 5 \\ x + 2 \\ \hline +2x - 10 \\ x^2 - 5x \\ \hline x^2 - 3x - 10 \end{array}$$

$x^2 - 4x - 96 =$
$(x + 8)(x - 12)$

ANSWER CHECK

$$\begin{array}{r} x + 8 \\ x - 12 \\ \hline -12x - 96 \\ x^2 + 8x \\ \hline x^2 - 4x - 96 \end{array}$$

$x^2 - x - 12 =$
$(x - 4)(x + 3)$

ANSWER CHECK

$$\begin{array}{r} x - 4 \\ x + 3 \\ \hline +3x - 12 \\ x^2 - 4x \\ \hline x^2 - x - 12 \end{array}$$

Lesson 9

1. $\sqrt{8} = \sqrt{4} \cdot \sqrt{2} = \mathbf{2\sqrt{2}}$
2. $\sqrt{75} = \sqrt{25} \cdot \sqrt{3} = \mathbf{5\sqrt{3}}$
3. $\sqrt{18} = \sqrt{9} \cdot \sqrt{2} = \mathbf{3\sqrt{2}}$
4. $\sqrt{24} = \sqrt{4} \cdot \sqrt{6} = \mathbf{2\sqrt{6}}$
5. $\sqrt{72} = \sqrt{36} \cdot \sqrt{2} = \mathbf{6\sqrt{2}}$
6. $\sqrt{54} = \sqrt{9} \cdot \sqrt{6} = \mathbf{3\sqrt{6}}$
7. $\sqrt{500} = \sqrt{100} \cdot \sqrt{5} = \mathbf{10\sqrt{5}}$
8. $\sqrt{45} = \sqrt{9} \cdot \sqrt{5} = \mathbf{3\sqrt{5}}$
9. $\sqrt{96} = \sqrt{16} \cdot \sqrt{6} = \mathbf{4\sqrt{6}}$
10. $\sqrt{128} = \sqrt{64} \cdot \sqrt{2} = \mathbf{8\sqrt{2}}$

Lesson 10

1. $x^2 + 7x + 10 = 0$
$(x + 2)(x + 5) = 0$

$$\begin{array}{ll} x + 2 = 0 & x + 5 = 0 \\ \underline{-2 \quad -2} & \underline{-5 \quad -5} \\ x \quad = \mathbf{-2} \text{ and } x \quad = \mathbf{-5} \end{array}$$

ANSWER CHECK

$(-2)^2 + 7(-2) + 10 = 0$ and $(-5)^2 + 7(-5) + 10 = 0$
$4 - 14 + 10 = 0$ $25 - 35 + 10 = 0$
$-10 + 10 = 0$ $-10 + 10 = 0$

$x^2 + 10x + 9 = 0$
$(x + 1)(x + 9) = 0$

$$\begin{array}{ll} x + 1 = 0 & x + 9 = 0 \\ \underline{-1 \quad -1} & \underline{-9 \quad -9} \\ x \quad = \mathbf{-1} \text{ and } x \quad = \mathbf{-9} \end{array}$$

ANSWER CHECK

$(-1)^2 + 10(-1) + 9 = 0$ and $(-9)^2 + 10(-9) + 9 = 0$
$1 - 10 + 9 = 0$ $81 - 90 + 9 = 0$
$-9 + 9 = 0$ $-9 + 9 = 0$

$x^2 + 15x + 56 = 0$
$(x + 8)(x + 7) = 0$

$$\begin{array}{ll} x + 8 = 0 & x + 7 = 0 \\ \underline{-8 \quad -8} & \underline{-7 \quad -7} \\ x \quad = \mathbf{-8} \text{ and } x \quad = \mathbf{-7} \end{array}$$

ANSWER CHECK

$(-8)^2 + 15(-8) + 56 = 0$ and $(-7)^2 + 15(-7) + 56 = 0$
$64 - 120 + 56 = 0$ $49 - 105 + 56 = 0$
$-56 + 56 = 0$ $-56 + 56 = 0$

2. $x^2 - 7x + 12 = 0$
$(x - 3)(x - 4) = 0$

$$\begin{array}{ll} x - 3 = 0 & x - 4 = 0 \\ \underline{+3 \quad +3} & \underline{+4 \quad +4} \\ x \quad = \mathbf{+3} \text{ and } x \quad = \mathbf{+4} \end{array}$$

ANSWER CHECK

$(3)^2 - 7(3) + 12 = 0$ and $(4)^2 - 7(4) + 12 = 0$
$9 - 21 + 12 = 0$ $16 - 28 + 12 = 0$
$-12 + 12 = 0$ $-12 + 12 = 0$

$x^2 - 17x + 60 = 0$
$(x - 5)(x - 12) = 0$

$$\begin{array}{ll} x - 5 = 0 & x - 12 = 0 \\ \underline{+5 \quad +5} & \underline{+12 \quad +12} \\ x \quad = \mathbf{5} \text{ and } x \quad = \mathbf{12} \end{array}$$

ANSWER CHECK

$(5)^2 - 17(5) + 60 = 0$ and $(12)^2 - 17(12) + 60 = 0$
$25 - 85 + 60 = 0$ $144 - 204 + 60 = 0$
$-60 + 60 = 0$ $-60 + 60 = 0$

$x^2 - 12x + 27 = 0$
$(x - 9)(x - 3) = 0$

$$\begin{array}{ll} x - 9 = 0 & x - 3 = 0 \\ \underline{+9 \quad +9} & \underline{+3 \quad +3} \\ x \quad = \mathbf{9} \text{ and } x \quad = \mathbf{3} \end{array}$$

ANSWER CHECK

$(9)^2 - 12(9) + 27 = 0$ and $(3)^2 - 12(3) + 27 = 0$
$81 - 108 + 27 = 0$ $9 - 36 + 27 = 0$
$-27 + 27 = 0$ $-27 + 27 = 0$

3. $x^2 + 2x - 24 = 0$

$(x + 6)(x - 4) = 0$

$$\begin{array}{rcl rcl} x + 6 &=& 0 & x - 4 &=& 0 \\ -6 & & -6 & +4 & & +4 \\ \hline x &=& -6 \text{ and } x & &=& 4 \end{array}$$

ANSWER CHECK

$(-6)^2 + 2(-6) - 24 = 0$ and $(4)^2 + 2(4) - 24 = 0$

$\quad 36 - 12 - 24 = 0 \qquad\qquad 16 + 8 - 24 = 0$

$\qquad\qquad 36 - 36 = 0 \qquad\qquad\qquad 24 - 24 = 0$

$x^2 + x - 72 = 0$

$(x - 8)(x + 9) = 0$

$$\begin{array}{rcl rcl} x - 8 &=& 0 & x + 9 &=& 0 \\ +8 & & +8 & -9 & & -9 \\ \hline x &=& 8 \text{ and } x & &=& -9 \end{array}$$

ANSWER CHECK

$(8)^2 + 8 - 72 = 0$ and $(-9)^2 + (-9) - 72 = 0$

$64 + 8 - 72 = 0 \qquad\qquad 81 - 9 - 72 = 0$

$\quad 72 - 72 = 0 \qquad\qquad\qquad 81 - 81 = 0$

$x^2 + 7x - 30 = 0$

$(x + 10)(x - 3) = 0$

$$\begin{array}{rcl rcl} x + 10 &=& 0 & x - 3 &=& 0 \\ -10 & & -10 & +3 & & +3 \\ \hline x &=& -10 \text{ and } x & &=& 3 \end{array}$$

ANSWER CHECK

$(-10)^2 + 7(-10) - 30 = 0$ and $(3)^2 + 7(3) + 30 = 0$

$\quad 100 - 70 - 30 = 0 \qquad\qquad 9 + 21 - 30 = 0$

$\qquad\quad 30 - 30 = 0 \qquad\qquad\qquad 30 - 30 = 0$

4. $x^2 - 4x - 21 = 0$

$(x - 7)(x + 3) = 0$

$$\begin{array}{rcl rcl} x - 7 &=& 0 & x + 3 &=& 0 \\ +7 & & +7 & -3 & & -3 \\ \hline x &=& 7 \text{ and } x & &=& -3 \end{array}$$

ANSWER CHECK

$(7)^2 - 4(7) - 21 = 0$ and $(-3)^2 - 4(-3) - 21 = 0$

$\quad 49 - 28 - 21 = 0 \qquad\qquad 9 + 12 - 21 = 0$

$\qquad\quad 49 - 49 = 0 \qquad\qquad\qquad 21 - 21 = 0$

$x^2 - 5x - 6 = 0$

$(x + 1)(x - 6) = 0$

$$\begin{array}{rcl rcl} x + 1 &=& 0 & x - 6 &=& 0 \\ -1 & & -1 & +6 & & +6 \\ \hline x &=& -1 \text{ and } x & &=& 6 \end{array}$$

ANSWER CHECK

$(-1)^2 - 5(-1) - 6 = 0$ and $(6)^2 - 5(6) - 6 = 0$

$\quad 1 + 5 - 6 = 0 \qquad\qquad 36 - 30 - 6 = 0$

$\qquad\quad 6 - 6 = 0 \qquad\qquad\qquad 36 - 36 = 0$

$x^2 - 10x - 24 = 0$

$(x - 12)(x + 2) = 0$

$$\begin{array}{rcl rcl} x - 12 &=& 0 & x + 2 &=& 0 \\ +12 & & +12 & -2 & & -2 \\ \hline x &=& 12 \text{ and } x & &=& -2 \end{array}$$

ANSWER CHECK

$(12)^2 - 10(12) - 24 = 0$ and $(-2)^2 - 10(-2) - 24 = 0$

$\quad 144 - 120 - 24 = 0 \qquad\qquad 4 + 20 - 24 = 0$

$\qquad\quad 144 - 144 = 0 \qquad\qquad\qquad 24 - 24 = 0$

5. $x^2 - 49 = 0$

$(x + 7)(x - 7) = 0$

$$\begin{array}{rcl rcl} x + 7 &=& 0 & x - 7 &=& 0 \\ -7 & & -7 & +7 & & +7 \\ \hline x &=& -7 \text{ and } x & &=& 7 \end{array}$$

ANSWER CHECK

$(-7)^2 - 49 = 0$ and $(7)^2 - 49 = 0$

$\quad 49 - 49 = 0 \qquad\quad 49 - 49 = 0$

$x^2 - 1 = 0$

$(x + 1)(x - 1) = 0$

$$\begin{array}{rcl rcl} x + 1 &=& 0 & x - 1 &=& 0 \\ -1 & & -1 & +1 & & +1 \\ \hline x &=& -1 \text{ and } x & &=& 1 \end{array}$$

ANSWER CHECK

$(-1)^2 - 1 = 0$ and $(1)^2 - 1 = 0$

$\quad 1 - 1 = 0 \qquad\quad 1 - 1 = 0$

$x^2 - 144 = 0$

$(x + 12)(x - 12) = 0$

$$\begin{array}{rcl rcl} x + 12 &=& 0 & x - 12 &=& 0 \\ -12 & & -12 & +12 & & +12 \\ \hline x &=& -12 \text{ and } x & &=& 12 \end{array}$$

ANSWER CHECK

$(-12)^2 - 144 = 0$ and $(12)^2 - 144 = 0$

$\quad 144 - 144 = 0 \qquad\quad 144 - 144 = 0$

Level 2   Review

1.
$$15 = 9x - 3$$
$$\underline{+3 \qquad +3}$$
$$\frac{18}{9} = \frac{9x}{9}$$
$$\mathbf{2} = x$$

2.
$$4a + 2 + 3a = a + 20$$
$$7a + 2 = a + 20$$
$$\underline{-a \qquad\qquad -a}$$
$$6a + 2 = 20$$
$$\underline{\qquad -2 \qquad -2}$$
$$\frac{6a}{6} = \frac{18}{6}$$
$$a = \mathbf{3}$$

3.
$$7x - 3 = 5(x + 5)$$
$$7x - 3 = 5x + 25$$
$$\underline{-5x \qquad -5x}$$
$$2x - 3 = 25$$
$$\underline{\quad +3 \qquad\quad +3}$$
$$\frac{2x}{2} = \frac{28}{2}$$
$$x = \mathbf{14}$$

4.
$$6n - 5 > 4n + 21$$
$$\underline{-4n \qquad -4n}$$
$$2n - 5 > 21$$
$$\underline{\qquad +5 \qquad +5}$$
$$\frac{2n}{2} > \frac{26}{2}$$
$$n > \mathbf{13}$$

5. $(2x - 3)/5$, or $\dfrac{(2x - 3)}{5}$

6. $\mathbf{t - 8}$

7. $c^2 - 6c = \mathbf{c(c - 6)}$

8. $x^2 + 6x - 16 =$ $\mathbf{(x + 8)(x - 2)}$

9. $\sqrt{28} = \sqrt{4} \cdot \sqrt{7} = \mathbf{2\sqrt{7}}$

10. $x^2 - 6x - 27 = 0$
$$(x - 9)(x + 3) = 0$$
$$x - 9 = 0 \qquad x + 3 = 0$$
$$\underline{\quad +9 \quad +9} \qquad \underline{\quad -3 \quad -3}$$
$$x = \mathbf{9} \text{ and } x = \mathbf{-3}$$

Level 3   Algebra Problem Solving

1.
$$6x + 5 = 17$$
$$\underline{\quad -5 \quad -5}$$
$$\frac{6x}{6} = \frac{12}{6}$$
$$x = \mathbf{2}$$

2.
$$43 = 9x - 2$$
$$\underline{+2 \qquad +2}$$
$$\frac{45}{9} = \frac{9x}{9}$$
$$\mathbf{5} = x$$

3.
$$4(x - 1) = x + 2$$
$$4x - 4 = x + 2$$
$$\underline{-x \qquad -x}$$
$$3x - 4 = 2$$
$$\underline{\quad +4 \qquad\quad +4}$$
$$\frac{3x}{3} = \frac{6}{3}$$
$$x = \mathbf{2}$$

4. boys $= x$
girls $= x + 9$
$$x + x + 9 = 65$$
$$2x + 9 = 65$$
$$\underline{\quad -9 \qquad -9}$$
$$\frac{2x}{2} = \frac{56}{2}$$
$$x = \mathbf{28\ boys}$$
$$x + 9 = 28 + 9 = \mathbf{37\ girls}$$

5. Tom's hours $= x$
Bill's hours $= x + 10$
Fred's hours $= 2x$
$$x + x + 10 + 2x = 94$$
$$4x + 10 = 94$$
$$\underline{\quad -10 \qquad -10}$$
$$\frac{4x}{4} = \frac{84}{4}$$
$$x = 21$$
Fred's hours $= 2(21) = \mathbf{42\ hr.}$

Lesson 1

1.
$$x + 15 = 21$$
$$\underline{\quad -15 \quad -15}$$
$$x = \mathbf{6}$$

2.
$$\frac{6x}{6} = \frac{72}{6}$$
$$x = \mathbf{12}$$

3.
$$25 \cdot \frac{x}{25} = 6 \cdot 25$$
$$x = \mathbf{150}$$

4.
$$30 = x - 27$$
$$\underline{+27 \qquad +27}$$
$$57 = x$$

5.
$$7x - x = 78$$
$$\frac{6x}{6} = \frac{78}{6}$$
$$x = \mathbf{13}$$

6.
$$9x - 3 = 2x + 25$$
$$\underline{-2x \qquad -2x}$$
$$7x - 3 = 25$$
$$\underline{\quad +3 \qquad\quad +3}$$
$$\frac{7x}{7} = \frac{28}{7}$$
$$x = \mathbf{4}$$

7.
$$10x - 5 = 2x + 19$$
$$\underline{-2x \qquad -2x}$$
$$8x - 5 = 19$$
$$\underline{\quad +5 \qquad +5}$$
$$\frac{8x}{8} = \frac{24}{8}$$
$$x = \mathbf{3}$$

8.
$$12x = 3x + 45$$
$$\underline{-3x \quad -3x}$$
$$\frac{9x}{9} = \frac{45}{9}$$
$$x = \mathbf{5}$$

9.
$$5(x - 2) = 30$$
$$5x - 10 = 30$$
$$\underline{\quad +10 \qquad +10}$$
$$\frac{5x}{5} = \frac{40}{5}$$
$$x = \mathbf{8}$$

10.
$$7(x - 1) = 4(x + 2)$$
$$7x - 7 = 4x + 8$$
$$\underline{-4x \qquad -4x}$$
$$3x - 7 = 8$$
$$\underline{\quad +7 \qquad\quad +7}$$
$$\frac{3x}{3} = \frac{15}{3}$$
$$x = \mathbf{5}$$

Lesson 2

1.
$$d = rt$$
$$\frac{210}{42} = \frac{42t}{42}$$
$$\mathbf{5\ hr.} = t$$

2.
$$d = rt$$
$$\frac{1194}{3} = \frac{r \cdot 3}{3}$$
$$\mathbf{398\ mph} = r$$

3.
$$d = rt$$
$$\frac{27}{6} = \frac{r \cdot 6}{6}$$
$$\mathbf{4\tfrac{1}{2}\ mph} = r$$

4.
$$c = nr$$
$$\frac{528}{22} = \frac{n \cdot 22}{22}$$
$$\mathbf{24} = n$$

5.　$c = nr$

$$\frac{420}{12} = \frac{12r}{12}$$

$\mathbf{\$35} = r$

6.　$c = nr$

$$\frac{5625}{4.5} = \frac{4.5r}{4.5}$$

$\mathbf{\$125}\ 0 = r$

$$4.5\overline{)5625.0}$$
$$\underline{45}$$
$$112$$
$$\underline{90}$$
$$225$$
$$\underline{225}$$
$$0\ 0$$

7.　$c = nr$

$$\frac{3380}{6.50} = \frac{n \cdot 6.50}{6.50}$$

$\mathbf{5}\ 20 = n$

$$6.50\overline{)3380.00}$$
$$\underline{3250}$$
$$1300$$
$$\underline{1300}$$
$$0\ 0$$

8.　$i = prt$

$$\frac{51}{850} = \frac{850 \cdot r \cdot 1}{850}$$

$r = \quad .06 = \mathbf{6\%}$

$$850\overline{)51.00}$$

9. $8\% = .08$

$i = prt$

$14 = 350 \cdot .08 \cdot t$

$$\frac{14}{28} = \frac{28t}{28}$$

$\frac{1}{2}$ yr. $= t = \mathbf{6\ mo.}$

10. $12\% = .12$

$i = prt$

$$\frac{4800}{.12} = \frac{p \cdot .12 \cdot 1}{.12}$$

$\mathbf{\$40,0}\ 00 = p$

$$.12\overline{)480\ 0.00}$$

Lesson 3

1. small no. $= x$

　large no. $= 3x$

$3x - 10 = \quad x + 18$

$\underline{-x \qquad\qquad -x}$

$2x - 10 = \qquad\quad 18$

$\underline{+ 10 \qquad\qquad +10}$

$$\frac{2x}{2} = \frac{28}{2}$$

$x = \mathbf{14}$

$3x = 3(14) = \mathbf{42}$

2. small no. $= x$

　large no. $= x + 7$

$3x - 4 = 2(x + 7)$

$3x - 4 = \quad 2x + 14$

$\underline{-2x \qquad\quad -2x}$

$x - 4 = \qquad 14$

$\underline{+ 4 \qquad\qquad +4}$

$x = \qquad \mathbf{18}$

$x + 7 = 18 + 7 = \mathbf{25}$

3. daughter's age $= x$

　Louise's age $= x + 26$

$x + 26 = \quad 4x + 2$

$\underline{-x \qquad\qquad -x}$

$26 = \quad 3x + 2$

$\underline{-2 \qquad\qquad -2}$

$$\frac{24}{3} = \frac{3x}{3}$$

$\mathbf{8} = x$

$x + 26 = 8 + 26 = \mathbf{34}$

4. grandfather's age $= x$

　Douglas's age $= x - 46$

$3(x - 46) = \quad x - 8$

$3x - 138 = \quad x - 8$

$\underline{-x \qquad\qquad -x}$

$2x - 138 = \qquad - 8$

$\underline{+ 138 \qquad\quad +138}$

$$\frac{2x}{2} = \frac{130}{2}$$

$x = \mathbf{65}$

$x - 46 = 65 - 46 = \mathbf{19}$

5. deductions $= x$

　take-home pay $= 5x$

$x + 5x = 324$

$$\frac{6x}{6} = \frac{324}{6}$$

$x = 54$

$5x = 5(54) = \mathbf{\$270}$

6. no. of men $= x$

　no. of women $= x + 15$

$x + x + 15 = 47$

$2x + 15 = \quad 47$

$\underline{- 15 \qquad\quad -15}$

$$\frac{2x}{2} = \frac{32}{2}$$

$x = 16$

$x + 15 = 16 + 15 = \mathbf{31}$

7. Assistant's wages $= x$

　Joe's wages $= x + 3$

$10x + 10(x + 3) = 350$

$10x + 10x + 30 = 350$

$20x + 30 = 350$

$\underline{- 30 \qquad\quad -30}$

$$\frac{20x}{20} = \frac{320}{20}$$

$x = \mathbf{\$16}$

$x + 3 = 16 + 3 = \mathbf{\$19}$

8. Jim's hours $= x$

　Carmen's hours $= x + 5$

　George's hours $= 2(x + 5)$

$x + x + 5 + 2(x + 5) = 95$

$2x + 5 + 2x + 10 = 95$

$4x + 15 = 95$

$\underline{- 15 \qquad\quad -15}$

$$\frac{4x}{4} = \frac{80}{4}$$

$x = \mathbf{20\ hr.}$

$x + 5 = 20 + 5 = \mathbf{25\ hr.}$

$2(x + 5) = 2(20 + 5) = \mathbf{50\ hr.}$

9.

	Age Now	Age in 5 Years
Andy	x	x + 5
Chris	x + 3	x + 3 + 5 = x + 8

$3(x + 5) = 2(x + 8)$
$+ 6$

$3x + 15 = 2x + 16$
$+ 6$

$3x + 15 = \quad 2x + 22$

$\underline{-2x \qquad\qquad -2x}$

$x + 15 = \qquad 22$

$\underline{- 15 \qquad\qquad -15}$

$x = \qquad \mathbf{7}$

$x + 3 = 7 + 3 = \mathbf{10}$

10. food $= 3x$

car $= 2x$

rent $= 4x$

$3x + 2x + 4x = 648$

$$\frac{9x}{9} = \frac{648}{9}$$

$x = 72$

$4x = 4(72) = 288

Level 3   Review

1.
$$4x - 9 = 23$$
$$\underline{+9 \quad +9}$$
$$\frac{4x}{4} = \frac{32}{4}$$
$$x = \mathbf{8}$$

2.
$$8x = 5x + 12$$
$$\underline{-5x \quad -5x}$$
$$\frac{3x}{3} = \frac{12}{3}$$
$$x = \mathbf{4}$$

3.
$$2(x - 3) = x + 7$$
$$2x - 6 = x + 7$$
$$\underline{-x \qquad\quad -x}$$
$$x - 6 = 7$$
$$\underline{+6 \qquad +6}$$
$$x = \mathbf{13}$$

4.
$$d = rt$$
$$234 = r \cdot 4\tfrac{1}{2}$$
$$\tfrac{2}{9} \cdot 234 = r\tfrac{9}{2} \cdot \tfrac{2}{9}$$
$$\mathbf{52 \ mph} = r$$

5.
$$c = nr$$
$$\frac{5.70}{.95} = \frac{n \cdot .95}{.95}$$
$$\mathbf{6 \ gal.} = n$$
$$.95\overline{)5.70}$$

6.
$$15\% = .15$$
$$i = prt$$
$$135 = 1200 \cdot .15t$$
$$\frac{135}{180} = \frac{180t}{180}$$
$$\tfrac{3}{4} \ yr. = t$$
$$\tfrac{3}{4} \times \tfrac{12}{1} = \mathbf{9 \ mo.}$$

7. small no. $= x$

large no. $= 2x - 1$

$$7x - 12 = 3(2x - 1)$$
$$7x - 12 = 6x - 3$$
$$\underline{-6x \qquad\quad -6x}$$
$$x - 12 = -3$$
$$\underline{+12 \qquad +12}$$
$$x = \mathbf{9}$$
$$2x - 1 = 2(9) - 1 = \mathbf{17}$$

8. Ann's wages $= x$

Dorothy's wages $= x + 4$

$$25x + 25(x + 4) = 500$$
$$25x + 25x + 100 = 500$$
$$50x + 100 = 500$$
$$\underline{-100 \quad -100}$$
$$\frac{50x}{50} = \frac{400}{50}$$
$$x = 8$$
$$x + 4 = 8 + 4 = \mathbf{\$12}$$

9. Mrs. Nash $= x$

Sally $= x + 3$

Mr. Nash $= 2(x + 3)$

$$x + x + 3 + 2(x + 3) = 33$$
$$2x + 3 + 2x + 6 = 33$$
$$4x + 9 = 33$$
$$\underline{-9 \quad -9}$$
$$\frac{4x}{4} = \frac{24}{4}$$
$$x = 6$$
$$2(x + 3) = 2(6 + 3) = \mathbf{18 \ hr.}$$

10.

	Age Now	Age 10 Years Ago
Carlos	x	$x - 10$
Alvaro	$x - 3$	$x - 3 - 10 =$
		$x - 13$

$$3(x - 13) - 5 = 2(x - 10)$$
$$3x - 39 - 5 = 2x - 20$$
$$3x - 44 = 2x - 20$$
$$\underline{-2x \qquad\qquad -2x}$$
$$x - 44 = -20$$
$$\underline{+44 \qquad\quad +44}$$
$$x = 24$$
$$x - 3 = 24 - 3 = \mathbf{21 \ yr.}$$

Chapter 6   Quiz

1. **point A**

2. $(-16) + (-19) + (12) =$
$$-16 - 19 + 12 =$$
$$-35 + 12 = \mathbf{-23}$$

3. $(-40)\left(-\tfrac{3}{4}\right)(+2) = +60$

4. $(-8m) + (-7m) + (-m) =$
$$-8m - 7m - m = \mathbf{-16m}$$

5. $(-6a)(+3ac)(-4c) = \mathbf{+72a^2c^2}$

6. $\dfrac{-20m^2n^3}{-10mn} = \mathbf{2mn^2}$

7. $-3xy + 4x =$
$$-3(-3)(7) + 4(-3) =$$
$$+63 - 12 = \mathbf{+51}$$

8.
$$\tfrac{3}{4}m - 2 = 34$$
$$\underline{\phantom{\tfrac{3}{4}m}+2 \quad +2}$$
$$\tfrac{4}{3} \cdot \tfrac{3}{4}m = 36 \cdot \tfrac{4}{3}$$
$$m = \mathbf{48}$$

9.
$$7a - 13 = 2a + 27$$
$$\underline{-2a \qquad\quad -2a}$$
$$5a - 13 = 27$$
$$\underline{+13 \qquad +13}$$
$$\frac{5a}{5} = \frac{40}{5}$$
$$a = \mathbf{8}$$

10.
$$6(s - 3) = 5(s + 2)$$
$$6s - 18 = 5s + 10$$
$$\underline{-5s \qquad\quad -5s}$$
$$x - 18 = 10$$
$$\underline{+18 \qquad +18}$$
$$x = \mathbf{28}$$

11.
$$8p + 1 \leq p + 15$$
$$\underline{-p \qquad\quad -p}$$
$$7p + 1 \leq 15$$
$$\underline{-1 \qquad -1}$$
$$\frac{7p}{7} \leq \frac{14}{7}$$
$$p \leq \mathbf{2}$$

12. $\tfrac{2}{3}(x - 6)$

13.
$$\begin{array}{r} x + 7 \\ \underline{x - 3} \\ -3x - 21 \\ \underline{x^2 + 7x} \\ x^2 + 4x - 21 \end{array}$$

14. $\sqrt{48} = \sqrt{16} \cdot \sqrt{3} = \mathbf{4\sqrt{3}}$

15. $x^2 + 13x + 40 = 0$

$(x + 5)(x + 8) = 0$

$$\begin{array}{ll} x + 5 = 0 & x + 8 = 0 \\ \underline{-5} & \underline{-8 \quad -8} \\ x = \mathbf{-5} \text{ and } x = \mathbf{-8} \end{array}$$

16. $5(x - 2) = 4(x + 1)$

$$\begin{array}{rcl} 5x - 10 & = & 4x + 4 \\ \underline{-4x} & & \underline{-4x} \\ x - 10 & = & 4 \\ \underline{+10} & & \underline{+10} \\ x & = & \mathbf{14} \end{array}$$

17. 9 mo. $= \frac{9}{12} = \frac{3}{4}$ yr.

$i = prt$

$132 = 1600r \cdot \frac{3}{4}$

$\frac{132}{1200} = \frac{1200r}{1200}$

$\frac{11}{100} = r$

$\frac{11}{100} = \mathbf{11\%}$

18. small no. $= x$

large no. $= 3x + 2$

$$\begin{array}{rcl} 7x - 4 & = & 3x + 2 + 10 \\ 7x - 4 & = & 3x + 12 \\ \underline{-3x} & & \underline{-3x} \\ 4x - 4 & = & 12 \\ \underline{+4} & & \underline{+4} \\ \frac{4x}{4} & = & \frac{16}{4} \\ x & = & \mathbf{4} \end{array}$$

$3x + 2 = 3(4) + 2 = 12 + 2 = \mathbf{14}$

19. Maria's wage $= x$

Al's wage $= x + 5$

Lucy's wage $= x + 2$

$30(x + x + 5 + x + 2) = 750$

$30(3x + 7) = 750$

$$\begin{array}{rcl} 90x + 210 & = & 750 \\ \underline{-210} & & \underline{-210} \\ \frac{90x}{90} & = & \frac{540}{90} \\ x & = & 6 \end{array}$$

$x + 5 = 6 + 5 = \mathbf{\$11}$

20.

	Age Now	Age in 20 Years
Katherine	x	x + 20
David	2x + 5	2x + 5 + 20 = 2x + 25

$2(2x + 25) = 3(x + 20) + 11$

$4x + 50 = 3x + 60 + 11$

$$\begin{array}{rcl} 4x + 50 & = & 3x + 71 \\ \underline{-3x} & & \underline{-3x} \\ x + 50 & = & 71 \\ \underline{-50} & & \underline{-50} \\ x & = & 21 \end{array}$$

$2x + 5 = 2(21) + 5 = \mathbf{47}$

Chapter 7 Geometry

Level 1 Geometry Skills

1. **acute**

2.
$$\begin{array}{r} 90° \\ \underline{-67°} \\ \mathbf{23°} \end{array}$$

3. Since $\angle b$ and $\angle f$ are corresponding, $\angle f = \mathbf{52°}$.

4. $\angle f$

5.
$$\begin{array}{r} 46° \\ \underline{+54°} \\ 100° \end{array} \qquad \begin{array}{r} 180° \\ \underline{-100°} \\ 80° = \angle Y \end{array}$$

Since $\angle Y$ is the largest angle, **WX** is the longest side.

Lesson 1

1. a. **acute** b. **right** c. **obtuse** d. **reflex**
2. a. **straight** b. **reflex** c. **right** d. **acute**
3. a. **acute** b. **acute** c. **obtuse** d. **right**
4. a. **reflex** b. **obtuse** c. **straight** d. **acute**
5. a. **acute** b. **obtuse** c. **reflex** d. **obtuse**

Lesson 2

1.
$$\begin{array}{r} 90° \\ \underline{-22°} \\ \mathbf{68°} \end{array}$$

2.
$$\begin{array}{r} 180° \\ \underline{-22°} \\ \mathbf{158°} \end{array}$$

3.
$$\begin{array}{r} 180° \\ \underline{-106°} \\ \mathbf{74°} \end{array}$$

4. $\angle b$

5. $\angle c$

6. **106°**

7. **74°**

8.
$$\begin{array}{r} 180° \\ \underline{-63°} \\ 117° \end{array}$$

$\frac{3m}{3} = \frac{117}{3}$

$m = \mathbf{39°}$

9. small angle $= x$

large angle $= x + 20$

$$\begin{array}{rcl} x + x + 20 & = & 90 \\ 2x + 20 & = & 90 \\ \underline{-20} & & \underline{-20} \\ \frac{2x}{2} & = & \frac{70}{2} \\ x & = & \mathbf{35°} \end{array}$$

$x + 20 = 35 + 20 = \mathbf{55°}$

10. small angle = x

large angle = 4x

$$x + 4x = 180$$
$$\frac{5x}{5} = \frac{180}{5}$$
$$x = \mathbf{36°}$$
$$4x = 4(36) = \mathbf{144°}$$

Lesson 3

1. $\angle t$
2. $\angle s$
3. $\angle q$
4. $\angle p$
5. **118°**
6.   180°
      $\underline{-\ 49°}$
      131°

7. **vertical**
8. **alternate interior**
9. **alternate exterior**
10. **corresponding**

Lesson 4

1.   32°   180;°
   $\underline{+\ 74°}$ $\underline{-\ 106°}$
    106°   74° = $\angle M$

Since two angles are equal, the triangle is **isosceles**.

2.   35°   180°
   $\underline{+\ 55°}$ $\underline{-\ 90°}$
    90°   90° = $\angle F$

Since there is one right angle, the triangle is **right**.

3.   42°   180°
   $\underline{+\ 42°}$ $\underline{-\ 84°}$
    84°   **96°**

4.   180°   **51°**
   $\underline{-\ 78°}$ $2\overline{)102}$
    102°

5.   90°   180°
   $\underline{+\ 33°}$ $\underline{-\ 123°}$
    123°   **57°**

6.   30°   180°
   $\underline{+\ 60°}$ $\underline{-\ 90°}$
    90°   90° = $\angle C$

Since $\angle C$ is the largest angle, side **AB** is the longest side.

7.   6 in.   24 in.
   $\underline{+\ 8}$ $\underline{-\ 14}$
   14 in.   10 in. = DF

Since side DF is the longest, $\angle E$ is the largest angle.

8.   48°   180°
   $\underline{+\ 66°}$ $\underline{-\ 114°}$
    114°   66° = $\angle Z$

Since $\angle X$ is the smallest angle, side **YZ** is the shortest side.

9. Since side NO is across from the right angle, side **NO** is the hypotenuse.

10.   60°   180°
   $\underline{+\ 60°}$ $\underline{-\ 120°}$
    120°   60°

Since the three angles are equal, the triangle is **equiangular**.

Level 1   Review

1. **acute**
2. **reflex**
3.   180°
   $\underline{-\ 84°}$
    96°
4.   90°
   $\underline{-\ 36°}$
    54°

5. Since $\angle t$ and $\angle u$ are vertical, $\angle u = \mathbf{112°}$.
6. $\angle z$

7.   180°
   $\underline{-\ 112°}$
    68°

8.   33°   180°
   $\underline{+\ 57°}$ $\underline{-\ 90°}$
    90°   90° = $\angle C$

Since side AB is across from the 90° angle, side AB is called the **hypotenuse**.

9.   49°   180°
   $\underline{+\ 82°}$ $\underline{-\ 131°}$
    131°   49° = $\angle Z$

Since two angles are equal, the triangle is **isosceles**.

10.   55°   180°
   $\underline{+\ 47°}$ $\underline{-\ 102°}$
    102°   78° = $\angle R$

Since $\angle R$ is the largest angle, side **PQ** is the longest.

Level 2   Geometry Applications

1. $\dfrac{\text{width}}{\text{length}}$ $\dfrac{3}{5} = \dfrac{15}{x}$
$$\frac{3x}{3} = \frac{75}{3}$$
$$x = \mathbf{25\ in.\ long}$$

2. **(4) RT = UW**. This satisfies the SAS requirement.

3. $c^2 = a^2 + b^2$
$c^2 = (3.2)^2 + (2.4)^2$
$c^2 = 10.24 + 5.76$
$c^2 = 16$
$c = \sqrt{16}$
$c = \mathbf{4\ m}$

4. **(1) (−3, 18)**
For Choice (1), $y = (-3)^2 - 3(-3) = +9 + 9 = 18$.
   Point $(-3, 18)$ is on the graph.
For Choice (2), $y = (-2)^2 - 3(-2) = +4 + 6 = 10$.
   Point $(-2, 9)$ is not on the graph.
For Choice (3), $y = (1)^2 - 3(1) = 1 - 3 = -2$.
   Point $(1, -1)$ is not on the graph.
For Choice (4), $y = (2)^2 - 3(2) = 4 - 6 = -2$.
   Point $(2, -6)$ is not on the graph.
For Choice (5), $y = (4)^2 - 3(4) = 16 - 12 = 4$.
   Point $(4, 8)$ is not on the graph.

5. **(4)** $\dfrac{+3}{2}$

$$\frac{y_2 - y_1}{x^2 - x^1} = \frac{11 - 5}{7 - 3} = \frac{6}{4} = \frac{3}{2}$$

Lesson 1

1. **Yes**. $\angle A =$ 80°   180°
    $\angle B =$ $\underline{60°}$ $\underline{-140°}$
           140°   40° = $\angle C$

    $\angle D =$ 80°   180°
    $\angle F =$ $\underline{40°}$ $\underline{-120°}$
           120°   60 = $\angle E$

Since the angles in $\triangle ABC$ are the same as the angles in $\triangle DEF$, the triangles are **similar**.

2. **No.** $\angle G = 65°$ $180°$

      $\angle H = \underline{55°}$ $\underline{-120°}$

           $120°$ $60° = \angle I$

      $\angle J = 60°$ $180°$

      $\angle K = \underline{50°}$ $\underline{-110°}$

           $110°$ $70° = \angle L$

    Since the angles in $\triangle GHI$ are not the same as the angles in $\triangle JKL$, the triangles are **not similar**.

3. **Yes.** The ratio of the short side to the long side in rectangle $MNOP$ is 3:4.

    The ratio of the short side to the long side in rectangle $QRST$ is 9:12 = 3:4.

    Since the ratios of the sides are the same, the rectangles are **similar**.

4. **No.** The ratio of the short side to the long side in the small rectangle is 5:6.

    The ratio of the short side to the long side in the large rectangle is 10:11.

    Since the ratios of the sides are not the same, the rectangles are **not similar**.

5. width $\dfrac{4}{5} = \dfrac{20}{x}$

    length

       $\dfrac{4x}{4} = \dfrac{100}{4}$

          $x = $ **25 in. long**

6. short leg $\dfrac{18}{30} = \dfrac{x}{25}$

    long leg

       $\dfrac{30x}{30} = \dfrac{450}{30}$

          $x = $ **15**

7. height $\dfrac{5}{3} = \dfrac{x}{72}$

    shadow

       $\dfrac{3x}{3} = \dfrac{360}{3}$

          $x = $ **120 ft.**

8. long leg $\dfrac{10}{3} = \dfrac{x}{24}$

    short leg

       $\dfrac{3x}{3} = \dfrac{240}{3}$

          $x = $ **80 ft.**

9. height $\dfrac{6}{3} = \dfrac{7}{x}$

    width

       $\dfrac{6x}{6} = \dfrac{21}{6}$

          $x = $ **$3\frac{1}{2}$ ft.**

10. short leg $\dfrac{5}{8} = \dfrac{x}{48}$

    long leg

       $\dfrac{8x}{8} = \dfrac{240}{8}$

          $x = $ **30 cm**

Lesson 2

1. **Yes.** The conditions **satisfy** the SAS requirement.
2. **Yes.** The conditions **satisfy** the ASA requirement.
3. **No.** Perimeter of triangle at the left is $6 + 8 + 10 = 24$ in. Missing side of triangle at right is $23 - 6 - 10 = 7$ in. The conditions **fail to satisfy** the SSS requirement.
4. **No.** In the triangle at the left, one leg is 4 inches. In the triangle at the right, the hypotenuse is 4 inches. Since corresponding sides are not equal, the conditions **fail to satisfy** the SAS requirement.
5. **Yes.** The conditions satisfy the ASA requirement.
6. **Yes.** Perimeter of triangle at the left is $4 + 4 + 6 = 14$. Missing side of triangle at right is $14 - 4 - 6 = 4$. Since the sides are the same, the conditions satisfy the SSS requirement.
7. **(4)** $\angle B = \angle E$. This condition satisfies the ASA requirement.
8. **(1)** $GI = JL$. This condition satisfies the SSS requirement.
9. **(2) B only.** This condition satisfies the SSS requirement.
10. **(5) A or C.** Condition A satisfies the SAS requirement. Condition C satisfies the ASA requirement.

Lesson 3

1. $c^2 = a^2 + b^2$

   $c^2 = 24^2 + 32^2$

   $c^2 = 576 + 1024$

   $c^2 = 1600$

   $c = \sqrt{1600}$

   $c = $ **40 ft.**

2. $c^2 = a^2 + b^2$

   $c^2 = 36^2 + 48^2$

   $c^2 = 1296 + 2304$

   $c^2 = 3600$

   $c = \sqrt{3600}$

   $c = $ **60 in.**

3. $c^2 = a^2 + b^2$

   $34^2 = a^2 + 30^2$

   $1156 = a^2 + 900$

   $\underline{-900 \quad -900}$

     $256 = a^2$

   $\sqrt{256} = a$

     16 $= a$

4. $c^2 = a^2 + b^2$

   $13^2 = 5^2 + b^2$

   $169 = 25 + b^2$

   $\underline{-25 \quad -25}$

     $144 = b^2$

   $\sqrt{144} = b$

   12 ft. $= b$

5. $c^2 = a^2 + b^2$

   $c^2 = 18^2 + 24^2$

   $c^2 = 324 + 576$

   $c^2 = 900$

   $c = \sqrt{900}$

   $c = $ **30 ft.**

6. $c^2 = a^2 + b^2$

   $c^2 = (1.2)^2 + (1.6)^2$

   $c^2 = 1.44 + 2.56$

   $c^2 = 4$

   $c = \sqrt{4}$

   $c = $ **2 cm**

7. $c^2 = a^2 + b^2$

   $26^2 = a^2 + 24^2$

   $676 = a^2 + 576$

   $\underline{-576 \quad -576}$

     $100 = a^2$

   $\sqrt{100} = a$

     $10 = a$

   base $= 2 \cdot 10 = $ **20 in.**

8. $c^2 = a^2 + b^2$

   $75^2 = 45^2 + b^2$

   $5625 = 2025 + b^2$

   $\underline{-2025 \quad -2025}$

     $3600 = b^2$

     $3600 = b^2$

   $\sqrt{3600} = b$

     60 $= b$

9. $c^2 = a^2 + b^2$

   $17^2 = a^2 + 15^2$

   $289 = a^2 + 225$

   $\underline{-225 \quad -225}$

     $64 = a^2$

   $\sqrt{64} = a$

   8 ft. $= a$

10. $c^2 = a^2 + b^2$

   $c^2 = 15^2 + 20^2$

   $c^2 = 225 + 400$

   $c^2 = 625$

   $c = \sqrt{625}$

   $c = $ **25 mi.**

Lesson 4

1. $I = (+6, +3)$ $N = (-7, 0)$
    $J = (+2, +7)$ $P = (-5, -8)$
    $K = (0, +5)$ $Q = (-3, -5)$
    $L = (-4, +6)$ $R = (0, -8)$
    $M = (-6, -1)$ $S = (+7, -4)$

2.

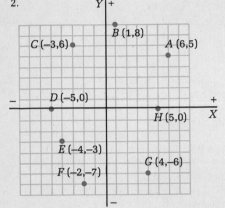

3. $A = (4, 3)$ $F = (-6, 0)$
    $B = (2, 7)$ $G = (-4, -4)$
    $C = (8, 8)$ $H = (-1, -7)$
    $D = (-1, 6)$ $I = (4, -3)$
    $E = (2, 5)$ $J = (0, -4)$

4.

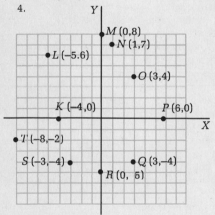

5. $A = (-1, -3)$ $F = (5, 0)$
    $B = (-5, -3)$ $G = (5, 7)$
    $C = (-7, 0)$ $H = (0, 4)$
    $D = (-5, 5)$ $I = (-1, 6)$
    $E = (2, -5)$ $J = (7, -7)$

Lesson 5

1. A is 5 spaces above the x-axis.
   B is 3 spaces below the x-axis.
   Distance AB is $5 + 3 = $ **8.**

2. B is 2 spaces left of the y-axis.
   C is 4 spaces right of the y-axis.
   Distance BC is $2 + 4 = $ **6.**

3. C is 3 spaces below the x-axis.
   D is 1 space above the x-axis.
   Distance CD is $3 + 1 = $ **4.**

4. $E = (x_1, y_1) = (+1, +2)$
   $F = (x_2, y_2) = (+13, +11)$
   $d = \sqrt{(x_2 - x_1)^2 + (y_2 - y_1)^2}$
   $= \sqrt{(13 - 1)^2 + (11 - 2)^2}$
   $= \sqrt{(12)^2 + (9)^2}$
   $= \sqrt{144 + 81}$
   $= \sqrt{225}$
   $= $ **15**

5. $G = (x_1, y_1) = (+5, +15)$
   $H = (x_2, y_2) = (+21, +3)$
   $d = \sqrt{(x_2 - x_1)^2 + (y_2 - y_1)^2}$
   $= \sqrt{(21 - 5)^2 + (3 - 15)^2}$
   $= \sqrt{(16)^2 + (-12)^2}$
   $= \sqrt{256 + 144}$
   $= \sqrt{400}$
   $= $ **20**

6. $I = (x_1, y_1) = (-5, -4)$
   $J = (x_2, y_2) = (+3, +2)$
   $d = \sqrt{(x_2 - x_1)^2 + (y_2 - y_1)^2}$
   $= \sqrt{(3 - (-5))^2 + (2 - (-4))^2}$
   $= \sqrt{(3 + 5)^2 + (2 + 4)^2}$
   $= \sqrt{(8)^2 + (6)^2}$
   $= \sqrt{64 + 36}$
   $= \sqrt{100}$
   $= $ **10**

7. $K = (x_1, y_1) = (-2, +7)$
   $L = (x_2, y_2) = (+3, -5)$
   $d = \sqrt{(x_2 - x_1)^2 + (y_2 - y_1)^2}$
   $= \sqrt{(3 - (-2))^2 + (-5 - (7))^2}$
   $= \sqrt{(3 + 2)^2 + (-5 - 7)^2}$
   $= \sqrt{(5)^2 + (-12)^2}$
   $= \sqrt{25 + 144}$
   $= \sqrt{169}$
   $= $ **13**

8. $P = (x_2, y_2) = (+1, +6)$
   $N = (x_1, y_1) = (+3, +8)$
   $M = \left(\dfrac{x_1 + x_2}{2}, \dfrac{y_1 + y_2}{2}\right)$
   $= \left(\dfrac{1 + 3}{2}, \dfrac{6 + 8}{2}\right)$
   $= \left(\dfrac{4}{2}, \dfrac{14}{2}\right)$
   $= $ **(2, 7)**

9. $Q = (x_1, y_1) = (-5, +2)$
   $R = (x_2, y_2) = (+3, +6)$
   $M = \left(\dfrac{x_1 + x_2}{2}, \dfrac{y_1 + y_2}{2}\right)$

   $= \left(\dfrac{-5 + 3}{2}, \dfrac{+2 + 6}{2}\right)$

   $= \left(\dfrac{-2}{2}, \dfrac{+8}{2}\right)$

   (−1, 4)

10. $S = (x_1, y_1) = (-3, -3)$
    $T = (x_2, y_2) = (+7, +5)$
    $M = \left(\dfrac{x_1 + x_2}{2}, \dfrac{y_1 + y_2}{2}\right)$

    $= \left(\dfrac{-3 + 7}{2}, \dfrac{-3 + 5}{2}\right)$

    $= \left(\dfrac{+4}{2}, \dfrac{+2}{2}\right)$

    $= \textbf{(2, 1)}$

Lesson 6

1. $y = x + 4$
   When $x = 3$, $y = 3 + 4 = \textbf{7}$.
   When $x = -2$, $y = -2 + 4 = \textbf{2}$.
   When $x = -5$, $y = -5 + 4 = \textbf{−1}$.

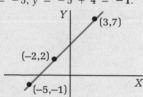

x	y
3	7
−2	2
−5	−1

2. $y = \dfrac{x}{2} + 1$
   When $x = 8$, $y = \dfrac{8}{2} + 1 = 4 + 1 = \textbf{5}$.

   When $x = 4$, $y = \dfrac{4}{2} + 1 = 2 + 1 = \textbf{3}$.

   When $x = -6$, $y = \dfrac{-6}{2} + 1 = -3 + 1 = \textbf{−2}$.

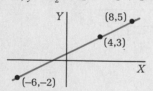

x	y
8	5
4	3
−6	−2

3. $y = -3x + 4$
   When $x = 3$, $y = -3(3) + 4 = -9 + 4 = \textbf{−5}$.
   When $x = 1$, $y = -3(1) + 4 = -3 + 4 = \textbf{1}$.
   When $x = -2$, $y = -3(-2) + 4 = +6 + 4 = \textbf{10}$.

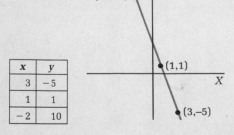

x	y
3	−5
1	1
−2	10

4. $y = -2x - 3$
   When $x = 2$, $y = -2(2) - 3 = -4 - 3 = \textbf{−7}$.
   When $x = -3$, $y = -2(-3) - 3 = 6 - 3 = \textbf{3}$.
   When $x = -4$, $y = -2(-4) - 3 = 8 - 3 = \textbf{5}$.

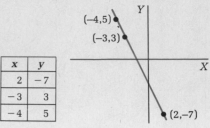

x	y
2	−7
−3	3
−4	5

5. **Yes**, when $x = 1$, $y = 5(1) + 3 = 5 + 3 = 8$.
   Point (1, 8) **is** on the graph.

6. **No**, when $x = 2$, $y = -3(2) + 1 = -6 + 1 = -5$.
   Point (2, −4) **is not** on the graph.

7. **No**, when $x = 3$, $y = -(3) + 6 = -3 + 6 = +3$.
   Point (3, 5) **is not** on the graph.

8. **Yes**, when $x = -8$, $y = \dfrac{3}{4}(-8) - 2 = -6 - 2 = -8$.
   Point (−8, −8) **is** on the graph.

9. **(3) (2, −3)**
   For Choice (1), when $x = 4$, $y = 4 - 5 = -1$.
   Point (4, 0) is not on the graph.
   For Choice (2), when $x = 3$, $y = 3 - 5 = -2$.
   Point (3, −4) is not on the graph.
   For Choice (3), when $x = 2$, $y = 2 - 5 = -3$.
   Point (2, −3) is on the graph.
   For Choice (4), when $x = -1$, $y = -1 - 5 = -6$.
   Point (−1, −5) is not on the graph.
   For Choice (5), when $x = -2$, $y = -2 - 5 = -7$.
   Point (−2, −8) is not on the graph.

10. **(1) (3, −9)**
    For Choice (1), when $x = 3$, $y = -2(3) - 3 = -6 - 3$
    $= -9$.
    Point (3, −9) is on the graph.
    For Choice (2), when $x = 2$, $y = -2(2) - 3 = -4 - 3$
    $= -7$.
    Point (2, −4) is not on the graph.
    For Choice (3), when $x = -1$, $y = -2(-1) - 3 = +2$
    $- 3 = -1$.
    Point (−1, +1) is not on the graph.
    For Choice (4), when $x = -3$, $y = -2(-3) - 3 = +6$
    $- 3 = +3$.
    Point (−3, +4) is not on the graph.
    For Choice (5) when $x = -5$, $y = -2(-5) - 3 = 10 -$
    $3 = 7$.
    Point (−5, 9) is not on the graph.

Lesson 7

1. $y = x^2 + 2$
   $y = 2^2 + 2 = 4 + 2 = \mathbf{6}$
   $y = 1^2 + 2 = 1 + 2 = \mathbf{3}$
   $y = 0^2 + 2 = 0 + 2 = \mathbf{2}$
   $y = (-1)^2 + 2 = 1 + 2 = \mathbf{3}$
   $y = (-2)^2 + 2 = 4 + 2 = \mathbf{6}$

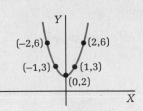

x	y
2	6
1	3
0	2
-1	3
-2	6

2. $y = x^2 - 2x$
   $y = 4^2 - 2(4) = 16 - 8 = \mathbf{8}$
   $y = 3^2 - 2(3) = 9 - 6 = \mathbf{3}$
   $y = 2^2 - 2(2) = 4 - 4 = \mathbf{0}$
   $y = 1^2 - 2(1) = 1 - 2 = \mathbf{-1}$
   $y = 0^2 - 2(0) = 0 - 0 = \mathbf{0}$
   $y = (-1)^2 - 2(-1) = +1 + 2 = \mathbf{3}$
   $y = (-2)^2 - 2(-2) = +4 + 4 = \mathbf{8}$

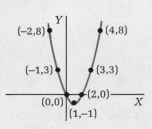

x	y
4	8
3	3
2	0
1	-1
0	0
-1	3
-2	8

3. $y = x^2 + x - 2$
   $y = 2^2 + 2 - 2 = 4 + 2 - 2 = \mathbf{4}$
   $y = 1^2 + 1 - 2 = 1 + 1 - 2 = \mathbf{0}$
   $y = 0^2 + 0 - 2 = 0 + 0 - 2 = \mathbf{-2}$
   $y = (-1)^2 + (-1) - 2 = 1 - 1 - 2 = \mathbf{-2}$
   $y = (-2)^2 + (-2) - 2 = 4 - 2 - 2 = \mathbf{0}$
   $y = (-3)^2 + (-3) - 2 = 9 - 3 - 2 = \mathbf{4}$

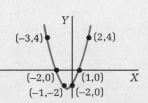

x	y
2	4
1	0
0	-2
-1	-2
-2	0
-3	4

4. $y = x^2 + x - 4$
   $y = 2^2 + 2 - 4 = 4 + 2 - 4 = \mathbf{2}$
   $y = 1^2 + 1 - 4 = 1 + 1 - 4 = \mathbf{-2}$
   $y = 0^2 + 0 - 4 = 0 + 0 - 4 = \mathbf{-4}$
   $y = (-1)^2 + (-1) - 4 = 1 - 1 - 4 = \mathbf{-4}$
   $y = (-2)^2 + (-2) - 4 = 4 - 2 - 4 = \mathbf{-2}$
   $y = (-3)^2 + (-3) - 4 = 9 - 3 - 4 = \mathbf{+2}$

x	y
2	2
1	-2
0	-4
-1	-4
-2	-2
-3	+2

5. **No,** when $x = 5$, $y = 5^2 - 4(5) = 25 - 20 = \mathbf{5}$.
   Point (5, 4) **is not** on the graph.
6. **Yes,** when $x = 4$, $y = 4^2 + 4 + 3 = 16 + 4 + 3 = \mathbf{23}$.
   Point (4, 23) **is** on the graph.
7. **Yes,** when $x = -3$, $y = (-3)^2 - (-3) - 1 = 9 + 3 - 1 = \mathbf{11}$.
   Point $(-3, 11)$ **is** on the graph.
8. **Yes,** when $x = -2$, $y = (-2)^2 + 3(-2) - 2 = 4 - 6 - 2 = \mathbf{-4}$.
   Point $(-2, -4)$ **is** on the graph.
9. **(2) (3, 15)**
   For Choice (1), $y = 4^2 + 2(4) = 16 + 8 = \mathbf{24}$.
   Point (4, 20) is not on the graph.
   For Choice (2), $y = 3^2 + 2(3) = 9 + 6 = \mathbf{15}$.
   Point (3, 15) is on the graph.
   For Choice (3), $y = 1^2 + 2(1) = 1 + 2 = \mathbf{3}$.
   Point (1, 2) is not on the graph.
   For Choice (4), $y = (-2)^2 + 2(-2) = 4 - 4 = \mathbf{0}$.
   Point $(-2, 1)$ is not on the graph.
   For Choice (5), $y = (-4)^2 + 2(-4) = 16 - 8 = \mathbf{8}$.
   Point $(-4, 6)$ is not on the graph.
10. **(4) (-2, 17)**
    For Choice (1), $y = 3^2 - 5(3) + 3 = 9 - 15 + 3 = -3$.
    Point (3, 3) is not on the graph.
    For Choice (2), $y = 2^2 - 5(2) + 3 = 4 - 10 + 3 = -3$.
    Point $(2, -4)$ is not on the graph.
    For Choice (3), $y = (-1)^2 - 5(-1) + 3 = 1 + 5 + 3 = 9$.
    Point $(-1, 8)$ is not on the graph.
    For Choice (4), $y = (-2)^2 - 5(-2) + 3 = 4 + 10 + 3 = 17$.
    Point $(-2, 17)$ is on the graph.
    For Choice (5), $y = (-3)^2 - 5(-3) + 3 = 9 + 15 + 3 = 27$.
    Point $(-3, 25)$ is not on the graph.

Lesson 8

1. $C = (x_1, y_1) = (4, 5)$
   $D = (x_2, y_2) = (6, 9)$
   $m = \dfrac{y_2 - y_1}{x_2 - x_1} = \dfrac{9 - 5}{6 - 4} = \dfrac{4}{2} = \mathbf{2}$
2. $E = (x_1, y_1) = (2, 1)$
   $F = (x_2, y_2) = (10, 7)$
   $m = \dfrac{y_2 - y_1}{x_2 - x_1} = \dfrac{7 - 1}{10 - 2} = \dfrac{6}{8} = \mathbf{\dfrac{3}{4}}$

3. $G = (x_1, y_1) = (4, 8)$
   $H = (x_2, y_2) = (9, 3)$
   $m = \dfrac{y_2 - y_1}{x_2 - x_1} = \dfrac{3 - 8}{9 - 4} = \dfrac{-5}{5} = \mathbf{-1}$

4. $I = (x_1, y_1) = (-6 + 10)$
   $J = (x_2, y_2) = (+8, +3)$
   $m = \dfrac{y_2 - y_1}{x_2 - x_1} = \dfrac{3 - (10)}{8 - (-6)} = \dfrac{3 - 10}{8 + 6} = \dfrac{-7}{14} = \mathbf{-\dfrac{1}{2}}$

5. **Line b has positive slope.**

6. $y = 2x - 3$
   $y = 2(0) - 3$
   $y = 0 - 3 = -3$
   y-intercept = **(0, -3)**

7. $y = -2x + 1$
   $y = -2(0) + 1$
   $y = 0 + 1 = 1$
   y-intercept = **(0, 1)**

8. $y = 5x - 7$
   $y = 5(0) - 7$
   $y = 0 - 7 = -7$
   y-intercept = **(0, -7)**

9. $y = 3x - 9$
   $0 = 3x - 9$
   $\underline{+9 \qquad\quad + 9}$
   $\dfrac{9}{3} = \dfrac{3x}{3}$
   $3 = x$
   x-intercept = **(3, 0)**

10. $y = 8x + 4$
    $0 = 8x + 4$
    $\underline{-4 \qquad\quad -4}$
    $\dfrac{-4}{8} = \dfrac{8x}{8}$
    $\dfrac{-1}{2} = x$
    x-intercept = $\left(\dfrac{-1}{2}, 0\right)$

Level 2 Review

1. $\begin{array}{l} \text{height} \\ \text{base} \end{array} \dfrac{9}{5} = \dfrac{x}{8}$
   $\dfrac{5x}{5} = \dfrac{72}{5}$
   $c = \mathbf{14\tfrac{2}{5}}$ ft.

2. $\dfrac{12}{20} = \dfrac{16}{x}$
   $\dfrac{12x}{12} = \dfrac{320}{12}$
   $x = \mathbf{26\tfrac{2}{3}}$

3. **(2)** $\angle D = \angle G$. This satisfies the ASA requirement.

4. $\begin{aligned} c^2 &= a^2 + b^2 \\ 35^2 &= 28^2 + b^2 \\ 1225 &= 784 + b^2 \\ \underline{-784} \quad &\underline{-784} \\ 441 &= b^2 \\ \sqrt{441} &= b \\ \mathbf{21} &= b \end{aligned}$

5. S is 1 space left of y-axis. T is 2 spaces right of y-axis. Distance ST is $1 + 2 = \mathbf{3}.$

6. $S = (x_1, y_1) = (-1, +3)$
   $U = (x_2, y_2) = (+2, -1)$
   $\begin{aligned} d &= \sqrt{(x_2 - x_1)^2 + (y_2 - y_1)^2} \\ &= \sqrt{(2 - (-1))^2 + (-1 - (+3))^2} \\ &= \sqrt{(2 + 1)^2 + (-1 - 3)^2} \\ &= \sqrt{(3)^2 + (-4)^2} \\ &= \sqrt{9 + 16} \\ &= \sqrt{25} \\ &= \mathbf{5} \end{aligned}$

7. **(4) (4, 8)**
   For Choice (1), $y = 3(-3) - 4 = -9 - 4 = -13$.
     Point $(-3, -10)$ is not on the graph.
   For Choice (2), $y = 3(-1) - 4 = -3 - 4 = -7$.
     Point $(-1, -5)$ is not on the graph.
   For Choice (3), $y = 3(2) - 4 = 6 - 4 = 2$.
     Point $(2, 3)$ is not on the graph.
   For Choice (4), $y = 3(4) - 4 = 12 - 4 = 8$.
     Point $(4, 8)$ is on the graph.
   For Choice (5), $y = 3(6) - 4 = 18 - 4 = 14$.
     Point $(6, 16)$ is not on the graph.

8. **(2) (3, 14)**
   For Choice (1), $y = (5)^2 + 2(5) - 1 = 25 + 10 - 1 = 34$.
     Point $(5, 36)$ is not on the graph.
   For Choice (2), $y = (3)^2 + 2(3) - 1 = 9 + 6 - 1 = 14$.
     Point $(3, 14)$ is on the graph.
   For Choice (3), $y = (-2)^2 + 2(-2) - 1 = 4 - 4 - 1 = -1$.
     Point $(-2, -2)$ is not on the graph.
   For Choice (4), $y = (-4)^2 + 2(-4) - 1 = 16 - 8 - 1 = 7$.
     Point $(-4, 6)$ is not on the graph.
   For Choice (5), $y = (-5)^2 + 2(-5) - 1 = 25 - 10 - 1 = 14$.
     Point $(-5, 12)$ is not on the graph.

9. **(3) -1**
   $C = (x_1, y_1) = (-2, 7)$
   $D = (x_2, y_2) = (3, 2)$
   $m = \dfrac{y_2 - y_1}{x_2 - x_1} = \dfrac{2 - 7}{3 - (-2)}$
   $ m = \dfrac{-5}{3 + 2}$
   $ m = \dfrac{-5}{5} = \mathbf{-1}$

10. **(2) (0, 3)**
    When $x = 0$, $\begin{aligned} y &= 5(0) + 3 \\ &= 0 + 3 = 3. \end{aligned}$
    Coordinates of the y-intercept are **(0, 3)**.

Level 3 Geometry Problem Solving

1. $\begin{aligned} P &= a + b + c \\ 48 &= a + 12 + 16 \\ 48 &= a + 28 \\ \underline{-28} \quad &\underline{-28} \\ 20 &= a \\ a &= \mathbf{20}\ \textbf{in.} \end{aligned}$

2. $\begin{aligned} P &= 2l + 2w \\ 40 &= 2l + 2 \cdot 9 \\ 40 &= 2l + 18 \\ \underline{-18} \quad &\underline{-18} \\ \dfrac{22}{2} &= \dfrac{2l}{2} \\ 11 &= l \\ l &= \mathbf{11}\ \textbf{ft.} \end{aligned}$

3. $\begin{aligned} A &= lw \\ \dfrac{95}{5} &= \dfrac{l \cdot 5}{5} \\ 19 &= l \\ l &= \mathbf{19}\ \textbf{ft.} \end{aligned}$

4. $A = lw$ $A = s^2$

    $A = 12.5 \cdot 8$ $100 = s^2$

    $A = 100 \text{ m}^2$ $\sqrt{100} = s$

                  $10 = s$

                  $s = \textbf{10 m}$

5. $w = x$ $P = 2l + 2w$

    $l = x + 8$ $100 = 2(x + 8) + 2x$

            $100 = 2x + 16 + 2x$

            $100 = 4x + 16$

           $\underline{-16 \quad\quad -16}$

            $\dfrac{84}{4} = \dfrac{4x}{4}$

            $21 = x$

            $w = \textbf{21 in.}$

            $l = 21 + 8 = \textbf{29 in.}$

6. $l = 3x$ $A = lw$

   $w = 2x$ $294 = 3x \cdot 2x$

           $\dfrac{294}{6} = \dfrac{6x^2}{6}$

           $49 = x^2$

          $\sqrt{49} = x$

           $7 = x$

           $l = 3 \cdot 7 = \textbf{21 ft.}$

           $w = 2 \cdot 7 = \textbf{14 ft.}$

7. $b = x$ $A = \frac{1}{2}bh$

    $h = 2x$ $64 = \frac{1}{2} \cdot x + 2x$

          $64 = x^2$

        $\sqrt{64} = x$

          $8 = x$

          $b = \textbf{8 m}$

          $h = 2 \cdot 8 = \textbf{16 m}$

8. $l = x$ $V = lwh$

    $h = \frac{1}{3}x$ $162 = x \cdot 6 \cdot \frac{1}{3}x$

          $\dfrac{162}{2} = \dfrac{2x^2}{2}$

          $81 = x^2$

        $\sqrt{81} = x$

          $9 = x$

          $l = \textbf{9 ft.}$

          $h = \frac{1}{3} \cdot 9 = \textbf{3 ft.}$

9. $c^2 = a^2 + b^2$

     $= 6^2 + 6^2$

     $= 36 + 36$

     $= 72$

   $c = \sqrt{72}$

     $= \sqrt{36} \cdot \sqrt{2}$

     $= \textbf{6}\sqrt{\textbf{2}}$

10. $c^2 = a^2 + b^2$

     $= 4^2 + 8^2$

     $= 16 + 64$

     $= 80$

   $c = \sqrt{80}$

     $= \sqrt{16} \cdot \sqrt{5}$

     $= \textbf{4}\sqrt{\textbf{5}} \textbf{ mi.}$ 8 mi.   4 mi.

Lesson 1

1. $A = lw$ 2. $P = 2l + 2w$

   $\dfrac{132}{22} = \dfrac{22w}{22}$ $124 = 2l + 2 \cdot 16$

    $6 = w$ $124 = 2l + 32$

   $w = \textbf{6 ft.}$ $\underline{-32 \quad\quad -32}$

                    $\dfrac{92}{2} = \dfrac{2l}{2}$

                    $46 = l$

                    $l = \textbf{46 ft.}$

3. $P = a + b + c$ 4. $C = \pi d$

   $55 = 17 + 17 + c$ $\dfrac{15.7}{3.14} = \dfrac{3.14d}{3.14}$

   $55 = 34 + c$ $5 = d$

  $\underline{-34 \quad\quad -34}$ $3.14\overline{)15.70}$

   $21 = c$

    $c = \textbf{21 in.}$

5. $A = s^2$ 6. $P = 4s$

   $81 = s^2$ $\dfrac{150}{4} = \dfrac{4s}{4}$

  $\sqrt{81} = s$ $37.5 = s$

    $9 = s$ $s = \textbf{37.5 ft.}$

    $s = \textbf{9 yd.}$

7. $A = \frac{1}{2}bh$ 8. $V = lwh$

  $108 = \frac{1}{2} \cdot 12 \cdot h$ $180 = l \cdot 10 \cdot \frac{1}{2}$

   $\dfrac{108}{6} = \dfrac{6h}{6}$ $\dfrac{180}{5} = \dfrac{l \cdot 5}{5}$

   $18 = h$ $36 = l$

   $h = \textbf{18 ft.}$ $l = \textbf{36 ft.}$

9. $P = 2l + 2w$ 10. $A = \frac{1}{2}bh$

   $46 = 2 \cdot 13.4 + 2w$ $84 = \frac{1}{2}b \cdot 28$

   $46.0 = 26.8 + 2w$ $\dfrac{84}{14} = \dfrac{14b}{14}$

  $\underline{-26.8 \quad -26.8}$ $6 = b$

   $\dfrac{19.2}{2} = \dfrac{2w}{2}$ $b = \textbf{6 in.}$

   $9.6 = w$

   $w = \textbf{9.6 cm}$

Lesson 2

1. $w = x$ $P = 2l + 2w$

    $l = x + 6$ $48 = 2(x + 6) + 2x$

             $48 = 2x + 12 + 2x$

             $48 = 4x + 12$

          $\underline{-12 \quad\quad -12}$

           $\dfrac{36}{4} = \dfrac{4x}{4}$

             $9 = x$

            $w = \textbf{9 ft.}$

             $l = 9 + 6 = \textbf{15 ft.}$

2. $w = x$ $P = 2l + 2w$

    $l = 3x$ $64 = 2(3x) + 2x$

            $64 = 6x + 2x$

           $\dfrac{64}{8} = \dfrac{8x}{8}$

            $8 = x$

          $w = \textbf{8 ft.}$

          $l = 3 \cdot 8 = \textbf{24 ft.}$

3. $a = x$ $P = a + b + c$
    $b = x + 2$ $36 = x + x + 2 + x + 4$
    $c = x + 4$ $36 = 3x + 6$
$$\underline{-6 -6}$$
$$\frac{30}{3} = \frac{3x}{3}$$
$$10 = x$$

$$a = \textbf{10 in.}$$
$$b = 10 + 2 = \textbf{12 in.}$$
$$c = 10 + 4 = \textbf{14 in.}$$

4. $a = 2x$ $P = a + b + c$
    $b = 2x$ $45 = 2x + 2x + x$
    $c = x$ $\frac{45}{5} = \frac{5x}{5}$
               $9 = x$

long side $= 2 \cdot 9 = \textbf{18 m}$

5. $w = x$ $A = lw$
    $l = 2x$ $98 = 2x \cdot x$
              $\frac{98}{2} = \frac{2x^2}{2}$
              $49 = x^2$
              $\sqrt{49} = x$
              $7 = x$

              $w = \textbf{7 yd.}$
              $l = 2 \cdot 7 = \textbf{14 yd.}$

6. $l = x$ $A = lw$
    $w = \frac{x}{2}$ $50 = x \cdot \frac{x}{2}$
              $2 \cdot 50 = \frac{x^2}{2} \cdot 2$
              $100 = x^2$
              $\sqrt{100} = x$
              $10 = x$

              $l = \textbf{10 in.}$
              $w = \frac{10}{2} = \textbf{5 in.}$

7. $l = 4x$ $A = lw$
    $w = 3x$ $300 = 4x \cdot 3x$
              $\frac{300}{12} = \frac{12x^2}{12}$
              $25 = x^2$
              $\sqrt{25} = x$
              $5 = x$

              $l = 4 \cdot 5 = \textbf{20 ft.}$
              $w = 3 \cdot 5 = \textbf{15 ft.}$

8. $b = 2x$ $A = \frac{1}{2}bh$
    $h = 3x$ $108 = \frac{1}{2} \cdot 2x \cdot 3x$
              $\frac{108}{3} = \frac{3x^2}{3}$
              $36 = x^2$
              $\sqrt{36} = x$
              $6 = x$

              $b = 2 \cdot 6 = \textbf{12 in.}$
              $h = 3 \cdot 6 = \textbf{18 in.}$

9. $b = x$ $A = \frac{1}{2}bh$
    $h = \frac{1}{3}x$ $24 = \frac{1}{2}x \cdot \frac{1}{3}x$
              $6 \cdot 24 = \frac{1}{6}x^2 \cdot 6$
              $144 = x^2$
              $\sqrt{144} = x$
              $12 = x$

              $b = \textbf{12 m}$
              $h = \frac{1}{3} \cdot 12 = \textbf{4 m}$

10. $l = x$ $V = lwh$
     $h = 8x$ $1440 = x \cdot 5 \cdot 8x$
              $\frac{1440}{40} = \frac{40x^2}{40}$
              $36 = x^2$
              $\sqrt{36} = x$
              $6 = x$

              $l = \textbf{6 in.}$
              $h = 8 \cdot 6 = \textbf{48 in.}$

Lesson 3

1. $c^2 = a^2 + b^2$
   $c^2 = 3^2 + 3^2$
       $= 9 + 9$
       $= 18$
   $c = \sqrt{18}$
       $= \sqrt{9} \cdot \sqrt{2}$
       $= \mathbf{3\sqrt{2}}$

2. $c^2 = a^2 + b^2$
   $c^2 = 2^2 + 3^2$
       $= 4 + 9$
       $= 13$
   $c = \mathbf{\sqrt{13}}$

3. $c^2 = a^2 + b^2$
   $c^2 = 4^2 + 4^2$
       $= 16 + 16$
       $= 32$
   $c = \sqrt{32}$
       $= \sqrt{16} \cdot \sqrt{2}$
       $= \mathbf{4\sqrt{2}}$

4. $c^2 = a^2 + b^2$
   $c^2 = 3^2 + 6^2$
       $= 9 + 36$
       $= 45$
   $c = \sqrt{45}$
       $= \sqrt{9} \cdot \sqrt{5}$
       $= \mathbf{3\sqrt{5}}$

3 mi.
6 mi.

5. $c^2 = a^2 + b^2$
   $c^2 = 8^2 + 8^2$
       $= 64 + 64$
       $= 128$
   $c = \sqrt{128}$
       $= \sqrt{64} \cdot \sqrt{2}$
       $= \mathbf{8\sqrt{2}\ mi.}$

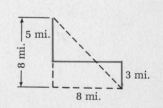
5 mi.   8 mi.   3 mi.   8 mi.

Level 3   Review

1. $P = 2l + 2w$
   $39 = 2l + 2 \cdot 7.5$
   $39 = 2l + 15$
$$\underline{-15 -15}$$
$$\frac{24}{2} = \frac{2l}{2}$$
$$12 = l$$
$$l = \textbf{12 ft.}$$

2. $A = lw$

$\dfrac{17}{3.4} = \dfrac{l \cdot 3.4}{3.4}$

$5 = l$

$3.4)\overline{17.0}$

$l = 5$ m

3. $V = lwh$

$420 = 7 \cdot w \cdot 10$

$\dfrac{420}{70} = \dfrac{70w}{70}$

$6 = w$

$w = \textbf{6 ft.}$

4. $A = s^2$ $A = lw$

$A = 6^2$ $\dfrac{36}{8} = \dfrac{8 \cdot w}{8}$

$A = 36$ $4\frac{4}{8} = w$

$w = \textbf{4}\frac{1}{2}\textbf{ in.}$

5. $a = x$ $P = a + b + c$

$b = x$ $28 = x + x + x + 4$

$c = x + 4$ $28 = 3x + 4$

$\dfrac{-4}{} \quad \dfrac{-4}{}$

$\dfrac{24}{3} = \dfrac{3x}{3}$

$8 = x$

$c = 8 + 4 = \textbf{12 yd.}$

6. $l = x$ $A = lw$

$w = \frac{x}{3}$ $75 = x \cdot \frac{x}{3}$

$3 \cdot 75 = \frac{x^2}{3} \cdot 3$

$225 = x^2$

$\sqrt{225} = x$

$15 = x$

$l = \textbf{15 ft.}$

$w = \frac{15}{3} = \textbf{5 ft.}$

7. $b = 3x$ $A = \frac{1}{2}bh$

$h = 4x$ $150 = \frac{1}{2} \cdot 3x \cdot 4x$

$\dfrac{150}{6} = \dfrac{6x^2}{6}$

$25 = x^2$

$\sqrt{25} = x$

$5 = x$

$b = 3 \cdot 5 = \textbf{15 yd.}$

$h = 4 \cdot 5 = \textbf{20 yd.}$

8. $w = x$ $V = lwh$

$l = 2x$ $216 = 2x \cdot x \cdot 12$

$\dfrac{216}{24} = \dfrac{24x^2}{24}$

$9 = x^2$

$\sqrt{9} = x$

$3 = x$

$w = \textbf{3 in.}$

$l = 2 \cdot 3 = \textbf{6 in.}$

9. $c^2 = a^2 + b^2$

$c^2 = 7^2 + 7^2$

$= 49 + 49$

$= 98$

$c = \sqrt{98}$

$= \sqrt{49} \cdot \sqrt{2}$

$= \textbf{7}\sqrt{\textbf{2}}$

10. $c^2 = a^2 + b^2$

$c^2 = 2^2 + 6^2$

$= 4 + 36$

$= 40$

$c = \sqrt{40}$

$= \sqrt{4} \cdot \sqrt{10}$

$= \textbf{2}\sqrt{\textbf{10}}$

Chapter 7   Quiz

1. $132°$ is **obtuse**.

2. $\begin{array}{r} 90° \\ -17° \\ \hline \textbf{73°} \end{array}$

3. $\angle d$ is alternate interior with $\angle e$.

4. $\begin{array}{r} 180° \\ -111° \\ \hline \textbf{69°} \end{array}$

5. $\begin{array}{r} 21° \\ +69° \\ \hline 90° \end{array}$ $\begin{array}{r} 180° \\ -90° \\ \hline 90° \end{array} = \angle T$

Since there is one right angle, $\triangle RST$ is a **right** triangle.

6. $\angle K = \begin{array}{r} 55° \\ \angle L = +63° \\ \hline 118° \end{array}$ $\begin{array}{r} 180° \\ -118° \\ \hline 62° \end{array} = \angle J$

Since $\angle L$ is largest, side JK is the longest.

7. $\dfrac{\text{height}}{\text{shadow}}$ $\dfrac{10}{7} = \dfrac{x}{35}$

$\dfrac{7x}{7} = \dfrac{350}{7}$

$x = \textbf{50}$

ft. high

8. $\dfrac{24}{32} = \dfrac{15}{x}$

$\dfrac{24x}{24} = \dfrac{480}{24}$

$x = \textbf{20 ft.}$

9. **(5) B or C.** Choice B and Choice C each satisfy the ASA requirement.

10. $c^2 = a^2 + b^2$

$39^2 = a^2 + 36^2$

$1521 = a + 1296$

$\dfrac{-1296 \qquad -1296}{225 = a^2}$

$\sqrt{225} = a$

$\textbf{15} = a$

11. A is 5 spaces below x-axis. B is 11 spaces above y-axis. Distance AB is $5 + 11 = \textbf{16}$.

12. $A = (x_1, y_1) = (-4, -5)$

$C = (x_2, y_2) = (+8, +11)$

$d = \sqrt{(x_2 - x_1)^2 + (y_2 - y_1)^2}$

$\sqrt{(8 - (-4))^2 + (11 - (-5))^2}$

$= \sqrt{(8 + 4)^2 + (11 + 5)^2}$

$= \sqrt{(12)^2 + (16)^2}$

$= \sqrt{144 + 256}$

$= \sqrt{400}$

$= \textbf{20}$

13. **(2) (4, 8)**
    For Choice (1), $y = 3(3) - 4 = 9 - 4 = 5$.
        Point (3, 6) is not on the graph.
    For Choice (2), $y = 3(4) - 4 = 12 - 4 = 8$.
        Point (4, 8) is on the graph.
    For Choice (3), $y = 3(-1) - 4 = -3 - 4 = -7$.
        Point $(-1, -6)$ is not on the graph.
    For Choice (4), $y = 3(-3) - 4 = -9 - 4 = -13$.
        Point $(-3, -12)$ is not on the graph.
    For Choice (5), $y = 3(-4) - 4 = -12 - 4 = -16$.
        Point $(-4, -10)$ is not on the graph.

14. When $x = 0$, $y = 3(0) - 4 = 0 - 4 = -4$. Coordinates of y-intercept are **(0, −4)**.

15. $P = (x_1, y_1) = (2, 3)$
    $Q = (x_2, y_2) = (8, 7)$
    $m = \dfrac{y_2 - y_1}{x_2 - x_1} = \dfrac{7 - 3}{8 - 2} = \dfrac{4}{6} = \mathbf{\dfrac{2}{3}}$

16. $V = lwh$
    $120 = 8 \cdot 6 \cdot h$
    $\dfrac{120}{48} = \dfrac{48h}{48}$
    2.5 m $= h$

17. $w = x$ $P = 2l + 2w$
    $l = 3x$ $56 = 2(3x) + 2x$
                $56 = 6x + 2x$
                $\dfrac{56}{8} = \dfrac{8x}{8}$
                $7 = x$
                $w = 7$ ft.
                $l = 3 \cdot 7 =$ **21 ft.**

18. $b = 6x$ $A = \frac{1}{2}bh$
    $h = 5x$ $135 = \frac{1}{2} \cdot 6x \cdot 5x$
                $\dfrac{135}{15} = \dfrac{15x^2}{15}$
                $9 = x^2$
                $\sqrt{9} = x$
                $3 = x$
                $b = 6 \cdot 3 =$ **18 ft.**
                $h = 5 \cdot 3 =$ **15 ft.**

19. $w = x$ $V = lwh$
    $h = 6x$ $1500 = 10 \cdot x \cdot 6x$
                $\dfrac{1500}{60} = \dfrac{60x^2}{60}$
                $25 = x^2$
                $\sqrt{25} = x$
                $5 = x$
                $w =$ **5 in.**
                $h = 6 \cdot 5 =$ **30 in.**

20. **(4) $10\sqrt{2}$**
    $$c^2 = a^2 + b^2$$
    $$= 10^2 + 10^2$$
    $$= 100 + 100$$
    $$= 200$$
    $$c = \sqrt{200}$$
    $$= \sqrt{100} \cdot \sqrt{2}$$
    $$= 10\sqrt{2}$$

Unit Test

1. **(2)** *Algebra/Problem Solving.*

   $x - y =$ the difference between hourly pay and hourly day-care charge.
   Multiply by 35 to find the weekly amount.

2. **12** *Geometry/Problem Solving.*

   Use the Pythagorean relationship:
   $c^2 = a^2 + b^2$, so $b^2 = c^2 - a^2$.
   $b^2 = 13^2 - 5^2$
   $ = 169 - 25$
   $ = 144$
   $b = \mathbf{12}$

3. **$6050** *Algebra/Problem Solving.*

   Let $h =$ Mike's salary before the GED.
   $3h = 3$ times his salary; 150 less that amount equals his current salary.
   $3h - 150 = 18,000$
   $3h = 18,150$

   $h = \mathbf{6050}$

4. **(5)** *Arithmetic/Problem Solving.*

   $9 \text{ mo} = \frac{9}{12} = \frac{3}{4} \text{ yr.}$
   $i = prt$
   $ = 21,000 \times \frac{8}{100} \times \frac{3}{4}$

   $ = \mathbf{1260}$

5. **(5)** *Arithmetic/Applications.*

   The amount of their mortgage is not given.

6. **(3)** *Arithmetic/Applications.*

   $i = prt$, so $r = i/pt$.
   $r = 450,000/(300,000 \times 25)$

   $ = .06$, or **6%**

7. **(5)** *Arithmetic/Applications.*

   Each year the number of plantings increases by a power of 16.

8. **15,900** *Arithmetic/Problem Solving.*

   Subtract the area of the house from the area of the lot minus the area of the walkway.
   $A = lw$
   $ = 150 \times 120$
   $ = 18,000$ (area of the lot without walkway)
   $A = lw$
   $ = 70 \times 30$
   $ = 2100$ (area of the house)

   $18,000 - 2100 = \mathbf{15,900}$

9. $\frac{1}{24,986}$ *Arithmetic/Problem Solving.*

   Between 75,000 and 50,001, there are 25,000 possible whole numbers.

   $25,000 - 14 = \mathbf{24,986}$ possible numbers after 14 tries.

10. **$375** *Algebra/Problem Solving.*

    Let $f =$ the price of a regular ticket.
    $\frac{3f}{8} =$ the price of a supersaver ticket
    $32(\frac{3f}{8}) =$ the amount of money collected for supersaver tickets
    $50f =$ the amount of money collected for regular tickets
    $32(\frac{3f}{8}) + 50f = 23,250$
    $12f + 50f = 23,250$
    $62f = 23,250$

    $f = \mathbf{375}$

11. **180** *Algebra/Problem Solving.*

    Let $a =$ attempts.
    $6:9 = 120:a$
    $6a = 1080$

    $a = \mathbf{180}$

12. **20** *Arithmetic/Applications.*

    The decimal would be moved 23 places; 6023 has the decimal already moved 3 places.

13. **(5)** *Arithmetic/Problem Solving.*

    The information in the graph represents a national average and cannot be applied to an individual company.

14. **(3)** *Arithmetic/Problem Solving.*

    $39.1 + 10.4 + 2.7 = 52.2\%$ of workers age 35 and over; $52.2\% = .522$.

    $110 \times .522 = \mathbf{57.42}$

15. **(5)** *Algebra/Problem Solving.*

    $.95b =$ Ed's salary
    $b + .95b =$ their combined salaries

    $4(b + .95b) =$ amount to spend

16. **(3)** *Algebra/Problem Solving.*

    $(e - d) =$ profit on the house

17. **$663** *Arithmetic/Applications.*

    Tax = $2339

    $3002 - 2339 = \mathbf{663}$

18. **(2)** *Geometry/Skills.*

    Lines perpendicular to the same line are parallel to each other.

19. **4 or 1** *Algebra/Applications.*

    Use substitution method or factor to $(n - 4)(n - 1)$.

20. **$4yz^4$** *Algebra/Applications.*

    48 divided by 12 = 4.
    Subtract exponents.

21. **(5)** *Geometry/Applications.*

Use points E $(-6, 11)$ and F $(15, -10)$.

$$m = \frac{y_2 - y_1}{x_2 - x_1}$$

$$m = \frac{-10 - 11}{15 - (-6)}$$

$$= \frac{-21}{21}$$

$$= -1$$

22. **(3)** *Geometry/Applications.*

$B = (-8, 3)$
$E = (-6, 11)$

$$d = \sqrt{(x_2 - x_1)^2 + (y_2 - y_1)^2}$$

$$= \sqrt{(-6 - (-8))^2 + (11 - 3)^2}$$

$$= \sqrt{(+2)^2 + (+8)^2}$$

$$= \sqrt{4 + 64}$$

$$= \sqrt{68}$$

23. **7.2** *Geometry/Applications.*

If s = side, then $12:20 = s:12$.

$20s = 144$

$s = 7.2$

24. **49** *Geometry/Skills.*

The bracing forms an isosceles triangle; therefore, both base angles are equal
$180 - 82 = 98$, the sum of the base angles.

25. **2599.92** *Arithmetic/Problem Solving.*

$27 - 4 = 23$

$V = \pi r^2 h$

$$= 3.14 \times 6 \times 6 \times 23 = \mathbf{2599.92}$$

26. **(3)** *Algebra/Problem Solving.*

27. **(4)** *Arithmetic/Applications.*

a. $\frac{1}{4} = \frac{8}{32}$

b. $\frac{5}{8} = \frac{20}{32}$

c. $\frac{9}{16} = \frac{18}{32}$

d. $\frac{1}{2} = \frac{16}{32}$

e. $\frac{3}{4} = \frac{24}{32}$

f. $\frac{11}{32} = \frac{11}{32}$

28. **1 hour 40 minutes** *Arithmetic/Applications.*

10,000 m = 10 km

$d = rt$, so $t = d/r$.

$t = \frac{10}{6} = 1\frac{2}{3}$ hr. = **1 hr. 40 min.**

Practice Items

1. **(2)** *Arithmetic/Applications.*

$$\begin{array}{r} 150 \\ -\ 85 \\ \hline 65 \end{array}$$ strike:not strike = 85:65 = **17:13**

2. **(5)** *Arithmetic/Applications.*

$\left(\frac{2}{3}\right)^3 = \frac{2}{3} \times \frac{2}{3} \times \frac{2}{3} = \frac{8}{27}$

3. **(3)** *Arithmetic/Applications.*

$$\begin{array}{r} 20{,}320 \\ 18{,}008 \\ +\ 17{,}400 \\ \hline 55{,}728 \end{array} \qquad \begin{array}{l} \mathbf{18{,}576\ ft.} \\ 3\overline{)55{,}728} \end{array}$$

4. **(2)** *Arithmetic/Applications.*

$\dfrac{\text{Smiths}}{\text{total}}$ $\dfrac{10}{500} = \dfrac{1}{50}$

5. **(3)** *Arithmetic/Applications.*

$$\begin{array}{r} 8 \\ 12 \\ +10 \\ \hline 30 \end{array} \qquad \begin{array}{l} \text{3 families} \\ \text{total} \end{array} \qquad \dfrac{30}{500} = \dfrac{3}{50}$$

6. **(4)** *Arithmetic/Applications.* **9 × 10 × 10 × 10 × 10**

7. **(2)** *Arithmetic/Problem solving.*

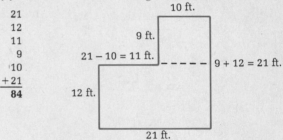

$$\begin{array}{r} 21 \\ 12 \\ 11 \\ 9 \\ 10 \\ +21 \\ \hline \mathbf{84} \end{array}$$

8. **(4)** *Arithmetic/Problem solving.*

Area of top rectangle: $A = lw$
$A = 10 \times 9 = 90$ ft.^{2}

Area of bottom rectangle: $A = lw$
$A = 21 \times 12 = 252$ ft.^{2}

Sum: $90 + 252 = \mathbf{342\ ft.^2}$

9. **(2)** *Arithmetic/Applications.*

$A = s^2$
$A = (4\frac{1}{2})^2$
$A = 4\frac{1}{2} \times 4\frac{1}{2}$
$A = \frac{9}{2} \times \frac{9}{2} = \frac{81}{4} = \mathbf{20\frac{1}{4}\ in.^2}$

10. **(1)** *Arithmetic/Applications.*

$$\begin{array}{r} 4.1\ \text{cm} \\ -\ 1.6 \\ \hline \mathbf{2.5\ cm} \end{array}$$

11. **(5)** *Arithmetic/Applications.*

$V = lwh$
$V = 6.5 \times 4 \times 20 = \mathbf{520\ cm^3}$

12. **(2)** *Arithmetic/Applications.* **B, A, D, E, C**
    A = 0.850 m
    B = 0.095 m
    C = 1.200 m
    D = 0.900 m
    E = 1.070 m

13. **(1)** *Graphs.* 26.5% is close to 25% = $\frac{25}{100} = \frac{1}{4}$

14. **(4)** *Graphs.*

    $$\begin{array}{r} 11.9\% \\ +26.5 \\ \hline 38.4\% \end{array} \qquad \begin{array}{r} 61.6\% \\ -38.4 \\ \hline \mathbf{23.2\%} \end{array}$$

15. **(5)** *Graphs.* **Insufficient data is given to solve the problem.** You do not know what percent of the population is 18 to 40.

16. **(3)** *Graphs.*

    $$26.5\% = .265 \qquad \begin{array}{r} .265 \\ \times 300{,}000 \\ \hline \mathbf{79{,}500.000} \end{array}$$

17. **(1)** *Arithmetic/Problem solving.* **600 + 60 × 80**

18. **(4)** *Arithmetic/Problem solving.*

    $$\begin{array}{r} \mathbf{20} \\ 150\overline{)3000} \end{array}$$

19. **(2)** *Arithmetic/Problem solving.*

    $$5\% = .05 \qquad \begin{array}{r} 120{,}000 \\ \times .05 \\ \hline 6000.00 \end{array} \qquad \frac{1}{3} \times \frac{6000}{1} = \mathbf{2000}$$

20. **(5)** *Arithmetic/Problem solving.* **Insufficient data is given to solve the problem.** You do not know the relationship between the new jobs and the change in population.

21. **(3)** *Arithmetic/Problem solving.* In two years, Elk Electronics will have 2 × 2000 = 4000 jobs.

    $$2\% = .02 \qquad \begin{array}{r} 120{,}000 \\ \times .02 \\ \hline 2400.00 \end{array} \text{ new jobs at Paulson's}$$

    $$\begin{array}{r} 4{,}000 \text{ jobs at Elk} \\ 2{,}400 \text{ jobs at Paulson's} \\ +120{,}000 \text{ existing jobs} \\ \hline \mathbf{126{,}400} \text{ total} \end{array}$$

22. **(2)** *Arithmetic/Problem solving.*

    $$\begin{array}{r} 2000 \text{ new jobs at Elk Electronics} \\ -\ 700 \text{ lost jobs} \\ \hline \mathbf{1300 \text{ more jobs}} \text{ is the net change.} \end{array}$$

23. **(5)** *Arithmetic/Problem solving.* $\dfrac{\mathbf{20 \times 6}}{\mathbf{3}}$

24. **(4)** *Arithmetic/Problem solving.*

    $$80\% = .08 \qquad \begin{array}{r} 20\ 0 \\ .8\overline{)160.0} \end{array} \qquad \begin{array}{r} \text{old weight} \quad 200 \\ \text{current weight} \ -160 \\ \hline \mathbf{40 \text{ lb. lost}} \end{array}$$

25. **(2)** *Arithmetic/Problem solving.*

    $$\begin{array}{l} \text{won} \quad 5 \\ \text{lost} \ \underline{+2} \\ \text{played} \ \ 7 \end{array} \qquad \begin{array}{l} \text{won} \quad \frac{5}{7} = \frac{x}{49} \\ \text{played} \end{array}$$

    $$\frac{7x}{7} = \frac{245}{7}$$
    $$x = \mathbf{35 \text{ won}}$$
    $$49 - 35 = \mathbf{14 \text{ lost}}$$

    (3) *Arithmetic/Problem solving.* **4(40) + 5(55)**

    (3) *Arithmetic/Applications.*

    $$\begin{array}{lr} \text{first oz.} & = \$.22 \\ \text{next 3 oz.} = 3 \times \$.17 = & .51 \\ \hline \text{total} & \mathbf{\$.73} \end{array}$$

28. **(2)** *Arithmetic/Applications.*

    $$\begin{array}{lr} \text{first oz.} & = \$\ .22 \\ \text{second oz.} & = \ \ \ .17 \\ \text{certified fee} & = \ \ \ .75 \\ \hline \text{total} & \mathbf{\$1.14} \end{array}$$

29. **(3)** *Algebra/Skills.* **C**

30. **(4)** *Algebra/Applications.*

    $$\begin{array}{rcl} 8a - 2 &=& 6 \\ +\ 2 & & +2 \\ \hline \frac{8a}{8} &=& \frac{8}{8} \\ a &=& \mathbf{1} \end{array}$$

31. **(1)** *Algebra/Applications.* **18m + 24 = 6(3m + 4)**

32. **(2)** *Algebra/Problem solving.* **4x + 2 = 3x + 11**

33. **(5)** *Algebra/Applications.*

    $$\begin{array}{rcl} 5y - 8 &=& 4y + 1 \\ -4y & & -4y \\ \hline y - 8 &=& 1 \\ +\ 8 & & +8 \\ \hline y &=& \mathbf{9} \end{array}$$

34. **(3)** *Algebra/Skills.*

    $$\frac{-8a^2}{-12a} - \frac{2a}{3}$$

35. **(5)** *Algebra/Applications.* **2x + 1**

36. **(4)** *Algebra/Applications.* **x² + 12x + 35 = (x + 7) (x + 5)**

37. **(2)** *Algebra/Applications.* **2(x − 6)**

38. **(3)** *Algebra/Applications.*

    $$\begin{array}{rcl} -2 &=& 4(3d - 2) \\ -2 &=& 12d - 8 \\ +8 & & +\ 8 \\ \hline \frac{6}{12} &=& \frac{12d}{12} \\ \frac{1}{2} &=& d \end{array}$$

39. **(1)** *Algebra/Problem solving.* $\dfrac{x}{2} - 1 = \dfrac{x}{3} + 5$

40. **(3)** *Algebra/Skills.* 12x − (−3x) = 12x + 3x = **15x**

41. **(5)** *Algebra/Problem Solving.*

    $$\begin{array}{rcl} 3x - 1 &=& 2(x + 5) \\ 3x - 1 &=& 2x + 10 \\ -2x & & -2x \\ \hline x - 1 &=& 10 \\ +\ 1 & & +1 \\ \hline x &=& \mathbf{11} \end{array}$$

42. **(4)** *Algebra/Applications.*

    $$\begin{array}{rcl} 3p - 2 &>& p + 8 \\ -p & & -p \\ \hline 2p - 2 &>& 8 \\ +\ 2 & & +2 \\ \hline \frac{2p}{2} &>& \frac{10}{2} \\ p &>& \mathbf{5} \end{array}$$

43. **(4)** *Algebra/Problem solving.*

    $$\begin{array}{ll} \text{Deborah's hours} = x & x + x + 4 + 2x = 28 \\ \text{Jeff's} \qquad\quad = x + 4 & \\ \text{Caroline's} \qquad = 2x & 4 + 4x = 28 \\ & \ \underline{-4 \qquad\quad -4} \\ & \frac{4x}{4} = \frac{24}{4} \\ & x = 6 \end{array}$$

    $$\text{Caroline's hours} \qquad = 2(6) = \mathbf{12}$$

44. **(3)** *Algebra/Applications.*

$$x^2 - 36 = 0$$
$$(x + 6)(x - 6) = 0$$

$$\begin{array}{ll} x + 6 = 0 & x - 6 = 0 \\ \underline{-6 \quad -6} & \underline{+6 \quad +6} \\ x = -6 \text{ and } x = +6 \end{array}$$

45. **(5)** *Geometry/Skills.*

$$\begin{array}{r} 180° \\ -\ 84° \\ \hline \mathbf{96°} \end{array}$$

46. **(2)** *Geometry/Skills.*

$$\begin{array}{rr} 62° & 180° \\ +\ 79° & -141° \\ \hline 141° & \mathbf{39°} = \angle BOC \end{array}$$

47. **(4)** *Geometry/Skills.*

$$\begin{array}{rrl} \angle P = & 59° & 180° \\ \angle Q = & \underline{64°} & \underline{-123°} \\ & 123° & \mathbf{57°} = \angle R \end{array}$$

Since $\angle Q$ is largest, side **PR** which is opposite is the longest side.

48. **(3)** *Geometry/Applications.*

$$c^2 = a^2 + b^2$$
$$c^2 = (3.2)^2 + (2.4)^2$$
$$c^2 = 10.24 + 5.76$$
$$c^2 = 16$$
$$c = \sqrt{16}$$
$$c = \mathbf{4}$$

49. **(4)** *Geometry/Applications.*

$$\begin{array}{l} \text{height} \\ \text{base} \end{array} \quad \frac{5}{2} = \frac{x}{30}$$
$$\frac{2x}{2} = \frac{150}{2}$$
$$x = \mathbf{75}$$

50. **(3)** *Geometry/Skills.*

$$\begin{array}{rr} 180° & \mathbf{66°} \\ -\ 48° & 2\overline{)132°} \\ \hline 132° & \end{array}$$

51. **(5)** *Geometry/Applications.*

$$F = (x_1, y_1) = (-2, -1)$$
$$G = (x_2, y_2) = (4, 2)$$
$$m = \frac{y_2 - y_1}{x_2 - x_1} = \frac{2 - (-1)}{4 - (-2)} = \frac{2 + 1}{4 + 2} = \frac{3}{6} = \frac{\mathbf{+1}}{\mathbf{2}}$$

52. **(2)** *Geometry/Applications.*

A is 2 spaces left of y-axis.
B is 4 spaces right of y-axis.
Distance AB is $2 + 4 = \mathbf{6.}$

53. **(4)** *Geometry/Applications.*

$$A = (x_1, y_1) = (-2, 5)$$
$$C = (x_2, y_2) = (4, -3)$$
$$M = \left(\frac{x_1 + x_2}{2}, \frac{y_1 + y_2}{2}\right)$$
$$M = \left(\frac{-2 + 4}{2}, \frac{5 - 3}{2}\right) = \left(\frac{+2}{2}, \frac{+2}{2}\right) = \mathbf{(1, 1)}$$

54. **(1)** *Geometry/Applications.*

Let $x = 0$ in $y = -x + 3$
$$y = -0 + 3$$
$$y = +3$$

Coordinates of the y-intercept are **(0, 3).**

55. **(3)** *Geometry/Problem solving.*

$$\begin{array}{ll} w = x & P = 2l + 2w \\ l = x + 7 & 66 = 2(x + 7) + 2x \\ & 66 = 2x + 14 + 2x \\ & 66 = 4x + 14 \\ & \underline{-14 \qquad -14} \\ & \frac{52}{4} = \frac{4x}{4} \\ & 13 = x \\ & l = 13 + 7 = \mathbf{20} \end{array}$$

56. **(2)** *Geometry/Problem solving.*

$$\begin{array}{ll} A = lw & A = s^2 \\ A = 20 \times 7.2 & 144 = s^2 \\ A = 144 & \sqrt{144} = s \\ & \mathbf{12} = s \end{array}$$

Practice Test

1. **(5)** *Arithmetic/Applications.*

   1 yd. = 36 in.   21 in.: 36 in. = **7:12**

2. **(4)** *Algebra/Skills.* **D**

3. **(3)** *Arithmetic/Applications.* In order: 23, 29, 32, 41, 56. The middle value is **32.**

4. **(1)** *Geometry/Skills.*

$$\begin{array}{rr} 72° & 180° \\ \times\ 2 & -144° \\ \hline 144° & \mathbf{36°} \end{array}$$

5. **(2)** *Algebra/Applications.* $\mathbf{2n - 7}$

6. **(5)** *Arithmetic/Applications.* $\mathbf{7 \times 10 \times 10 \times 10}$

7. **(4)** *Geometry/Skills.*

$$\begin{array}{r} 180° \\ -121° \\ \hline \mathbf{59°} \end{array}$$

8. **(3)** *Arithmetic/Problem solving.*

$$\begin{array}{r} 12.5 \\ 60\overline{)750.0} \\ \underline{60} \\ 150 \\ \underline{120} \\ 30\ 0 \\ \underline{30\ 0} \end{array}$$

9. **(1)** *Arithmetic/Problem solving.* $\mathbf{12(1800 - 400)}$

10. **(2)** *Arithmetic/Problem solving.*

$$P = 2l + 2w$$
$$P = 2(45) + 2(30)$$
$$P = 90 + 60 = \mathbf{150 \text{ ft.}}$$

11. (2) *Arithmetic/Problem solving.*

Area of large rectangle: $A = lw = 75 \times 50 = 3750$ ft.
Area of pool: $A = lw = 45 \times 30 = 1350$ ft.
Difference = area of walkway   = **2400** ft.

12. (3) *Arithmetic/Problem solving.*

$$\begin{array}{r} 13.4 \\ -10.7 \\ \hline \mathbf{2.7} \end{array}$$

13. (4) *Arithmetic/Problem solving.* **United States and Australia**

14. (4) *Arithmetic/Problem solving.*

voters   3   voters $\dfrac{3}{5} = \dfrac{5400}{x}$
nonvoters $+2$ registered
registered $\overline{5}$

$$\dfrac{3x}{3} = \dfrac{27,000}{3}$$
$$x = \textbf{9000 registered}$$

15. (5) *Arithmetic/Problem solving.* **2(4.5) + 4(3.5)**

16. (4) *Arithmetic/Problem solving.*

$8\% = .08$

$$\begin{array}{r} \$1350 \\ \times .08 \\ \hline \$108.00 \end{array} \quad \begin{array}{r} \$1350 \\ +108 \\ \hline \$1458 \end{array} \quad \begin{array}{r} \$1,458 \\ \times 12 \\ \hline 2\,916 \\ 14\,58 \\ \hline \mathbf{\$17,496} \end{array}$$

17. (5) *Arithmetric/Problem solving.*

$10\% = .10$

$$\begin{array}{r} \mathbf{\$1350} \\ \times .10 \\ \hline \mathbf{\$135.00} \end{array}$$

18. (5) *Arithmetic/Problem solving.* **Insufficient data is given to solve the problem.** You do not know the amount that goes into the pension plan.

19. (2) *Arithmetic/Problem solving.*

$$\begin{array}{r} \$1350 \\ +135 \\ \hline \$1485 \end{array} \quad \begin{array}{r} \$1,485 \\ \times 12 \\ \hline 2\,970 \\ 14\,85 \\ \hline \$17,820 \end{array} \quad \begin{array}{r} 12 \\ \times \$40 \\ \hline \$480 \end{array} \quad \begin{array}{r} \$17,820 \\ -480 \\ \hline \mathbf{\$17,340} \end{array}$$

20. (3) *Algebra/Skills.* $4ab - 6ab + 7ab = 11ab - 6ab = $ **5ab**

21. (3) *Geometry/Skills.*

$$\begin{array}{r} 90° \\ -23° \\ \hline \mathbf{67°} \end{array}$$

22. (4) *Algebra/Applications.*

$$\begin{array}{rl} \frac{3}{5}c + 1 =& 13 \\ -1 & -1 \\ \hline \frac{5}{3} \cdot \frac{3}{5}c =& 12 \cdot \frac{5}{3} \\ c =& \mathbf{20} \end{array}$$

23. (4) *Arithmetic/Applications.*

$A = \frac{1}{2} bh$
$A = \frac{1}{2} \times 2\frac{3}{4} \times 1\frac{1}{2}$
$A = \frac{1}{2} \times \frac{11}{4} \times \frac{3}{2} = \frac{33}{16} = \mathbf{2\frac{1}{16}}$ **in.²**

24. (2) *Algebra/Applications.* $\dfrac{\mathbf{5x + 1}}{\mathbf{4}}$

25. (5) *Algebra/Skills.*

$\dfrac{-4xy}{12x} = \dfrac{-y}{3}$

26. (4) *Graphs.* **1983**

27. (5) *Graphs.* **Insufficient data is given to solve the problem.** You do not know the value of the cassettes that were sold.

28. (1) *Graphs.* **The number of cassettes sold in 1984 was about three times the number sold in 1980.** The number of cassettes sold in 1980 was a little over 100 million. The number sold in 1984 was a little over 300 million. $\frac{300}{100} = \mathbf{3}.$

29. (4) *Graphs.*

$$\begin{array}{r} 320 \text{ million} \\ -110 \\ \hline 210 \text{ million, which is about } \mathbf{200 \text{ million}} \end{array}$$

30. (1) *Arithmetic/Applications.*

$$\begin{array}{rl} 3\frac{1}{4} =& 3\frac{2}{8} = 2\frac{2}{8} + \frac{8}{8} = 2\frac{10}{8} \\ -1\frac{7}{8} & = 1\frac{7}{8} \\ \hline & \mathbf{1\frac{3}{8}} \textbf{ in.} \end{array}$$

31. (2) *Algebra/Applications.* $24y - 32 = 8(3y - 4)$

32. (4) *Arithmetic/Applications.*

$V = \pi r^2 h$
$= 3.14 \,(2.5)^2 \times 4$
$= 3.14 \times 2.5 \times 2.5 \times 4$
$= \mathbf{78.5 \; m^3}$

33. (3) *Algebra/Problem solving.* $2(x - 5) = 12$

34. (4) *Arithmetic/Problem solving.*

right   48   right $\dfrac{48}{60} = \dfrac{4}{5}$ $\dfrac{4}{5} \times \dfrac{100\%}{1} = \mathbf{80\%}$
wrong $+12$ total
total $\overline{60}$

35. (2) *Algebra/Applications.*

$$\begin{array}{rl} 3(e + 4) =& 6 \\ 3e + 12 =& 6 \\ -12 =& -12 \\ \hline \frac{3e}{3} =& \frac{-6}{3} \\ e =& \mathbf{-2} \end{array}$$

36. (4) *Geometry/Applications.*

height $\dfrac{8}{3} = \dfrac{x}{48}$
shadow

$$\dfrac{3x}{3} = \dfrac{384}{3}$$
$$x = \mathbf{128 \text{ ft.}}$$

37. (5) *Algebra/Applications.* $x^2 - 100 = (x + 10)(x - 10)$

38. (2) *Algebra/Applications.*

$$\begin{array}{rl} 5(t + 1) <& 20 \\ 5t + 5 <& 20 \\ -5 & -5 \\ \hline \frac{5t}{5} <& \frac{15}{5} \\ t <& \mathbf{3} \end{array}$$

39. (1) *Arithmetic/Applications.* **D, B, E, A, C**

$A = 1\frac{15}{16}$ lb.

$B = 2\frac{7}{16}$ lb.

$C = 1\frac{3}{4} = 1\frac{12}{16}$ lb.

$D = 2\frac{1}{2} = 2\frac{8}{16}$ lb.

$E = 2\frac{3}{8} = 2\frac{6}{16}$ lb.

40. **(3)** *Arithmetic/Applications.*

$(.4)^2 = .4 \times .4 = .16$

$(.01)^2 = .01 \times .01 = .0001$

$(.4)^2 + (.01)^2 = .16 + .0001 = \textbf{.1601}$

41. **(4)** *Arithmetic/Applications.*

$\begin{array}{r} 60 \\ 48 \\ +\,36 \\ \hline 144 \end{array}$
tomato $\dfrac{60}{\text{total}\ 144} = \dfrac{5}{12}$

42. **(3)** *Arithmetic/Applications.*

now: $36 - 6 = 30$ cans of chicken soup

$144 - 12 = 132$ total

$\dfrac{\text{chicken}\ 30}{\text{total}\ 132} = \dfrac{5}{22}$

43. **(4)** *Algebra/Problem solving.* $3 + 6 = 4$

44. **(1)** *Geometry/Problem solving.*

$w = x \qquad\qquad A = lw$

$l = 2x \qquad\qquad 800 = 2x \cdot x$

$\qquad\qquad\qquad \dfrac{800}{2} = \dfrac{2x^2}{2}$

$\qquad\qquad\qquad 400 = x^2$

$\qquad\qquad\qquad \sqrt{400} = x$

$\qquad\qquad\qquad \textbf{20} = x$

45. **(2)** *Algebra/Applications.*

$\begin{array}{rcl} 7s + 1 &=& 3s + 3 \\ -3s && -3s \\ \hline 4s + 1 &=& 3 \\ -1 && -1 \\ \hline 4s &=& 2 \\ \hline 4 && 4 \\ s &=& \tfrac{1}{2} \end{array}$

46. **(2)** *Geometry/Skills.* Since side *XZ* is longest, ∠**Y**, which is opposite side *XZ*, is the largest angle.

47. **(3)** *Algebra/Problem solving.*

$\begin{array}{rcl} 5x + 2 &=& x + 30 \\ -x && -x \\ \hline 4x + 2 &=& 30 \\ -2 && -2 \\ \hline 4x &=& 28 \\ \hline 4 && 4 \\ x &=& 7 \end{array}$

48. **(3)** *Geometry/Applications.*

$c^2 = a^2 + b^2$

$17^2 = a^2 + 8^2$

$289 = a^2 + 64$

$\underline{-64 = -64}$

$225 = a^2$

$\sqrt{225} = a$

$\textbf{15} = a$

49. **(1)** *Geometry/Applications.* **A only.** This satisfies the SSS requirement.

50. **(2)** *Algebra/Applications.*

$y^2 + y - 12 = 0$

$(y + 4)(y - 3) = 0$

$\begin{array}{rcl} y + 4 = 0 & & y - 3 = 0 \\ \underline{-4 \quad -4} & & \underline{+3 \quad +3} \\ y = -4 & \text{and} & y = +3 \end{array}$

51. **(2)** *Geometry/Applications.*

$A = (x_1, y_1) = (-1, 7) \qquad m = \dfrac{y_2 - y_1}{x_2 - x_1}$

$B = (x_2, y_2) = (4, 2)$

$m = \dfrac{2 - 7}{4 - (-1)} = \dfrac{-5}{4 + 1} = \dfrac{-5}{5} = \textbf{-1}$

52. **(4)** *Arithmetic/Problem solving.* $\dfrac{3000}{15 \times 60}$

53. **(4)** *Algebra/Problem solving.*

car payments $= x$		$x + 2x + 3x = 696$
food $= 2x$		$\dfrac{6x}{6} = \dfrac{696}{6}$
mortgage $= 3x$		$x = 116$

mortgage $= 3(116) = \textbf{348}$

54. **(5)** *Geometry/Applications.*

$G = (x_1, y_1) = (-5, -3) \qquad H = (x_2, y_2) = (7, 6)$

$d = \sqrt{(x_2 - x_1)^2 + (y_2 - y_1)^2}$

$d = \sqrt{(7 - (-5))^2 + (6 - (-3))^2}$

$d = \sqrt{(7 + 5)^2 + (6 + 3)^2}$

$d = \sqrt{(12)^2 + (9)^2}$

$d = \sqrt{144 + 81}$

$d = \sqrt{225}$

$d = \textbf{15}$

55. **(3)** *Geometry/Applications.*

$G = (x_1, y_1) = (-5, -3) \qquad M = \left(\dfrac{x_1 + x_2}{2}, \dfrac{y_1 + y_2}{2}\right)$

$H = (x_2, y_2) = (7, 6)$

$M = \left(\dfrac{-5 + 7}{2}, \dfrac{-3 + 6}{2}\right) = \left(\dfrac{2}{2}, \dfrac{3}{2}\right) = \textbf{(1, } 1\tfrac{1}{2}\textbf{)}$

56. **(2)** *Geometry/Applications.*

H is 6 spaces above the x-axis.

I is 3 spaces below the x-axis.

Distance *HI* is $6 + 3 = \textbf{9}$.

Simulated Test

1. **(4)** *Arithmetic/Applications.*

$10^4 = 10 \times 10 \times 10 \times 10 = 10,000$

$5^2 = 5 \times 5 = 25$

$10^4 - 5^2 = 10,000 - 25 = \textbf{9975}$

2. **(2)** *Arithmetic/Problem solving.*

$\begin{array}{r} \textbf{25 wk.} \\ 180\overline{)4500} \\ \underline{360} \\ 900 \\ \underline{900} \end{array}$

3. **(2)** *Algebra/Skills.* **B**

4. **(5)** *Arithmetic/Problem solving.*

green	3	green	$\dfrac{3}{7}$	$= \dfrac{x}{56}$
white	$+4$	total		
total	7		$\dfrac{7x}{7}$	$= \dfrac{168}{7}$

$x = \textbf{24}$ gallons of green paint

5. **(1)** *Algebra/Applications.* $\dfrac{3x + 8}{6}$

6. **(4)** *Arithmetic/Applications.*

$15	6% = .06	$45
$\times 3$		$\times .06$
$45		**$2.70**

7. **(3)** *Algebra/Applications.* $\frac{1}{2}c - 20$

8. **(5)** *Algebra/Applications.*

$$\frac{2}{3}b - 2 = 8$$
$$\underline{+ 2 \quad +2}$$
$$\frac{3}{2} \cdot \frac{2}{3}b = 10 \cdot \frac{3}{2}$$
$$b = \mathbf{15}$$

9. **(3)** *Geometry/Skills.*

$\angle M =$	61°	180°
$\angle N =$	52°	$-113°$
	113°	$67° = \angle O$

Since $\angle O$ is largest, side **MN**, which is opposite $\angle O$, is the longest side.

10. **(4)** *Algebra/Skills.*

$\dfrac{15c^2d}{-25cd} = \dfrac{-3c}{5}$

11. **(2)** *Algebra/Skills.* $(-5m^3)(-2m) = \mathbf{+10m^4}$

12. **(3)** *Algebra/Problem solving.*

$$2x - 7 = x + 3$$
$$\underline{-x \qquad -x}$$
$$x - 7 = 3$$
$$\underline{+ 7 \qquad +7}$$
$$x = \mathbf{10}$$

13. **(3)** *Arithmetic/Applications.*

3.60
2.45
4.63
+ 3.84
14.52

3.63 in.
4)14.52

14. **(2)** *Arithmetic/Applications.*

15	large	$\dfrac{20}{60} = \dfrac{1}{3}$
25	total	
+20		
60		

15. **(3)** *Arithmetic/Applications.*

now: 15 small   small $\dfrac{15}{55} = \dfrac{3}{11}$
23 medium   total
+17 large
55

16. **(2)** *Arithmetic/Problem solving.* **7(6) + 3(9)**

17. **(2)** *Geometry/Skills.*

180°
− 55°
125°

62.5°
2)125.0°

18. **(4)** *Arithmetic/Problem solving.* **1980**

19. **(1)** *Arithmetic/Problem solving.* **While the number of newspapers gradually decreased, the number of periodicals increased.**

20. **(3)** *Arithmetic/Applications.* **4 × 10 × 10 × 10 × 10 × 10 × 10**

21. **(2)** *Arithmetic/Problem solving.* $\dfrac{12,000}{8 \times 100}$

22. **(3)** *Arithmetic/Problem solving.*

$24	$384
$\times 16$	$+100$
144	**$484**
24	
$384	

23. **(1)** *Arithmetic/Problem solving.*

$36	$432	$492
$\times 12$	$+60$	-439
72	$492	**$ 53**
36		
$432		

24. **(5)** *Arithmetic/Problem solving.* **Insufficient data is given to solve the problem.** You do not know the cost of the guarantee.

25. **(1)** *Arithmetic/Problem solving.*

5% = .05	$389	$389.00
	$\times .05$	19.45
	$19.45	+ 10.00
		$418.45

26. **(4)** *Algebra/Applications.*

$$2n - 7 \geq 5$$
$$\underline{+ 7 \quad +7}$$
$$\frac{2n}{2} \geq \frac{12}{2}$$
$$n \geq \mathbf{6}$$

27. **(1)** *Arithmetic/Applications.* **D, C, A, E, B**

A = 0.650 kg
B = 0.500 kg
C = 1.090 kg
D = 1.450 kg
E = 0.505 kg

28. **(3)** *Algebra/Problem solving.* $\dfrac{x - 5}{2} = 7$

29. **(4)** *Arithmetic/Problem solving.* **2(35) + 3(55)**

30. **(5)** *Algebra/Applications.*

$$8z - 1 = 6z + 4$$
$$\underline{-6z \qquad -6z}$$
$$2z - 1 = 4$$
$$\underline{+ 1 \qquad +1}$$
$$\frac{2z}{2} = \frac{5}{2}$$
$$z = 2\frac{1}{2}$$

31. **(2)** *Graphs.* **$270**

32. **(5)** *Graphs.* **Insufficient data is given to solve the problem.**

You do not know what part of earnings went to taxes and social security.

33. **(3)** *Graphs.* **Gross earnings more than doubled from 1970 to 1985.**

In 1970, the current dollar earnings were almost $200 per week. In 1985, the current dollar earnings were more than $400 per week. $\frac{400}{200} = 2$.

34. **(1)** *Graphs.* **1985 earnings were less than 1970 earnings.**

35. **(1)** *Arithmetic/Applications.*

5.2 cm
− 2.9
2.3 cm

36. **(1)** *Arithmetic/Problem solving.*

$$90 \text{ ft.} = \frac{90}{3} = 30 \text{ yd.} \qquad P = 2l + 2w$$
$$36 \text{ ft.} = \frac{36}{3} = 12 \text{ yd.} \qquad P = 2(30) + 2(12)$$
$$P = 60 + 24 = \textbf{84 yd.}$$

37. **(2)** *Arithmetic/Problem solving.*

$$A = lw$$
$$A = 90 \times 36$$
$$A = 3240 \text{ ft.}^2$$

$$\begin{array}{r} 10.8 \\ 300\overline{)3240.0} \\ \underline{300} \\ 240 \\ \underline{0} \\ 240\ 0 \\ \underline{240\ 0} \end{array}$$

38. **(3)** *Algebra/Applications.*

$$5(t - 3) = 2(t + 6)$$
$$5t - 15 = 2t + 12$$
$$\underline{-2t} \qquad \underline{-2t}$$
$$3t - 15 = 12$$
$$\underline{+ 15} \qquad \underline{+15}$$
$$\frac{3t}{3} = \frac{27}{3}$$
$$t = \textbf{9}$$

39. **(5)** *Geometry/Skills.* **106°** Alternate exterior angles are equal.

40. **(2)** *Arithmetic/Applications.*

$$V = s^3 = (1.5)^3 =$$

$$\begin{array}{r} 1.5 \\ \times 1.5 \\ \hline 7\ 5 \\ 1\ 5 \\ \hline 2.2\ 5 \\ \times\ 1.5 \\ \hline 1\ 1\ 2\ 5 \\ 2\ 2\ 5 \\ \hline 3.3\ 7\ 5 \end{array}$$

to the nearest tenth = **3.4** m^3

41. **(4)** *Arithmetic/Applications.*

$$\begin{array}{l} \$40,000 \\ \underline{-\ 16,000} \\ \$24,000 \text{ owed} \end{array} \qquad \begin{array}{l} \text{paid:owed} = 16,000{:}24,000 \\ = \textbf{2:3} \end{array}$$

42. **(5)** *Geometry/Applications.*

$$c^2 = a^2 + b^2$$
$$c^2 = 48^2 + 36^2$$
$$c^2 = 2304 + 1296$$
$$c^2 = 3600$$
$$c \quad \sqrt{3600}$$
$$c = \textbf{60 m}$$

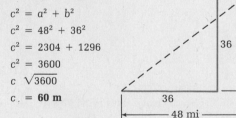

43. **(4)** *Algebra/Applications.* $x^2 + 2x - 48 =$ **(x + 8) (x − 6)**

44. **(3)** *Arithmetic/Applications.*

$$A = \pi r^2$$
$$= 3.14\ (6.5)^2$$
$$= 3.24 \times 6.5 \times 6.5$$
$$= 132.665 \text{ to the nearest tenth} = \textbf{132.7 m}^2$$

45. **(2)** *Geometry/Skills.*

$$\begin{array}{r} 27° \\ \underline{+29°} \\ 56° \end{array} \qquad \begin{array}{r} 90° \\ \underline{-56°} \\ \textbf{34°} = \angle ROS \end{array}$$

46. **(5)** *Geometry/Applications.*

$$\begin{array}{l} \text{short leg} \\ \text{long leg} \end{array} \quad \frac{3}{7} = \frac{42}{x}$$
$$\frac{3x}{3} = \frac{294}{3}$$
$$x = \textbf{98}$$

47. **(1)** *Algebra/Applications.* $a^2 - 12a = \textbf{\textit{a}(\textit{a} − 12)}$

48. **(2)** *Geometry/Applications.*

$$S = (x_1, y_1) = (-4, -3) \qquad T = (x_2, y_2) = (2, 7)$$
$$m = \frac{y_2 - y_1}{x_2 - x_1}$$
$$m = \frac{7 - (-3)}{2 - (-4)} = \frac{7 + 3}{2 + 4} = \frac{10}{6} = \frac{\textbf{+5}}{\textbf{3}}$$

49. **(4)** *Geometry/Problem solving.*

$$b = x \qquad A = \tfrac{1}{2}bh$$
$$h = \tfrac{x}{3} \qquad 54 = \tfrac{1}{2} \cdot x \cdot \tfrac{x}{3}$$
$$6 \cdot 54 = \tfrac{x^2}{6} \cdot 6$$
$$324 = x^2$$
$$\sqrt{324} = x \qquad \text{Guess 20.}$$
$$18 = x$$

$$\begin{array}{r} 16 \\ 20\overline{)324} \\ \underline{20} \\ 124 \\ 120 \end{array} \qquad \begin{array}{r} 16 \\ +20 \\ \hline 36 \end{array} \qquad \begin{array}{r} 18 \\ 2\overline{)36} \end{array}$$

50. **(4)** *Geometry/Applications.* **both A and B**

Condition A satisfies the ASA requirement.
Condition B satisfies the SAS requirement.

51. **(3)** *Algebra/Problem solving.* **4(x − 1) = 2(x + 9)**

52. **(5)** *Geometry/Applications.*

$$\text{When } x = 0,\ y = +3(0) - 4$$
$$y = 0 - 4$$
$$y = -4$$

The coordinates of the y-intercept are **(0, −4)**.

53. **(2)** *Algebra/Applications.*

$$c^2 + 3c - 18 = 0$$
$$(c + 6)\,(c - 3) = 0$$
$$c + 6 = 0 \qquad c - 3 = 0$$
$$\underline{-\ 6} \quad \underline{-6} \qquad \underline{+\ 3} \quad \underline{+3}$$
$$c = \textbf{−6} \text{ and } c = \textbf{+3}$$

54. **(4)** *Geometry/Applications.*

S is 6 spaces left of the y-axis.
T is 10 spaces right of the y-axis.
Distance ST is 6 + 10 = **16.**

55. **(5)** *Geometry/Applications.*

$$R = (x_1, y_1) = (-6, -4) \qquad T = (x_2, y_2) = (10, 8)$$
$$d = \sqrt{(x_2 - x_1)^2 + (y_2 - y_1)^2}$$
$$d = \sqrt{(10 - (-6))^2 + (8 - (-4))^2}$$
$$d = \sqrt{(10 + 6)^2 + (8 + 4)^2}$$
$$d = \sqrt{(16)^2 + (12)^2}$$
$$d = \sqrt{256 + 144}$$
$$d = \sqrt{400}$$
$$d = \textbf{20}$$

56. **(1)** *Algebra/Problem solving.* **Fred gets \$12, and Gordon gets \$7.**

Fred's wage $= x$

Gordon's wage $= x - 5$

$40x + 40(x - 5) = 760$

$40x + 40x - 200 = 760$

$80x - 200 = 760$

$\underline{\quad\ \ + 200 \qquad + 200}$

$\dfrac{80x}{80} = \dfrac{960}{80}$

$x = 12$

$x - 5 = 12 - 5 = 7$

Description	Formula
AREA (A) of a:	
square	$A = s^2$; where s = side
rectangle	$A = lw$; where l = length, w = width
parallelogram	$A = bh$; where b = base, h = height
triangle	$A = \frac{1}{2} bh$; where b = base, h = height
circle	$A = \pi r^2$; where π = 3.14, r = radius
PERIMETER (P) of a:	
square	$P = 4s$; where s = side
rectangle	$P = 2l + 2w$; where l = length, w = width
triangle	$P = a + b + c$; where, a, b, and c are the sides
circumference (C) of a circle	$C = \pi d$; where π = 3.14, d = diameter
VOLUME (V) of a:	
cube	$V = s^3$; where s = side
rectangular container	$V = lwh$; where l = length, w = width, h = height
cylinder	$V = \pi r^2 h$; where π = 3.14, r = radius, h = height
Pythagorean relationship	$C^2 = a^2 + b^2$; where c = hypotenuse, a and b are legs of a right triangle
distance (d) between the two points in a plane	$d = \sqrt{(x_2 - x_1)^2 + (y_2 - y_1)^2}$; where (x_1, y_1) and (x_2, y_2) are two points in a plane
slope of a line (m)	$m = \dfrac{y_2 - y_1}{x_2 - x_1}$; where (x_1, y_1) and (x_2, y_2) are two points in a plane
mean	$mean = \dfrac{x_1 + x_2 \cdots + x_n}{n}$; where the x's are the values for which a mean is desired, and n = number of values in the series
median	$median$ = the point in an ordered set of numbers at which half of the numbers are above and half of the numbers are below this value
simple interest (i)	$i = prt$, where p = principal, r = rate, t = time
distance (d) as function of rate and time	$d = rt$; where r = rate, t = time
total cost (c)	$c = nr$; where n = number of units, r = cost per unit